MATHEMATICAL BIOECONOMICS

"I'm worried, Charlie! A catch like this could knock the bottom out of the market."

—Reprinted from *Punch*

MATHEMATICAL BIOECONOMICS: THE OPTIMAL MANAGEMENT OF RENEWABLE RESOURCES

COLIN W. CLARK

A WILEY-INTERSCIENCE PUBLICATION

JOHN WILEY & SONS, New York • London • Sydney • Toronto

Library of Congress Cataloging in Publication Data

Clark, Colin Whitcomb, 1931–
  Mathematical bioeconomics.

  (Pure and applied mathematics)
  "Wiley-Interscience publication."
  Includes index.
  1. Biology, Economic—Mathematical models.
2. Fishery management—Mathematical models. 3. Forestry management—Mathematical models. 4. Renewable natural resources—Mathematical models. I. Title.
[DNLM: 1. Models, Theoretical. 2. Economics. 3. Environment. 4. Mathematics. 5. Fisheries. 6. Trees.
HC103.7 C593m]

QH705.C4   333.9'5'0184   76-16473
ISBN 0-471-15856-9

Printed in the United States of America

10 9 8 7 6 5 4 3 2

PREFACE

Although an immense body of both popular and scientific literature covers the many aspects of biological conservation, as yet the subject can hardly be said to have developed a sound theoretical basis. The purpose of this book is to provide an introduction to such a theory of conservation. To preclude the possibility of confusion, it should be stated at the outset that our subject matter is the conservation of productive resources, rather than the preservation of natural environments. This distinction is important, because the term "conservation" is frequently employed in both senses. Thus the applications in this book are largely to the commercial fishery and forestry industries, rather than to the recreational uses of natural resource stocks.

An emphasis on the productive aspects of renewable-resource management obviously implies that both economics and biology play important roles in the theory. But resource conservation is also largely a problem of the optimal use of resource stocks *over time.* Conservation theory therefore must be established on explicit dynamic mathematical models of biological processes and must concern itself with the problem of dynamic optimization.

Undoubtedly the lack of a viable theory of conservation is to a large degree a consequence of the mathematical difficulties that are considered inherent in dynamic optimization theory. These difficulties are certainly real (especially in any practical situation), but one of the main purposes of this book is to demonstrate that the mathematical calculations in conservation theory are not as complex as is commonly supposed. For example, Chapters 1–3 in the book describe a dynamic optimization model of the fishery, which encompasses much of the existing theoretical literature on the subject, but which—even for the rigorous solution—only requires a single application of Green's theorem in the plane (i.e., integration by parts). Even this technicality can be omitted by the reader, because the optimal solution is described in terms that render its validity almost self-evident.

More advanced optimization techniques (namely the maximum principle) are employed in later chapters, but no previous knowledge of these methods is assumed. In fact the mathematical prerequisites consist only of a basic familiarity with calculus and simple differential equations. No stochastic models are employed, so that a knowledge of probability and statistics is not required.

Insofar as the disciplines of biology and economics are concerned, there are no prerequisites. All the necessary concepts and models are carefully defined and described before they are introduced into the theory. However, interested readers with a limited background in these subjects may wish to do some additional reading. To aid these readers, a list of basic background references precedes the Bibliography.

A brief outline of the book is as follows. Chapters 1–3 discuss a basic (one-dimensional) dynamic fishery model, and introduce various necessary concepts from biology and economics. In particular Chapter 3 concentrates on the capital-theoretic aspects of resource exploitation. Chapter 4 discusses the mathematical techniques of optimal control theory, which are extensively applied later in the book. Chapter 6 examines the phase-plane analysis of dynamical systems, which is also employed later in the book. A somewhat deeper discussion of the economic aspects of resource management is given in Chapter 5, which also contains a brief survey of the theory of exhaustible resources. The first six chapters of the book utilize continuous-time models of biological processes. Chapter 7 examines the study of discrete-time models, which in many ways are more flexible than continuous-time models. Finally, Chapters 8 and 9 extend the theory to more complex biological models that involve age structure (Chapter 8) and multispecies systems (Chapter 9). These last two chapters indicate that despite the complexities of realistic biological models, the basic bioeconomic theorems described in earlier chapters remain valid and provide great insight into more general problems.

The methods employed in this book are entirely analytic to provide a firm theoretical basis for renewable-resource management. Although many practical examples are mentioned, they serve as illustrations rather than as fully proved applications of the theory. Biometric and econometric methods, numerical optimization techniques, and computer simulation modeling—important as they are in the practice of biological resource management—are not discussed in this book.

COLIN W. CLARK

Vancouver, Canada
February 1976

ACKNOWLEDGMENTS

I acknowledge my indebtedness to the many friends and colleagues with whom I have had the pleasure of discussing the ideas in this book. The influence of my good friend and colleague, Gordon R. Munro (who has kindly contributed Section 3.4), is much deeper than may be apparent. Thanks to Gordon's tireless assistance, I have some confidence that the economic aspects of this book are reasonably close to a professional standard; any blunders still remaining can only be attributed to my own incompetence. I have also benefited greatly from my association with other economists, including A. D. Scott, P. H. Pearse, P. G. Bradley, P. H. Neher, J. A. Crutchfield, and M. Spence. The following biologists have also been most helpful: P. A. Larkin, N. J. Wilimovsky, and S. J. Holt. On the mathematical side, I am indebted to J. de Pree, P. L. Katz, and F. H. Clarke, and to my students, M. Friedlaender and W. J. Reed. I would also like to express my appreciation to my former department head, R. D. James, whose continual encouragement was most welcome.

CONTENTS

MATHEMATICAL BIOECONOMICS

INTRODUCTION

The management of renewable resources, where it has been practiced at all, has generally been based on the concept of maximum sustainable yield (commonly abbreviated MSY). This is perhaps the simplest possible management objective that accounts for the fact that a biological resource stock cannot be exploited too heavily without an ultimate loss of productivity.

The concept of MSY itself is based on a model of biological growth (see Figure 1) that assumes that at any given population level less than a certain level K, a surplus production exists that can be harvested in perpetuity without altering the stock level. (See Chapter 1 for a further discussion of this model.) If the surplus is not harvested, on the other hand, this causes a corresponding increase in the stock level, which ultimately approaches the environmental carrying capacity K, where surplus production is reduced to zero.

Since surplus production equals sustainable yield at each population level, it follows that MSY is achieved at the population level where surplus production is greatest (i.e., at the level where the growth rate of the population is maximized). For most populations to which this model applies, the MSY level is found to lie between 40% and 60% of the environmental carrying capacity.

In recent years it has become apparent that the MSY concept is, in many respects, far too simplistic to serve as a valid operational objective for the management of most living resource stocks. Severe objections have been raised on both biological and socioeconomic grounds. On the biological side, the word "yield" may often be ambiguous, especially in cases in which several ecologically interdependent species are harvested. In such cases the maximization of yield for each species separately is clearly impossible, so that some method of forming a weighted sum must

1

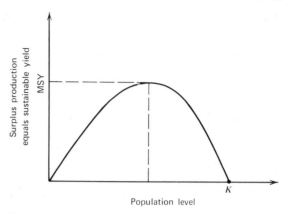

Figure 1

be employed. The outcome will then depend on the weighting system, whether in terms of gross weight, protein or caloric content, or economic value. The maximum yield concept clearly must be modified if these complications are to be incorporated.

The word "sustainable" is equally problematic biologically. Many resource stocks, particularly marine fish populations, are subject to large and unpredictable variations. Large yields that may be quite feasible when population levels are high can obviously not be sustained at lower population levels. Modification of the MSY concept is also required to encompass such fluctuations.

Significant as these biological objections to MSY may be, the economic shortcomings of the concept are even more important. The MSY concept is clearly addressed solely to the benefits of resource exploitation (even here the concept is notably superficial) and competely ignores the cost side of cost-benefit considerations. This fundamental flaw means that the MSY concept is virtually useless for descriptive theories of renewable resource exploitation. It can be stated unequivocally that any commercial resource industry will weigh the costs of exploitation equally with the benefits. As is explained soon it is extremely unlikely that in any particular case an MSY harvest policy will prove to be optimal in an economic sense.

The question remains as to whether MSY possesses any normative justification. The position taken in this book is that if taken as a *constraint* on rather than as a desideratum of exploitation, MSY may indeed possess desirable aspects. As we will see, in various situations commercial exploitation may lead to serious overexploitation of biological resources in the sense that the population becomes reduced to a level far below MSY.

There are convincing reasons why such an outcome—which may actually be "optimal" from the point of view of resource exploiters—may be far from optimal from the social viewpoint. When this is the case, public intervention may be justifiable, and MSY may provide a good rallying point. Even so, it is important to recognize that management policies based solely upon the achievement of MSY will almost inevitably lead to severe difficulties arising from the economic irrelevance of the concept.

Recognition of the inadequacy of the MSY concept has resulted in a trend to replace it with some concept of optimum sustainable yield (OSY). Unfortunately, this concept is often referred to without any clear understanding as to what objective is to be "optimized." The purpose of this book, as its title suggests, is the accurate identification and analysis of optimal resource management policies. Although it certainly cannot be claimed that all important aspects of the question have been resolved, or even considered, it is hoped that the reader will find that some progress has been achieved and that directions for further advances have been clarified.

The concept of optimal resource management used in this book is based on the standard cost-benefit criterion of maximizing present values of net economic revenues. This criterion is relevant to both private and public management decisions, although the specification of costs and benefits is not necessarily the same in both cases. (Private management is normally concerned only with actual, internalized costs; public management is often concerned with social and external costs as well.)

It should be noted at the outset that few biological resource stocks, historically speaking, have been managed on the present-value criterion. Most marine fisheries, for example, have hardly been "managed" at all, in any strictly economic sense of the term. Several important fisheries have been subject to a certain degree of control by various international agencies, but there are probably no cases in which these agencies could be said to possess full management authority. Whether satisfactory results can be achieved in the absence of such authority remains an important question.

One agency that has been publicly prominent in recent years is the International Whaling Commission (IWC), established in 1946 to regulate whaling. Since the difficulties associated with the conservation of whales in many ways typify the general problem of biological resource conservation, we introduce some of the main concepts that are developed in this book, using the whaling industry as an example.

We begin by asking, hypothetically, how the whale resource might be managed by a single firm (or consortium of firms) that possessed complete rights to the exploitation of whales. First, we consider a single population, the Antarctic blue whales. We will use the simple surplus-production model shown in Figure 1, ignoring for the moment the many biological limitations of this model. Whaling scientists have estimated the environmental carrying capacity for Antarctic blue whales at $K = 150,000$ whales and MSY at approximately 2000 whales per annum. Assuming that the surplus-production curve is symmetric, we see that a standing population of 75,000 whales is required to produce MSY.

Imagine that through its past operations the industry has reduced the original 150,000 whales to the MSY level of 75,000. If $10,000 represents the market value of the products obtained from an average blue whale, then an MSY policy will result in an annual revenue of $20 million. The industry has other options, however. For example, it might decide to capture the remaining 75,000 whales immediately, overlooking the possibility of sustained production altogether. Assuming that this could be accomplished in one season and neglecting the problem of disposing of several million tons of whale oil in a short time (demand elasticity, to be discussed in Chapter 5), the industry would obtain a lump-sum revenue of $750 million. Invested at the rather conservative rate of 5% per annum, this sum would yield an annual return of $37.5 million. On the basis of these elementary calculations, extermination of the blue-whale population appears to constitute a considerably more profitable policy than MSY.

The preceding argument has been deliberately oversimplified; some of the implications of a more realistic analysis are discussed in the following paragraphs. Nevertheless, the argument illustrates one of the fundamental aspects of the economics of resource management. The owner of a resource stock tends to view the stock as a *capital asset;* this is equally true for exhaustible resources (see Section 5.2) and for renewable resources. He expects the asset to earn dividends at the "normal" rate of return; otherwise, the owner would attempt to dispose of the asset. This result, which can be thought of as the first fundamental theorem of resource economics, was developed by H. Hotelling (1931).

From this result we see why the conservation of whales is a particularly difficult problem. Growth rates of whale populations typically range from 5% to 10% per annum. These rates are of the same order of magnitude, if not below, the expected rates of return on alternative investments (the so-called opportunity-cost rate of return; see Chapter 3). This may mean that the whaling industry itself has little motivation to conserve its own resource base. Before a final conclusion can be drawn, however, many other aspects of the problem must be taken into consideration.

For example, what about the *cost* of catching whales? It is apparent that once the blue whale population has been greatly depleted, the cost of locating whales somewhere within the 25 million square kilometers of their Antarctic feeding grounds may become exorbitant. At their present population level, thought to be around 8000 blue whales, exploitation seems to be barely feasible (see Section 2.6).

A second basic aspect of renewable resource exploitation, particularly relevant to widespread populations such as whales and fishes, may now be discussed; as the population level is reduced, the cost efficiency of harvesting decreases. The result of this effect is to make it desirable to maintain a higher population level than would otherwise be the case.

Here, we see an interesting and most significant dichotomy regarding "optimal" resource harvesting policies. As is shown in detail in Chapter 2, the "discounting" effect associated with capital opportunity costs always has the effect of pushing the optimal population level below the MSY level, whereas cost-efficiency considerations have the opposite effect. When both effects are considered, the optimum may turn out to be on either side of MSY, depending on the relative strengths of the two effects [see Eq. (2.16)]. It is interesting to observe, however, that the clash between these two effects does not ordinarily lead to some form of oscillatory harvesting, in which one effect and then the other dominates. Although such harvesting policies are sometimes employed (e.g., in fisheries, where they are referred to as "pulse-fishing" policies), we mantain (see Sections 5.4 and 8.7) that this phenomenon arises from a different bioeconomic cause associated with efficiencies of scale in the harvesting process.

Returning to the blue-whale population, we see that (1) the "inferior-asset" problem may cause whalers to be disdainful of whale conservation, but that (2) the species may still survive, because complete extermination is not economically feasible. It should be emphasized that the inferior-asset problem is particularly important in whaling because of the *slow growth rates* of whale populations. Other resource stocks with much greater "biotic potential," such as tuna populations, would not be expected to suffer from overexploitation because of capital-asset effects. That such populations often *are* subject to overexploitation is due to other causes, associated with the *open-access* or *common-property* nature of these resources. This extremely important question is taken up after further discussion of the whaling problem.

Clearly, there are other important aspects to whaling than those we have mentioned thus far. For example, the price of whales may increase or decrease, according to the scarcity of whales or to exogenous changes in the demand for whale products. Such price changes, which may have a strong effect on commercial exploitation, are examined in Chapters 3 and 5. The rate of interest may also vary, for example, during inflationary periods. If the value of whales also increases (or decreases) at the inflationary (or deflationary) rate, the net effect will be canceled. If not, the result will be the same as a net price change and can be handled by the methods described in Chapter 3. Similarly, costs may vary over time, either as a result of technological change or due to the indirect political pressures imposed by conservationists. These effects can also be studied by utilizing the methods developed here, although we have not paid special attention to them at this point. (Refer to Section 3.4 for a more detailed discussion.)

Next, we must consider the existence of other whale species. Historically, exploitation of Antarctic whales concentrated mainly on the blue-whale population until the 1950s, when the developing shortage of blue whales led to increased harvests of fin whales, a smaller species about half the weight of the blue whale. Of course, blue whales were still harvested whenever they were sighted, but the mainstay of the industry for the next two decades was the fin whale. As fin-whale stocks became depleted, the even smaller sei and Minke whales were taken.

Under these circumstances the blue-whale population could be further reduced well below the level where exploitation based solely on blue whales would be profitable. As is shown in Chapter 9, it is possible that the blue whale species would ultimately be eliminated. Such considerations were instrumental in the decision of the International Whaling Commission (IWC) in 1965 to afford complete protection to blue-whale stocks and subsequently to extend this protection to other severely depleted whale stocks.

The joint exploitation of multispecies systems is similar to the situation, well known to ecologists, of a single predator with several prey species. Here, the less productive prey populations often tend to disappear under the predation pressure. To what extent a similar result may be true for the "optimal" exploitation of multiple-resource stocks depends partly on the feasibility and the cost of selective harvesting. This difficult question is discussed in Chapter 9.

To add a touch of realism to our analysis of the whaling industry, we now drop the assumption that a single firm or consortium controls whaling—although to a certain extent the IWC does perhaps play the role of a consortium of whaling firms. We now ask how the exploitation of whales would be expected to proceed in the absence of the IWC. Whales would then be an open-access resource, in the sense that any nation or firm would be free to enter the industry without any form of regulation. The economic theory of open-access resources (or common-property resources, as they are usually rather inaccurately referred to) was developed by H. S. Gordon (1954). This theory predicts an ultimate "bionomic equilibrium" between the resource and the exploiting industry, which occurs at a stock level where the revenue flow exactly equals the cost (opportunity cost; see Section 2.1) of exploitation. A simple argument supports this theory (see Chapter 2). Obviously equilibrium cannot be established when costs exceed revenues, for some exploiters would be forced to leave the industry and seek other employment. Conversely, equilibrium cannot be established when revenues exceed costs, for additional exploiters would be attracted from other areas of employment where remuneration is relatively lower. The only condition of this argument is the a priori assumption that an equilibrium will necessarily become established at some stock level. Interestingly, this assumption can be justified by other considerations, but only if the surplus production curve (Figure 1) is assumed to be convex (see Chapter 6, Section 5).

Gordon's results regarding "rent dissipation" in open-access resources may be considered the second fundamental theorem of resource economics, complementing Hotelling's theorem for individually owned resource stocks. Because most biological resources are not individually owned, Gordon's theorem is perhaps more important in this case. This theorem has many important consequences. Open-access exploitation is more intense than profit-maximizing management, and it is more likely to lead to adverse biological results, including the possibility of extinction. Increases in demand cause the open-access resource to be more heavily exploited, and once the MSY level is surpassed, lead to progressively lower levels of production. Technological progress that improves the efficiency of exploitation may then have a negative influence, producing further declines in productivity.

From the theoretical viewpoint open-access exploitation can be treated as the limiting case of privately optimal management, in which the rate of discount becomes infinitely large. This is intuitively clear, because the competitive aspect of open-access exploitation inhibits the motive of conservation for future yields. It follows, however, that many of the disadvantages of open-access exploitation may also arise under private ownership if high discount rates are employed. In the case of whales an annual discount rate of 10% is high enough to affect conservation severely. In forestry management also, discount rates as high as 10% have been known to produce devastating effects (see Chapter 8). The irreversibility of these effects implies that such overexploitation may be undesirable, even though calculations based on present conditions might seem to justify it.

The classical literature on the economics of fisheries has principally adopted the criterion of maximization of sustained economic rent (i.e., net economic revenue) as the ideal for fisheries management (see Chapter

2). This criterion should be discarded. It is defective because it overlooks the opportunity cost of capital—or, to be more precise, fixes it at zero. It is difficult to explain why such a zero-discounting criterion has been maintained for so long by economists in Western capitalistic countries. I hope that this book will establish the foolishness of supposing that discounting doesn't make any significant difference. It is the difference between an imaginary, static, utopian world and the real, dynamic world in which we live. High interest rates are admittedly deleterious to conservation, but a zero interest rate is a practical impossibility.

To conclude this introduction, let us return once again to our whaling example. What role has the IWC played in regulating whaling? The severe depletion of blue-whale and fin-whale stocks—and, to a lesser degree, of other stocks of whales—has transpired under its auspices. For many years, certainly prior to the appointment of a Scientific Committee in 1963, IWC activities consisted primarily of setting annual quotas for total whale harvests. These nonspecific quotas had the minimum effect of limiting open-access competition, which otherwise could have led to an extremely wasteful overextension of the capacity of whaling fleets. No doubt, the IWC was partially responsible for the profitability of whaling from 1946 through 1965—and beyond. But, as we have seen, the profitability of whaling does not imply the conservation of whales. Thus, the interests of whalers and of conservationists may be in direct conflict.

The conservationist movement, however, has had significant effects on recent IWC decisions. It seems that a compromise is currently being forged between the profit motives of the Japanese and Russian whaling industries and the expressed wishes of conservationists. Such a compromise is very much in the spirit of this book.

This is not a book about whaling. Many other resource-management examples will be examined, including fisheries, forestry, and wildlife management. But the real purpose of the book is to formulate a dynamic theory of renewable resource management. The theory is by no means complete: stochastic aspects are neglected, as are the problems of making decisions based on uncertain or incomplete information. All that can be hoped for is that this book will constitute the beginning of a scientific theory of conservation—a theory that is long overdue.

1

ELEMENTARY DYNAMICS
OF EXPLOITED
POPULATIONS

In this chapter we discuss some extremely simple mathematical models for the exploitation of biological resources. These models are based on an ordinary differential equation of the form

$$\frac{dx}{dt} = F(x) - h(t),$$ (1.1)

where $x = x(t)$ denotes the size of the resource population at time t, where $F(x)$ is a given function representing the natural growth rate of the population, and where $h(t)$ represents the rate of removal, or harvesting.

Such a model obviously involves a large number of simplifications and abstractions from the behavior of real-world biological populations. These simplifications are discussed in the final section of this chapter, and some of them are examined in greater detail in subsequent portions of the book.

Whenever the harvest rate $h(t)$ exceeds the natural growth rate $F(x)$, Eq. (1.1) implies that the population level will decline ($dx/dt < 0$), and vice versa. If $h(t) \equiv F(x)$, the population remains at a constant level. In other words, the natural growth rate $F(x)$ also equals the *sustainable yield* that can be harvested while maintaining a fixed population level x.

The fact that the sustainable yield depends on the stock level of a renewable resource is an elementary but fundamental principle of renewable resource management. Equally important, however, are the non-equilibrium situations that arise when $h(t) \neq F(x)$. Throughout this book

9

we are particularly concerned with the biological and economic implications of dynamic, nonequilibrium resource-harvesting models. In many (but not all) cases equilibrium solutions emerge from the dynamic theory and possess desirable optimality properties.

The differential equation adopted here [Eq. (1.1)] as a model of population growth and harvesting, has also been studied extensively by economists as a model of capital growth and consumption. The fact that these two problems can be described by a single mathematical model is not a mere coincidence. From the point of view of human needs, a resource stock is simply a particular form of capital that can either be consumed or conserved. What distinguishes a biological resource from a stock of traditional capital (such as buildings, machinery, human expertise, etc.), of course, is the mechanism of growth: biological resources grow "by the gift of nature"; traditional capital can only increase through human effort.

The capital-theoretic implications of renewable resource management will be discussed in some detail in Chapter 3. In this chapter we will concentrate primarily on the purely biological aspects of our model.

1.1 THE LOGISTIC GROWTH MODEL

Suppose that, in a certain population, both the birth rate b and the mortality rate m are proportional to the population size x. Writing $r = b - m$ for the *net proportional growth rate* of the population, we then obtain the differential equation

$$\frac{dx}{dt} = rx \tag{1.2}$$

as a continuous-time model of population growth. The solutions to this simple equation $x(t) = x(0)e^{rt}$ grow exponentially to infinity if $r > 0$ and decrease exponentially to zero if $r < 0$.

Under ideal conditions, where the availability of space and other resources does not inhibit growth, many biological populations are observed to grow at an approximately exponential rate initially. Clearly, however, such a process cannot proceed indefinitely. As the population level x increases, some environmental limitation must force the proportional growth rate to decline. To model this effect, Eq. (1.2) may be modified to the form

$$\frac{dx}{dt} = r(x) \cdot x, \tag{1.3}$$

where $r(x)$ is some decreasing function of x. The proportional growth rate

$$r(x) = \frac{F(x)}{x}$$

now depends on the population level x. If $r(x)$ is a *decreasing* function of x, this model is said to describe a process of feedback, or *compensation*, which controls the growth of the population as its level increases.

The simplest and perhaps the most useful example is obtained when $r(x) = r(1 - x/K)$, so that Eq. (1.3) becomes

$$\frac{dx}{dt} = rx\left(1 - \frac{x}{K}\right) = F(x). \tag{1.4}$$

This is the famous *logistic equation*, first proposed as a population model by P. F. Verhulst in 1838. The constant r, assumed positive, is called the *intrinsic growth rate*, since the proportional growth rate for small x approximately equals r. The positive constant K is usually referred to as the environmental *carrying capacity* or saturation level.

Although Eq. (1.4) can be easily solved explicitly for $x = x(t)$, the main features of the solution are apparent directly from the equation itself. First, we observe that the equation possesses two equilibrium solutions, namely $x \equiv 0$ and $x \equiv K$. Moreover, we have

$$0 < x < K \quad \text{implies} \quad \dot{x} > 0,$$

whereas

$$x > K \quad \text{implies} \quad \dot{x} < 0.$$

(The notation $\dot{x}$ is used interchangeably with dx/dt according to convenience.) It follows that K is a *stable* equilibrium, or, to be more precise, K is *globally asymptotically stable* for positive x, in the sense that

$$\lim_{t \to \infty} x(t) = K, \quad \text{provided} \quad x(0) > 0. \tag{1.5}$$

These facts are illustrated in Figure 1.1, where part (a) shows the growth function $F(x) = rx(1 - x/K)$ and the arrows indicate the direction of change of $x(t)$ with increasing t. Figure 1.1*b* illustrates two typical solution curves $x(t)$, approaching the equilibrium K from above and below. The lower curve, with its characteristic *ogive* shape, is usually referred to as a *logistic growth curve*.

The logistic differential equation [Eq. (1.4)] is easily solved by separation of variables. Write Eq. (1.4) in the form

$$\frac{dx}{x(K - x)} = \frac{r}{K}\, dt,$$

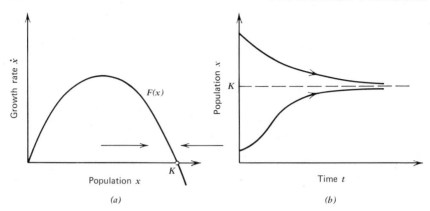

Figure 1.1 The logistic equation: (*a*) the logistic growth function $F(x) = rx(1 - x/K)$; (*b*) typical solution curves.

or

$$\left(\frac{1}{x} + \frac{1}{K - x}\right) dx = r\, dt,$$

so that, by integration,

$$\ln \frac{x}{K - x} = rt + \ln \frac{x_0}{K - x_0} \quad \text{where} \quad x_0 = x(0).$$

The solution may be rewritten in the form

$$x(t) = \frac{K}{1 + ce^{-rt}}, \quad \text{where} \quad c = \frac{K - x_0}{x_0}. \tag{1.6}$$

This equation may also be verified by direct substitution. The limit condition [Eq. (1.5)] follows directly from Eq. (1.6); moreover, we see that $x(t)$ converges towards K at an exponential rate as $t \to \infty$.

Harvesting

Now, let us suppose that the population described by the logistic equation [Eq. (1.4)] is subject to harvesting at a rate $h(t)$. Then Eq. (1.4) becomes

$$\frac{dx}{dt} = F(x) - h(t), \tag{1.7}$$

where $F(x) = rx(1 - x/K)$. What can be said about the dynamic behavior of the population?

An important special case arises when $h(t) = h \equiv \text{constant}$:

$$\frac{dx}{dt} = F(x) - h. \tag{1.8}$$

In case $h < \max F(x) = \frac{1}{4}rK$, Eq. (1.8) possesses two equilibria, x_1 and x_2 (see Figure 1.2a). Notice that $\dot{x} > 0$ when x lies between x_1 and x_2, while $\dot{x} < 0$ elsewhere. It follows that x_2 is a stable equilibrium and that x_1 is an unstable equilibrium. If the initial population is at $x = K$, for example, then $x(t)$ will converge asymptotically under constant harvesting to the equilibrium x_2. But if the initial population is less than x_1, then $x(t)$ will approach 0. In the latter case, the approach is not asymptotic [since 0 is not an actual equilibrium of the *equation* (1.3), although 0 is certainly a biological equilibrium]; instead, $x(t)$ is reduced to 0 in a finite time.

If $h > \max F(x)$, as it is in Figure 1.2b, the population approaches 0 for any initial level $x(0)$. Finally, in the special case when $h = \max F(x)$, there is a single equilibrium at $x_1 = K/2$, which is "semistable" in the sense that $x(t) \to x_1$ if $x(0) > x_1$ while $x(t) \to 0$ if $x(0) < x_1$.

In spite of various limitations, the model given by Eq. (1.8) provides a number of significant predictions concerning the harvesting of renewable resources. First, there exists a *maximum sustainable yield* MSY

$$h_{\text{MSY}} = \max F(x),$$

with the property that any larger harvest rate will lead to the depletion of the population (eventually to zero).

Second, the population level $x = x_{\text{MSY}}$ at which the productivity of the resource is maximized is *not* the natural equilibrium level K; in this model, it is only half that level. Indeed, there is no sustainable yield at the population level $x = K$.

These two simple phenomena, which appear to be typical of most practical resource harvesting situations, are often misunderstood by the general public and by inexperienced conservationists. Because it no

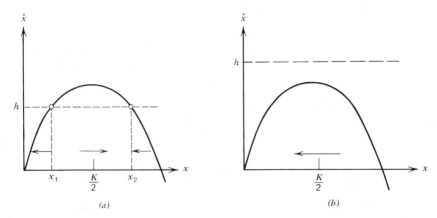

Figure 1.2 Logistic model with constant harvest rate h: (a) $h < \max F(x)$; (b) $h > \max F(x)$.

longer exists at its original level of abundance, a resource may be thought to be overexploited. Yet, significant harvesting, because it increases the mortality rate of the population, almost invariably leads to a decrease in the equilibrium population level. Most biologists consider a resource to be overexploited only when the population has been reduced to a level below the maximum sustainable yield x_{MSY}. In this book, we will adopt the phrase *biological overexploitation* to describe this situation.

Although the decreased abundance of a resource stock is not necessarily a sign of overexploitation, the layman's intuitive feeling that "things aren't what they used to be" has some justification. The fewer fish there are in a lake, for instance, the harder it is to catch a fish—all other things being equal. In economic terms, the *cost* of catching fish tends to rise as the population is reduced. When the costs as well as the benefits are taken into consideration, it might be argued that the optimal stock level should be *higher* than x_{MSY}. (This idea will be examined in greater detail in Chapter 2.)

Fishing Effort

In addition to data pertaining to the catch, fishery statistics normally include information under the heading *fishing effort*, measured in units appropriate for the fishery in question. In some cases, the unit of measurement is simply the total number of vessel-days per unit time; in other cases, more detailed information regarding the number of nets, lines, or traps hauled is available.

The ratio of catch divided by effort is almost always taken as at least a rough indication of the current stock level of the fish population. We will use the phrase *catch-per-unit-effort hypothesis* to describe an assumption that catch-per-unit-effort is proportional to the stock level, or that

$$h = qEx, \qquad (1.9)$$

where E denotes effort and q is constant, called the *catchability coefficient*. The catch-per-unit-effort hypothesis can be derived from a simple probabilistic fishing model, which will be discussed with alternative models in Chapter 7.

Since it is not necessary for our purposes to specify units of effort, we can normalize units by setting

$$q = 1.$$

We then substitute Eq. (1.9) into our basic harvesting model, Eq. (1.7):

$$\frac{dx}{dt} = F(x) - Ex = rx\left(1 - \frac{x}{K}\right) - Ex. \qquad (1.10)$$

We will first consider the solutions of this equation under the assumption that E is constant.

To obtain the equilibria of Eq. (1.10), we set $dx/dt = 0$. For any $E < r$, we find that the equation has a unique nonzero equilibrium x_1, given by

$$x_1 = K\left(1 - \frac{E}{r}\right). \tag{1.11}$$

Moreover, this equilibrium is always asymptotically stable (see Figure 1.3).

The equilibrium harvest or *sustainable yield* $Y = h$ corresponding to E is given by

$$Y = Ex_1 = KE\left(1 - \frac{E}{r}\right), \tag{1.12}$$

provided that $E < r$. The graph of this equation, a parabola (see Figure 1.4), is the *yield-effort curve* for the model given by Eq. (1.10). This model is commonly called the *Schaefer model* after biologist M. B. Schaefer, who studied it extensively and applied it to various fish populations. Later in this book, we will utilize the Schaefer model and examine some of Schaefer's data.

The smooth nature of the relationship between effort and sustainable yield in the Schaefer model should be observed. With increasing levels of effort, the sustainable yield rises smoothly to a maximum level (at $E = r/2$, $x_1 = K/2$) and then declines equally smoothly to zero (at $E = r$, $x_1 = 0$). A smooth relationship between effort and sustainable yield is highly desirable in actual fisheries, since it implies that "incremental" methods can be successfully applied to management policies.

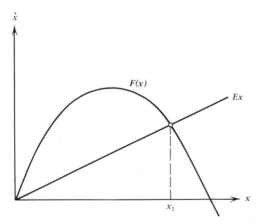

Figure 1.3 Logistic model with constant rate of effort E.

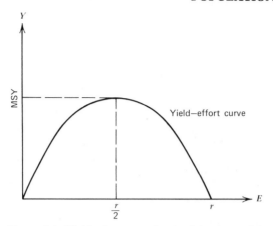

Figure 1.4 Yield–effort curve for the Schaefer model.

Unfortunately, while many fish populations appear to exhibit a smooth yield–effort relationship, others have suffered sudden collapses under the pressure of heavy fishing. We will now turn to the study of growth models that produce this kind of behavior.

1.2 GENERALIZED LOGISTIC MODELS: DEPENSATION

Many alternatives to the logistic growth model have been proposed. Some of these are described in more detail in the exercises that appear at the end of this chapter. In this section, we will discuss the class of models of the form

$$\frac{dx}{dt} = F(x) \tag{1.13}$$

where the growth curve has one of the forms shown in Figure 1.5. In each case, a stable equilibrium exists at $x = K$. Equation (1.13) is said to define a *generalized logistic growth model.*

The logistic model itself—and, more generally, any model with a growth function that appears in Figure 1.5a, such that the proportional growth rate $r(x) = F(x)/x$ is a decreasing function of x—is called a *pure compensation model.*

On the other hand, if $r(x)$ is an increasing function of x for certain values of x, a process of *depensation* is said to exist. Thus, the curve in Figure 5.1b exhibits depensation for $0 < x < K^*$ and compensation for $x > K^*$. Since all curves under discussion are assumed to be compensatory for large values of x, we will describe this curve simply as a *depensation curve.*

A depensation curve with the property $F(x) < 0$ for certain values of x near $x = 0$, as in Figure 1.5c, is called a *critical depensation curve.* In this

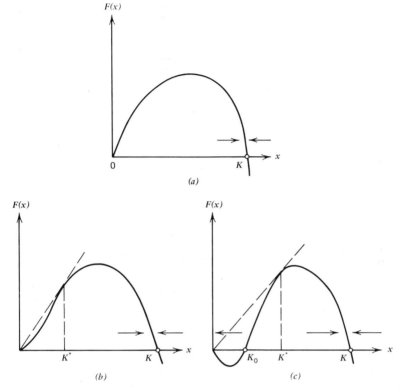

Figure 1.5 Growth curve types: (a) pure compensation; (b) depensation; (c) critical depensation.

case, an unstable equilibrium K_0 exists, such that

$$\lim_{t \to \infty} x(t) = 0 \quad \text{whenever} \quad x(0) < K_0.$$

The value $x = K_0$ is called the *minimum viable population level.*

The terms "compensation" and "depensation" are used by fishery biologists. Models of the life histories of fish populations that produce compensation and depensation processes are described in Chapter 7.

Yield–effort Curves

Now, suppose that the population is harvested by means of a constant effort E, so that Eq. (1.13) becomes

$$\frac{dx}{dt} = F(x) - Ex. \tag{1.14}$$

We wish to construct the yield–effort curve $Y = Y(E)$.

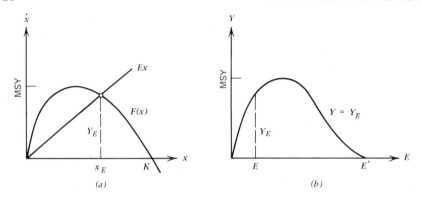

Figure 1.6 Pure compensation: (a) growth curve; (b) yield–effort curve.

In the case of pure compensation, each level of effort E produces a unique and stable equilibrium population x_E and a corresponding yield $Y_E = F(x_E)$—(see Figure 1.6). Thus, the yield-effort curve rises to a maximum MSY and then decreases smoothly as effort is further increased. The sustainable yield is zero for $E \geq E^*$, where

$$E^* = F'(0) = \max r(x) = r^*; \qquad (1.15)$$

that is, $E^* = r^*$ is the maximum proportional growth rate or the *intrinsic growth rate* of the population. Therefore, if the harvesting rate E exceeds the intrinsic growth rate r^*, the population will be driven towards extinction. For the logistic model given by Eq. (1.4), $r^* = r$.

Thus, pure compensation models are similar in most important respects to the special case of logistic growth discussed in the previous section.

Depensation

When the growth curve $F(x)$ exhibits depensation, the situation changes significantly (see Figure 1.7), for then Eq. (1.14) possesses multiple equilibrium solutions. For each level of effort $E < E^* = \max r(x)$ (see Figure 1.7), there exists a stable equilibrium population $_1x_E$ and a corresponding stable sustainable yield $_1Y_E$. But for $E > E^\dagger = F'(0)$, there also exists an unstable equilibrium population $_2x_E$. If the initial population level $x(0) > {_2x_E}$, the ultimate equilibrium is established at $x = {_1x_E}$; if $x(0) < {_2x_E}$, the ultimate equilibrium is established at $x = 0$ (assuming, of course, that E remains constant throughout).

As in the case of pure compensation, a critical effort level E^* exists such that $Y_E = 0$ when $E > E^*$. As before,

$$E^* = \max r(x).$$

The yield-effort curve in this case, however, is strikingly different from the compensation model, since the former now exhibits a *discontinuity* at $E = E^*$, with the sustainable yield suddenly jumping to zero as E exceeds this critical level.

System ecologists are familiar with many biological phenomena in which "large effects arise from small causes," and this will be a recurrent theme throughout the book. From the mathematical viewpoint, Eq. (1.14) is said to undergo a *bifurcation* at the critical parameter value $E = E^*$, since the solutions of the system change discontinuously as E passes through E^*. The theory of dynamical systems and resultant bifurcations will be taken up in Chapter 6.

The implications for management policymaking of a depensation yield-effort curve as shown in Figure 1.7b can be described in the following terms. First, the "incremental" approach suitable under the Schaefer model is no longer appropriate, since a small increase of effort beyond E^* leads to a population collapse. Second, the depensation model exhibits a certain *hysteresis effect*. Suppose that effort reaches a level $E > E^*$; and then that $x(t)$ approaches 0. Now, while x is still positive, assume that effort is reduced to some level below E^*. It does not necessarily follow that the system will return to the sustained-yield mode $_1Y_E$. Unless effort is reduced to a level for which $_2x_E < x$ (Figure 1.7a), the population will continue to decrease. In order to return to the sustained yield $_1Y_E$, it may be necessary to reduce effort to a level below $E^\dagger$ (Figure 1.7b).

The depensation model predicts actual extinction of the fish population if effort remains above a critical level. There are two reasons to anticipate a less extreme outcome in real-world fisheries: (1) effort will tend to be

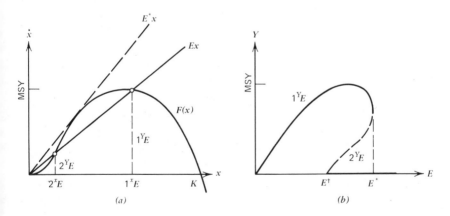

Figure 1.7 Noncritical depensation; (a) growth curve; (b) yield–effort curve.

reduced when catches decline to low levels, and (2) inevitably, some fish will avoid capture for one reason or another.

Although few if any important fisheries have been fished to biological extinction, various other animal populations have become extinct following heavy commercial exploitation. The critical depensation model allows for such a possibility.

Critical Depensation

The case of critical depensation (see Figure 1.8) exhibits all of the phenomena of the depensation model encountered above as well as an additional phenomenon: *irreversibility*. We note (Figure 1.8a) in this case, that every level of effort $E \geq 0$ gives rise to two equilibria, $_1x_E$ and $_2x_E$, and also that $x = 0$ is a locally stable equilibrium for every E.

If effort rises to a supercritical level, the population may be reduced to some level below the minimum viable population K_0. Once this has occurred, the ultimate extinction of the population is ensured, regardless of what happens to future effort levels. Whenever x falls below K_0, an irreversible extinction process begins.

In the early 1960s, the Antarctic blue-whale (*Balaenoptera musculus*) population had been reduced by whaling to a level from which recovery seemed impossible (see Small, 1971). But under IWC protection since 1965 the blue-whale population has apparently increased at a rate of approximately 4% per annum to a level of about 8000 whales (1975 estimate). Data on the fin whale (*B. phaesalus*) (see Allen, 1973) seem to suggest at least noncritical depensation in the growth curve for that population. In view of the fact that the Antarctic feeding grounds of these populations amount to some 10 million square miles, breeding might be expected to fall off considerably at low population levels.

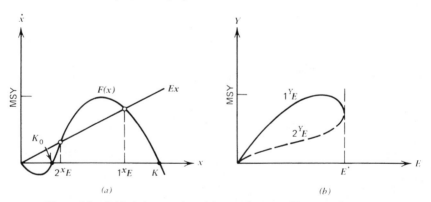

Figure 1.8 Critical depensation: (*a*) growth curve; (*b*) yield–effort curve

Another population whose growth curve may have exhibited critical depensation was the passenger pigeon (*Ectopistes migratorius*). The original population of some 7 billion birds was hunted to extinction during the nineteenth and early twentieth centuries (see Silverberg, 1967 for a chronicle of this and other modern extinctions). Whether hunting, forest removal, or "natural causes" ultimately caused the extinction of this species will never be known. The loss of the passenger pigeon did have one positive effect, however, in that it led to an upsurge of interest in resource conservation among Americans.

1.3 SUMMARY AND CRITIQUE

In this introductory chapter, we have discussed resource-harvesting models based on the elementary differential equation

$$\frac{dx}{dt} = F(x) - h(t).$$

Idealized as it may be, this model reflects several important aspects of the dynamics of renewable resource exploitation. Such a model implies that whenever the rate of harvest exceeds the natural rate of growth, the population biomass level will decline. Moreover, corresponding to each biomass level x is a certain rate of harvest $h = F(x)$ that just balances the natural rate of growth and thus maintains an equilibrium.

Some such relationship between natural productivity and harvestable "surplus" must obviously exist for any renewable resource stock. In practical applications, however, the simplicity of a model of this sort may be seriously misleading. Ideally, the predictive value of the model should be tested against experimental or field data. If found to be seriously deficient, the model should be modified or replaced.

One particular danger associated with the use of *any* mathematical model in renewable resource management is the problem of extrapolation from available data. A simple logistic model may be acceptable when exploitation rates are low, but can produce errors when extended to higher rates, as our depensation models clearly suggest.

Since field experiments can seldom be performed without endangering the resource stock itself, it becomes important to develop mathematical models that, in contrast to the ad hoc models of this chapter, incorporate known biological characteristics of the population. Such models are described in later chapters, particularly in Chapter 7.

The deterministic nature of the models discussed in this chapter also seriously limits their usefulness. The problem of introducing stochastic elements, however, is far from trivial, especially for the dynamic optimization problems that are the main object of study in this book. We shall largely (but not entirely) ignore stochastic problems in this book.

Other important aspects of realistic population models that need to be considered further include:

1. Delays and periodic (seasonal) effects (Chapter 7).
2. Age structure and related effects (Chapter 8).
3. Multispecies, ecosystem effects (Chapter 9).
4. Spatial effects and diffusion (Chapter 9).

Before examining these numerous difficulties, however, the next few chapters will be devoted to a study of the economic aspects of renewable resource exploitation. For simplicity, the basic biological model discussed in this chapter will be retained throughout the discussion.

EXERCISES

1. Given the logistic model with constant-effort harvesting:

$$\frac{dx}{dt} = rx\left(1 - \frac{x}{K}\right) - Ex, \quad x(0) = K,$$

   determine $x(t)$ explicitly. [This is readily obtained by using Eq. (1.6).] Verify from the form of the solution that if $E \leq r$, then $x(t) \to K(1 - E/r)$ as $t \to \infty$, whereas if $E > r$, then $x(t) \to 0$ exponentially as $t \to \infty$. In this model, how does the harvest rate vary over time?

2. Find x_{MSY} and the maximum sustainable yield $F(x_{MSY})$ for the "Gompertz law" of population growth:

$$\frac{dx}{dt} = rx \ln \frac{K}{x}.$$

   Also sketch the yield-effort curve. (What is E^*?)

3. Consider a modified logistic growth law:

$$\frac{dx}{dt} = rx^{\alpha}\left(1 - \frac{x}{K}\right) \quad (\alpha > 0).$$

   Sketch such growth curves for $\alpha \lesseqgtr 1$, and discuss the corresponding yield-effort curves.

4. The model

$$\frac{dx}{dt} = a - bx, \quad (a, b > 0)$$

was proposed by Schoener (1973). Determine the equation of the yield-effort curve and explain its peculiar feature.

5. Discuss the model

$$\frac{dx}{dt} = rx\left(\frac{x}{K_0} - 1\right)\left(1 - \frac{x}{K}\right),$$

where $0 < K_0 < K$.

6. The following "extreme" form of density-dependent growth has been discussed by Wiegert (1974):

$$\frac{dx}{dt} = F(x) = \begin{cases} rx & (x < K) \\ -\infty & (x > K). \end{cases}$$

Find the yield-effort relationship.

BIBLIOGRAPHICAL NOTES

The logistic growth model seems to have first been used by P. F. Verhulst (1838) as a model of human population growth. Reasonable fits with experimental data have been established by Pearl (1930) in *Drosophila melanogaster* (fruit-fly) populations, by Gause (1935) in *Paramecium* and *Tribolium* (flour-beetle) populations, and by others (see Pielou, 1969; Emlen, 1973; or almost any ecology text for further details). Feller (1940) has pointed out that almost any data for populations that increase to an asymptotic level will fit the logistic model to some degree, but a better fit can be obtained in most cases with alternative models.

Many alternative forms for the growth function $F(x)$ have been suggested (see May, 1973, for a brief summary; also Wiegert, 1974, for recent data). Constant-rate harvesting models have been studied by Brauer and Sanchez (1975).

Depensation is sometimes referred to as the "Allee effect" (Allee, 1931). Some of the consequences in terms of multiple equilibria are discussed by Watt (1968), Holling (1973), and, in the context of fisheries, by Larkin, Raleigh, and Wilimovsky (1964). The corresponding double-valued yield-effort curves were described by Clark (1974); also see Southey (1972). Various biological mechanisms that might be expected to give rise to depensation will be described in Chapter 7.

The concept of fishing effort, which plays a fundamental role in fisheries biology, is notoriously difficult to define and to quantify precisely. Further discussions can be found in Beverton and Holt (1957), pp. 172–77, and in Rothschild (1971).

2
ECONOMIC MODELS OF RENEWABLE RESOURCE HARVESTING

We come now to the fundamental issue of this book: how do economic forces affect the harvesting of renewable resources? In answering this question it is important to distinguish between *descriptive* and *normative* theories; that is, to separate the question of what *does* happen from what *should* happen. In this chapter we concentrate primarily on descriptive theory; normative theory is discussed in depth in Chapter 5.

2.1 THE OPEN-ACCESS FISHERY

The economic theory of the open-access or common-property fishery was developed by H. S. Gordon (1954). Although Gordon's model pertains specifically to fisheries, it can be applied with equal force to many other biological resource industries.

By definition, an *open-access resource* is one in which exploitation is completely uncontrolled: anyone can harvest the resource. Few present-day resource stocks satisfy this definition completely. Land-based resources are normally controlled at least in part by government or private owners. Most marine fisheries are subject to some form of international regulation. Frequently, however, resource control is quite minimal. Moreover, the *purpose* of establishing most regulatory agencies is to prevent the worst abuses of open-access exploitation. It is important, therefore, to define and to understand these abuses.

Gordon's model is a static or an equilibrium model based on the parabolic yield-effort curve described in Section 1.1 (in particular, see

Figure 1.4). [The Gordon model is sometimes referred to as the *Gordon–Schaefer* model, because the logistic growth model has been used extensively by fisheries biologist M. Schaefer (1957).] Recall that each point on this curve corresponds to the sustainable yield $Y(E)$, measured, for example, in terms of biomass, resulting from the application of a given level of fishing effort E. If we assume a constant *price* p per unit of harvested biomass, then the function

$$TR = pY(E)$$

represents the *total sustainable revenue* resulting from the effort E. This curve (shown in Figure 2.1) has the same parabolic shape as the yield-effort curve. Of course, the assumption of constant price is highly specialized; later in the book p becomes a variable.

Now we impose a total cost curve TC on the same figure, assuming that in the simplest case the costs of fishing are proportional to the effort expended:

$$TC = c \cdot E,$$

where c is constant. The difference between total sustained revenue TR and total cost TC is called the *sustainable economic rent* provided by the fishery resource at each given level of effort E:

$$\text{Sustainable economic rent} = TR - TC = pY(E) - cE. \qquad (2.1)$$

Gordon's principal result can now be stated as follows: *In the open-access fishery effort tends to reach an equilibrium* (bionomic equilibrium) *at the level $E = E_\infty$ at which total revenue TR equals total cost TC* (i.e., when the economic rent is completely "dissipated"). (The significance of the subscript ∞ here becomes apparent in Section 2.5.) This conclusion can be

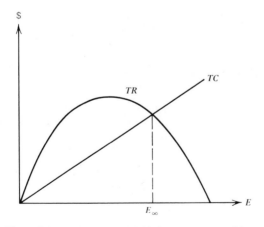

Figure 2.1 Gordon's model of the open-access fishery.

justified on the basis of two arguments:

1. No level of effort $E > E_\infty$ can be maintained indefinitely, for this would produce a situation in which the total costs of fishing would exceed the total revenues. At least some of the fishermen would lose money and would withdraw from the fishery, reducing the level of effort E.

2. No level of effort $E < E_\infty$ can be maintained indefinitely, because of the open-access condition: at such an effort level the fishermen would earn a profit, additional fishermen would be attracted to the fishery, and effort would increase.

Although this analysis is an oversimplification in several important respects, in summary it can be applied to explain many observable characteristics of the open-access exploitation of fisheries and other resource stocks.

Opportunity Cost

One oversimplification of the preceding analysis is its failure to define the term *cost*. For example, the condition that fishermen would leave the fishery when their costs exceeded their revenues must include the stipulation that the fishermen could earn higher incomes by working some other fishery (the case considered by Gordon) or by entering an entirely new occupational field.

In economics the term *opportunity cost* describes the cost of undertaking a particular activity, including the cost of *not* undertaking the most profitable alternative activity. A fisherman who is considering catching halibut, for example, must weigh the possibility of catching halibut against the possibility of catching salmon and against any other employment alternatives that are available. Elementary textbooks emphasize that in economic analysis, costs must always be understood in terms of opportunity costs.

It is most important to recognize the role of opportunity cost in Gordon's model of the open-access fishery. If $E > E_\infty$ (in an equilibrium situation) then opportunity costs exceed revenues and fishermen leave the fishery and seek other employment. Conversely, if $E < E_\infty$ then revenues exceed opportunity costs and effort tends to increase, because fishing is more profitable than other employment.

A related prediction of the Gordon theory is that the fishermen who remain in an open-access fishery tend to have the least number of alternative employment opportunities available (i.e., the lowest opportunity costs). In most cases, this seems to be true and can lead to significant social policy problems (see Section 3.4).

Economic Inefficiency

You may reasonably ask what (if anything) is wrong with a situation in which fishermen are earning their exact opportunity cost from fishing? Two possible responses to this question are:

1. The fishery resource, which is *capable* of producing positive economic rent, is actually producing zero rent, because an excessive level of effort is being utilized. Neither the fishermen, nor society at large are enjoying the benefits that could accrue if the fishery were under proper management. This situation may be described as *economic overfishing*.

2. The fishery may also suffer from *biological overfishing*, a situation in which the sustained biomass yield is less than the maximum sustainable yield MSY. In extreme cases biological productivity can be reduced to a near-zero level.

The important aspects of overfishing will be discussed in the following two sections.

Externalities

By definition an *externality* is a cost or benefit that is imposed on others as the result of a particular activity. External costs imposed on a large number of people are called *social costs*. The most familiar examples involve environmental pollution, and extensive literature concerning the economics of pollution has been produced. Although this material applies to certain aspects of this book, it is not dealt with directly here.

The open-access exploitation of common-property resources can also be considered a question of externalities. Thus the operation of one fishing vessel imposes costs on other vessels because it reduces the fish stock and thereby increases unit harvest costs for all vessels. External costs may also arise from crowding and from the interference of fishing gear. (On the other hand, external benefits can occur when one vessel locates a school of fish and other vessels "home in" on it.) The idea that open-access exploitation is a special problem of externalities is not pursued in this book, because this viewpoint does not seem to lend itself readily to analytic treatment.

Further Examples of Open-Access Resources

Natural resources other than fisheries are often subject to open-access exploitation. Wildlife populations are important examples of common-property resources for which conservation measures are often essential. Public grazing lands, particularly in arid areas, can become seriously depleted unless properly managed.

Oil fields are another well-known example of a resource area in which severe economic losses may result from uncontrolled exploitation. When several producers utilize a common pool of oil, unified control of drilling and extraction rates is essential to prevent wasteful competition (see Cummings, 1969). Similar considerations apply to groundwater supplies.

2.2 ECONOMIC OVERFISHING

Gordon's theory of the open-access fishery predicts an equilibrium in which economic rent is dissipated as effort expands to a level E_∞ at which revenues exactly equal opportunity costs. In terms of the logistic yield-effort model this equilibrium is determined by the equations

$$\frac{dx}{dt} = rx\left(1 - \frac{x}{K}\right) - Ex = 0,$$

$$TR - TC = pEx - cE = 0.$$

These equations can be readily solved for the equilibrium-effort level $E = E_\infty$:

$$E_\infty = r\left(1 - \frac{c}{pK}\right), \tag{2.2}$$

and for the corresponding stock level $x = x_\infty$:

$$x_\infty = \frac{c}{p}. \tag{2.3}$$

Since the equilibrium level of effort E_∞ is determined by both biological and economic parameters, Gordon's term *bionomic equilibrium* appropriately describes this situation.

If the biological parameters r and K are assumed given then the bionomic equilibrium effort level E_∞ becomes a function only of the *cost–price ratio c/p*. If fishing costs are sufficiently high relative to the price of fish, viz if

$$\frac{c}{K} > p,$$

as in curve TC_1 in Figure 2.2, the fishery will not be exploited at all. Thus many fish species do not support any commercial fishery, simply because their market value, even though it may be positive, does not offset the expense of catching the fish.

At somewhat higher price levels (or lower cost levels) the fishery becomes profitable and bionomic equilibrium is established at a level such as E_∞^2 in Figure 2.2. At this stage the effort is still below the level of maximum sustainable biological yield E_{MSY}, and biological overfishing

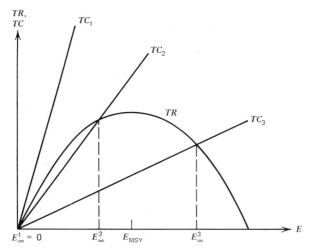

Figure 2.2 Bionomic equilibrium levels E_∞^1, E_∞^2, and E_∞^3 corresponding to progressively lower cost–price ratios c/p.

does not occur. But if the cost–price ratio becomes sufficiently low, equilibrium is established at a level such as $E_\infty^3 > E_{MSY}$, and biological overfishing occurs.

Notice that open-access exploitation in the Gordon model does not lead to the extinction of the fish population, provided the cost of fishing is positive. The degree to which this prediction remains valid in more general models is discussed in Section 2.3 and elsewhere in this book.

Now consider the case of an open-access fishery in bionomic equilibrium at some effort level $E_\infty > 0$ that yields zero economic rent. If effort can somehow be reduced, clearly positive levels of economic rent can be achieved. This is particularly evident when E_∞ exceeds E_{MSY}, because a reduction of effort has the double effect of increasing revenues *and* decreasing fishing costs. In any case the excess expenditure of effort characteristic of the open-access fishery results in economic inefficiency. (The *degree* of economic overfishing associated with the bionomic equilibrium of the open-access fishery is further delineated in Section 2.5.)

Of course, economic overfishing does not necessarily imply biological overfishing (see curve TC_2 in Figure 2.2). Gordon's analysis suggests, however, that as the price of fish rises or as technological innovations reduce the cost of fishing, the open-access fishery will eventually experience biological as well as economic overfishing. The history of many fisheries supports this prediction.

Knowing that the bionomic equilibrium level of effort E_∞ is inefficient, we can now ask (1) what is the optimum effort level, and (2) what is the optimum sustainable yield?

It is tempting to suggest (as Gordon and many subsequent authors have done) that the optimum level of fishing effort should be expected to maximize the sustainable economic rent $TR - TC$ (see Figure 2.3). However, this suggestion neglects an essential ingredient of the problem, namely the *dynamics* of both economic and biological processes. This is a crucial omission, the implications of which occupy our attention throughout most of this chapter and indeed of this book.

For example, imagine an open-access fishery in bionomic equilibrium with $E = E_\infty$, as in Figure 2.3. Now suppose that an agency is established to regulate the fishery to increase its sustained yield. To achieve this purpose it will be necessary to reduce the level of fishing effort. It is clear, however, that although a reduction of effort would *ultimately* lead to an increase in yield, its *immediate* effect would be to produce a decrease in yield.

To derive this result from our mathematical model, we note that

$$\text{yield} = Y = E \cdot x.$$

Even if effort changes discontinuously from E_∞ to some lower level $E < E_\infty$, the biomass level given by

$$\frac{dx}{dt} = F(x) - Ex$$

would increase in a continuous fashion. The resulting dynamics involving effort, population biomass, and yield are shown in Figure 2.4.

In biological conservation we cannot hope to "get something for nothing." If future yields are to be increased, current harvest levels must

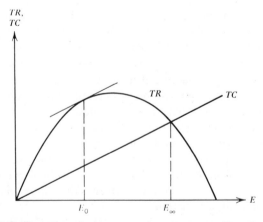

Figure 2.3 Maximization of sustainable economic rent occurs at $E = E_0$. This is a dubious optimum, however (see Section 2.4).

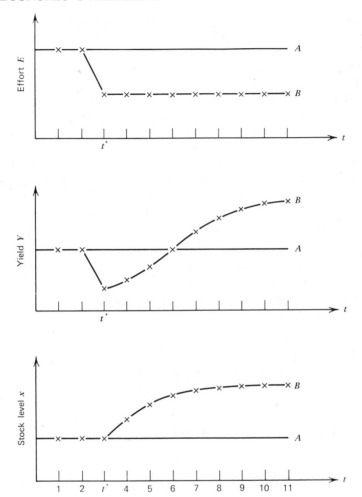

Figure 2.4 Effects of reducing fishing effort at time $t = t^*$. Curves B show result of the reduction; curves A correspond to no reduction in effort [Garrod (1973), p. 1982].

be reduced. The fundamental problem then becomes one of determining the optimal trade-off to be made between current and future harvests. This problem, which is the very essence of resource conservation, is an exceedingly difficult one, not from a mathematical viewpoint, perhaps, but certainly from a political and a philosophical viewpoint.

The standard device used to handle questions of intertemporal economic benefits is *time discounting*. Although there is considerable controversy as to the social justifiability of this concept (Solow, 1974), time discounting is normal practice in business management. If we are to develop a realistic descriptive theory of renewable resource exploitation, this procedure must be examined in detail (see Section 2.4).

It should be noted that the concept of maximizing sustained economic rent amounts to taking a particular and quite an extreme position regarding intertemporal trade-offs. Indeed, as is explained later, it amounts to setting the discount rate equal to zero. Consequently it is based on the underlying assumption that current sacrifices are immaterial if they ultimately lead to permanent increases in economic benefits. Until recently, the literature on fishery economics has largely failed to recognize this underlying assumption, although the necessity of reducing current harvests in order to rebuild an overfished stock has often been noted. [See Beverton and Holt (1957, pp. 396–404); Cushing (1968, p. 97).]

2.3 BIOLOGICAL OVERFISHING

In the previous section we saw that the unregulated open-access fishery tends to result in economic overfishing and to reach an equilibrium at which sustained economic benefits are in the ideal case reduced to zero. The historical development of many fisheries supports this theory, a particularly spectacular example being the Peruvian anchovy (*Engraulis ringens*) fishery (see Table 2.1).

TABLE 2.1. STATISTICAL HISTORY OF THE
PERUVIAN ANCHOVY FISHERY
(reference: Institut del Mar del Peru, 1974).

Year	Number of Boats	Number of Fishing Days	Catch (million tons)
1959	414	294	1.91
1960	667	279	2.93
1961	756	298	4.58
1962	1069	294	6.27
1963	1655	269	6.42
1964	1744	297	8.86
1965	1623	265	7.23
1966	1650	190	8.53
1967	1569	170	9.82
1968	1490	167	10.26
1969	1455	162	8.96
1970	1499	180	12.27
1971	1473	89	10.28
1972	1399	89	4.45
1973	1256	27	1.78

As a result of the rapidly increasing demand for fish meal, this fishery became the world's largest fishery during the 1960s. Its annual catch of approximately 10 million tons amounted to about 15% by weight of the total global catch of marine fisheries, including mammals and crustaceans. The capacity of its fishing fleets (total tonnage of vessels) increased steadily during the fishery's growth period. To prevent the destruction of the fishery, Peruvian authorities were forced to impose ever shorter fishing seasons. By 1972 the fishery's capacity was at least twice as high as the level necessary to achieve MSY.

Biologists warned that the Peruvian anchovy, although perhaps not actually overfished at the time, was in a precarious situation (Paulik, 1971). Every few years an incursion of warm tropical waters (called "El Niño" because of its occurrence during the Christmas season) could be anticipated to cause severe reductions in the anchovy population. If fishing were not carefully controlled, the fishery could collapse completely.

In 1973 these predictions came true. "El Niño" arrived but, against biologists' advice, the fishery was opened. By the end of March the anchovy had all but disappeared from Peruvian waters, precipitating a subsequent "anchovy crisis" (Idyll, 1973) that had a worldwide effect on food prices.

In view of serious attempts by Peruvian authorities to base the exploitation of the anchovy on scientific principles, the failure was particularly discouraging. Control of the situation seems to have been lost by 1973 at the political level. It is questionable whether the expert biological advice was matched economically. Fleet capacity was not properly controlled, and no "crop insurance" was established in preparation for the arrival of "El Niño."

However, the 1973 Peruvian anchovy crisis seems to have had the beneficial effect of emphasizing the need for economic control. Fleet capacity has reportedly been reduced substantially since 1973. Although catch statistics are no longer published (presumably to preclude speculation), the 1974 catch is believed to have exceeded four million tons. It is hoped that the Peruvian anchovy will recover to full productivity and that a viable, long-term industry will be possible in spite of occasional "failures."

Overfishing "Catastrophes"

We know from Chapter 1 that a fishery model based on the logistic growth curve does not predict rapid or sudden decreases in yield as effort levels increase. Particular care must therefore be taken when applying the

logistic model to extrapolate from existing data. If some form of depensation occurs at low population levels, the predictions of the logistic model
may become totally unreliable.

For example, consider the case of noncritical depensation (Figure 1.7)
that leads to a total sustainable revenue curve TR, as shown in Figure 2.5.
Suppose that the cost curve TC intersects the revenue curve TR on its
unstable branch (see Figure 2.5). To what extent is E_∞ a bionomic
equilibrium? Now consider a level of effort $E > E_\infty$. In Figure 2.5 two
equilibrium yields correspond to E: one at point P and one at point Q. If
the fishery were in equilibrium at P, TR would exceed TC and effort
would *expand* toward the critical level E^*. At this stage a rapid collapse
of the fish population could occur, causing the population level and the
corresponding yield to fall sharply toward the zero equilibrium level.

Once this occurs, we obtain $TC > TR$ and effort would then decrease
toward E_∞. But the fishery would then be operating near the zero branch
of the equilibrium yield curve, and the situation above would be reversed.
Consequently, effort could ultimately fall below E_∞ as a result of the
hysteresis effect (Chapter 1).

It is not clear whether the fishery would undergo a series of oscillations
or whether a rather tenuous equilibrium would eventually be established at
E_∞. We cannot determine this from our model unless we postulate some
particular reaction between the rate of increase or decrease of fishing
effort and the observed flow of economic rent. The simplest assumption

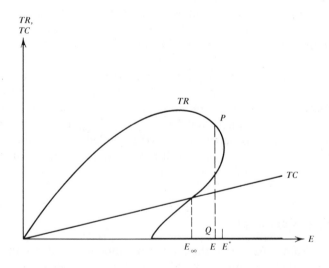

Figure 2.5   Bionomic equilibrium for noncritical depensation.

would be in the form

$$\frac{dE}{dt} = k(pxE - cE), \qquad (2.4)$$

where k is constant; that is, $\dot{E}$ is proportional to current economic rent. We would then have to study the system consisting of Eq. (2.4) and the biomass equation

$$\frac{dx}{dt} = F(x) - Ex. \qquad (2.5)$$

Equations (2.4) and (2.5) constitute a two-dimensional "dynamical system," to be examined in Chapter 6. There it is shown that for the logistic (Gordon) model the equilibrium point is (globally asymptotically) stable, but that for the case of depensation the equilibrium point is unstable, so that oscillations will occur.

Our analysis shows that the open-access fishery based on a depensatory growth curve does not lead to the extinction of the fish population, provided that the depensation is noncritical. This follows directly from Eq. (2.4), which implies that effort decreases whenever x falls below x_∞, where $px_\infty = c$. (It should be noted that the conclusion that open-access exploitation does not lead to extinction of the fish population except under critical depensation also depends on the assumed relationship between yield rate and the costs of fishing $Y = Ex$, with cost $\sim E$. These relationships imply that cost is proportional to Y/x, so that fishing costs approach infinity as $x \to 0$. Fishing therefore becomes infinitely unprofitable at low population levels. The possibility that extinction can occur at finite cost is discussed in section 2.8.)

In a critical depensation model, however, biological extinction is possible under open-access exploitation. A sufficient condition for this is

$$x_\infty < K_0,$$

where K_0 is the minimum viable population size (see Figure 1.8).

2.4 OPTIMAL FISHERY MANAGEMENT

Now imagine an open-access fishery that has achieved bionomic equilibrium with severe biological overfishing (point A in Figure 2.6). Suppose that a management agency is established to improve the operation of the fishery. What management objectives should be set? Once established, how can these objectives be achieved?

Clearly the level of fishing effort ought to be reduced somehow, and reducing effort has invariably been the objective of fishery regulatory

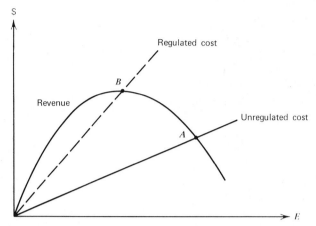

Figure 2.6   Fishery regulation by increasing the costs of fishing.

agencies. In most cases the stated aim of fishery regulation has been to achieve MSY by such methods as restricting the length of the fishing season, setting total catch limitations, and regulating the type of fishing gear used.

Restrictions of this kind generally increase the costs of fishing. For example, gear regulations usually reduce the efficiency of the fisherman. Closed seasons also decrease efficiency, because fishermen and equipment must remain idle during the closures. Total catch limitations force fishermen to compete vigorously to catch enough fish before the total limit is reached.

It is obvious, however, that increasing the costs of fishing can be an effective way to reduce effort and to increase the sustainable yield. The cost curve may be adjusted to meet the revenue curve at the MSY point, which then becomes the point of bionomic equilibrium (see point B in Figure 2.6). But economists studying the fishing industry have severely criticized such regulatory methods, stating that they achieve a primarily biologically oriented goal by introducing deliberate economic inefficiency. Neither the goal nor the means to achieve it are closely related to human needs. (Of course an increased supply of fish *at a lower price* would produce beneficial effects. Here, however, we are assuming a constant price independent of supply; the more general question of finite demand elasticity is examined in Chapter 5.)

Let us defer the question of choosing an optimal method of controlling effort (see Section 4.6) until we consider the problem of determining the optimal effort level. For this purpose we assume that methods of control can be found that do not increase the costs of fishing. We therefore assume fishing costs to be fixed.

The Sole Owner

In addressing the problem of optimal fishery management it is convenient to adopt the fiction of a "sole owner," who may be imagined as either a private firm or a government agency that owns complete rights to the exploitation of the given fish population. We adopt certain economic assumptions that imply that a private owner adopts the same objective as a public social manager attempting to maximize social welfare. In this way we sidestep the issue of descriptive versus normative theories by assuming that these theories are equivalent for the case of sole ownership. This assumption is generally untenable in several important ways, to be examined in detail in Chapter 5.

The Production Function

As before, we denote the natural growth rate of a given fish population (or other biological resource) by $F(x)$ and the rate of harvest by $h(t)$:

$$\frac{dx}{dt} = F(x) - h(t), \quad t \geq 0. \tag{2.6}$$

The harvest rate $h(t)$ is assumed to be determined by two quantities: the current size of the stock $x = x(t)$ and the rate of the harvesting effort $E = E(t)$:

$$h = Q(E, x). \tag{2.7}$$

The function $Q(E, x)$, which relates the "factors of production" E and x to the rate of production $h(t)$ is termed the *production function* for the given resource industry.

For econometric purposes it is often convenient to hypothesize that $Q(E, x)$ is in the form

$$Q(E, x) = aE^{\alpha}x^{\beta} \tag{2.8}$$

where a, α, and β are positive constants. (The form of Eq. (2.8) is sometimes called a "log-linear" relationship, because $\log Q$ depends linearly on $\log E$ and $\log x$. Thus techniques of linear regression can be used to estimate the coefficients a, α, β.) In this chapter we restrict Eq. (2.8) in one direction but generalize it in another. We set $\alpha = 1$, but we replace the function ax^{β} with an arbitrary, nondecreasing function $G(x)$:

$$h = Q(E, x) = G(x) \cdot E. \tag{2.9}$$

The reasons for this choice are primarily mathematical. Linearity in E will lead to a linear optimization model, which is much easier to handle than a nonlinear model. On the other hand, it does not complicate the theory if x^β is replaced by an arbitrary, nondecreasing function $G(x)$, which may often be more realistic. The assumption that the coefficient $G(x)$ is nondecreasing is clearly desirable, because it implies that the rate of harvest corresponding to a given effort E cannot decrease if the population level increases. Further discussion of the production function in fisheries can be found in Chapter 7.

The Objective Functional

Now assume that the price p of the harvested resource is a fixed constant; furthermore assume that the cost c of a unit of effort is also constant. Then the net economic revenue produced by an input of effort $E \, \Delta t$ is given by

$$
\begin{aligned}
R \, \Delta t = R(x, E) \, \Delta t &= [ph - cE] \, \Delta t \\
&= [pG(x) - c]E \, \Delta t \\
&= [p - c(x)]h \, \Delta t,
\end{aligned} \tag{2.10}
$$

where

$$
c(x) = \frac{c}{G(x)}.
$$

Since a unit harvest ($h \, \Delta t = 1$) incurs a cost equal to

$$
cE \, \Delta t = \frac{c}{G(x)} \, h \, \Delta t = c(x)h \, \Delta t = c(x),
$$

we see that $c(x)$ equals the *unit harvesting cost when the population level is x*. Since $G(x)$ is nondecreasing, the unit harvesting cost is a nonincreasing function of x.

Now assume that the sole owner's objective is the maximization of the *total discounted net revenues* derived from exploitation of the resource. If $\delta > 0$ is a constant denoting the (continuous) rate of discount, this objective may be expressed as

$$
\begin{aligned}
PV &= \int_0^\infty e^{-\delta t} R(x, E) \, dt \\
&= \int_0^\infty e^{-\delta t} \{p - c[x(t)]\}h(t) \, dt,
\end{aligned} \tag{2.11}
$$

where PV is present value. The economic significance of this expression and of time discounting in general are discussed in greater depth in Chapter 3.

According to our assumptions, then, the sole owner attempts to utilize a harvest rate $h = h(t)$ that leads to the largest possible value for the expression in Eq. (2.11). Note that Eq. (2.11) also depends on the population level $x(t)$, which is itself related to the harvest rate, according to Eq. (2.6). The variables $x(t)$ and $h(t)$ must also satisfy the constraints

$$x(t) \geq 0 \tag{2.12}$$

and

$$h(t) \geq 0. \tag{2.13}$$

In modern terminology, maximizing the expression in Eq. (2.11) subject to these conditions is a problem in *optimal control theory*. Fortunately it is a particularly simple problem, one that we can solve rigorously by using elementary methods. In later portions of the book, however, we utilize the more difficult mathematical techniques of optimal control theory to be described in Chapter 4.

2.5 THE OPTIMAL HARVEST POLICY

Several mathematical techniques can be used to determine the harvest policy $h(t)$ that maximizes Eq. (2.11). Here we use a formal approach. In Section 2.7 we introduce a completely elementary but rigorous method for solving optimization problems of this kind.

Substitution $h(t) = F(x) - \dot{x}$ into Eq. (2.11), we obtain

$$PV = \int_0^\infty e^{-\delta t}[p - c(x)][F(x) - \dot{x}]\, dt. \tag{2.14}$$

This integral has the form

$$\int \phi(t, x, \dot{x})\, dt,$$

and we may therefore apply the classical Euler necessary condition for a maximum:

$$\frac{\partial \phi}{\partial x} = \frac{d}{dt}\frac{\partial \phi}{\partial \dot{x}}. \tag{2.15}$$

[It should be emphasized that the use of the Euler equation at this point can be completely avoided (see Section 2.7). It is only used here to obtain a quick (but partial) solution to the problem.] In our case we have

$$\frac{\partial \phi}{\partial x} = \frac{\partial}{\partial x}\{e^{-\delta t}[p - c(x)][F(x) - \dot{x}]\}$$

$$= e^{-\delta t}\{-c'(x)[F(x) - \dot{x}] + [p - c(x)]F'(x)\}$$

$$\frac{d}{dt}\frac{\partial \phi}{\partial \dot{x}} = \frac{d}{dt}\frac{\partial}{\partial \dot{x}}\{\cdots\}$$

$$= \frac{d}{dt}\{-e^{-\delta t}[p - c(x)]\}$$

$$= e^{-\delta t}\{\delta[p - c(x)] + c'(x)\dot{x}\}.$$

Equating these expressions and simplifying, we obtain

$$-c'(x)F(x) + [p - c(x)]F'(x) = \delta[p - c(x)],$$

or

$$F'(x) - \frac{c'(x)F(x)}{p - c(x)} = \delta. \qquad (2.16)$$

Equation (2.16) is derived formally; this equation reappears throughout this chapter and the remainder of the book. Note that Eq. (2.16) is an implicit equation for the population x. Let us assume for now that Eq. (2.16) has a unique solution $x = x^*$ that we call the *optimal (equilibrium) population level*.

The following questions arise:

1. Under what circumstances is x^* an optimizing solution to our original problem?

2. What is the optimal method of approaching x^* from some initial population level $x(0) \neq x^*$?

3. How is x^* dependent on the parameters of the problem?

4. What is the economic significance of Eq. (2.16)?

Questions 1 and 2 can be answered easily, although some additional mathematical arguments are required to justify these answers. Question 3 is also fairly simple (see the following discussion). Question 4 is answered in Section 3.2.

Optimal Equilibrium and Approach

We continue to assume that x^* is the unique solution to Eq. (2.16). Given an initial population $x(0)$, the optimal harvest policy may then be described simply as follows: utilize the harvest rate $h^*(t)$ that drives the population level $x = x(t)$ toward x^* as rapidly as possible. If h_{max} denotes the maximum feasible harvest rate, we then have

$$h^*(t) = \begin{cases} h_{max} & \text{whenever} \quad x > x^* \\ F(x^*) & \text{whenever} \quad x = x^* \\ 0 & \text{whenever} \quad x < x^*. \end{cases} \tag{2.17}$$

The corresponding optimal population level $x = x(t)$ is shown in Figure 2.7. If $x(0)$ is at point A, then maximum-rate harvesting reduces x to x^*; if $x(0)$ is at point B, the fishery is closed ($h = 0$) until x increases to x^*.

This simple control policy may appear somewhat unrealistic. For example, closing down a fishery completely seems to be an extreme action, particularly if the closure is expected to be lengthy. Later in Chapter 5 it becomes apparent that the extreme form of the optimal control policy results from the *linearity* of our optimization model, which implies that no particular benefit or nonbenefit is associated with the rate of harvest but that only the amount of the catch is of importance.

It is interesting to observe that the optimal harvest policy, even in its present dynamic setting, results in an equilibrium yield rather than in some form of "pulse fishing" whereby the stock is fished heavily and then allowed to build up again. This is also a consequence of linearity to be discussed in Chapter 5.

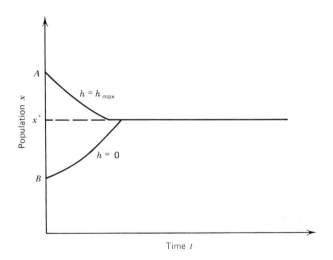

Figure 2.7 The optimal population $x = x(t)$.

The Effect of Discounting

How does the introduction of time discounting affect the optimal yield $Y^* = F(x^*)$? To discern this, we rewrite Eq. (2.16) in the form

$$F'(x)[p - c(x)] - c'(x)F(x) = \delta[p - c(x)],$$

or

$$\frac{d}{dx}\{[p - c(x)]F(x)\} = \delta[p - c(x)]. \qquad (2.18)$$

Notice that because sustained yield implies $h(t) \equiv F(x)$, the expression

$$\rho(x) = [p - c(x)]F(x)$$

represents the *sustainable economic rent* at population level x. Thus Eq. (2.18) asserts that

$$\frac{d\rho}{dx} = \delta[p - c(x)]. \qquad (2.19)$$

This equation has an interesting "marginal" interpretation. Let us consider the equilibrium population level x as a decision variable. Then a marginal decrease in x (i.e., $\Delta x = 1$) produces an immediate net revenue equal to $[p - c(x)]\Delta x = p - c(x)$. It also causes a decrease in the sustained rent equal to $\Delta\rho \approx \rho'(x)\Delta x = \rho'(x)$. The present value of this loss of rent is given by

$$\int_0^\infty e^{-\delta t}\rho'(x)\, dx = \frac{1}{\delta}\frac{d\rho}{dx}.$$

Equation (2.19) therefore asserts that when x is optimal, *the marginal immediate gain must equal the present value of the marginal future loss.* (See Section 3.2 for further discussion of the economic significance of this condition.)

Zero Discount Rate

Two special cases are of particular interest. First, if $\delta = 0$ we see that Eq. (2.19) implies the maximization of sustainable economic rent $\rho(x)$. In other words, when future revenues are not discounted relative to current revenues, the optimal harvest policy results in the maximization of sustainable economic yield, or rent. [Note that the case $\delta = 0$ must be treated as a limiting case, because the integral in Eq. (2.11) diverges for $\delta = 0$. Alternatively, we may pass to a finite time-horizon problem (see Section 2.7).] In this case the necessary sacrifice of current benefits does not affect the optimization decision, because the ultimate gain—no matter how small and no matter how long it is delayed—lasts forever and, not being discounted, outweighs any current loss.

Maximization of sustained economic rent, which (as noted earlier) is frequently recommended as an optimal management objective, is thus viewed as a particularly extreme proposal. For this objective is based on the supposition that society is, or should be, willing to make arbitrary current sacrifices to benefit future generations, the only provision being that total long-term economic benefits must be increased thereby.

Infinite Discount Rate

At the opposite end of the spectrum is the case $\delta \to +\infty$. It is safe to assume that $d\rho/dx$ is bounded, so Eq. (2.19) implies that as $\delta \to +\infty$ we have $x \to x_\infty$ where

$$p = c(x_\infty). \tag{2.20}$$

It therefore follows that

$$\rho(x_\infty) = F(x_\infty)[p - c(x_\infty)] = 0. \tag{2.21}$$

In other words, the sole owner who adopts an infinite discount rate will exploit the fishery at precisely the same level of exploitation that would be reached under open-access conditions.

This result is not unexpected. For the effect of an infinite rate of discount is simply to set a zero value on future revenues, which is precisely what fishermen in an open-access fishery are forced to do.

Our formulation of the optimal harvesting problem thus contains the two special limiting cases of the "most conservative" management policy (maximization of sustained economic rent) and the "least conservative" policy (dissipation of economic rent, as in the open-access fishery). The optimal solution for finite, positive discount rates always lies between these two extremes, as the following section indicates.

The Case $0 < \delta < +\infty$

How does the optimal population level x^*, as determined by Eq. (2.19), depend on the rate of discount δ? (The dependence of x^* on price and cost parameters is described later in this chapter.) Let us suppose that the left side of the equation $d\rho/dx$ is a decreasing function of x; this assumption (which obviously implies that the equation has a unique solution $x = x^*$) is valid for the Schaefer model, which we discuss in the next section. (The assumption is not valid in general; for example, it is invalid when $F(x)$ is depensatory. Such cases are discussed in Section

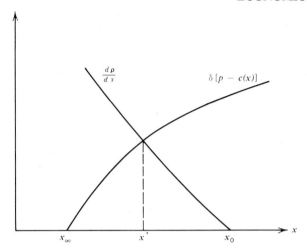

Figure 2.8   Graphical determination of the optimum population level x^*.

2.8.) We then conclude that the optimum population level x^* (see Figure 2.8) lies between x_∞ and x_0, where $\rho'(x_0) = 0$, and moreover that x^* decreases monotonically from x_0 to x_∞ as δ increases from 0 to ∞. (To visualize this, simply observe the effect in Figure 2.8 of increasing δ.)

Our results (under the assumption of uniqueness adopted above) may be summarized as follows:

1. There exists an optimal equilibrium population level $x = x^*$ that is determined by the fundamental Eq. (2.16) or, equivalently, by Eq. (2.19).

2. The optimal population level x^* is a function of the discount rate δ and of the other economic and biological parameters of the problem.

3. The value of x^* lies between the rent-maximizing level x_0 (corresponding to $\delta = 0$) and the rent-dissipating level x_∞ (corresponding to $\delta = \infty$). This value reflects the inevitable compromise between the desire for current versus future revenues.

4. When the initial population level $x(0) \neq x^*$, the optimal harvest policy is a "most-rapid approach" policy, driving the population x to the optimal level x^* as rapidly as possible.

The remainder of Chapter 2 is devoted to a further analysis of this solution. Perhaps it should be emphasized that the attractive simplicity of the solution is critically dependent on certain assumptions that are built into our present model. Subsequent chapters investigate the effects of altering these assumptions in realistically important ways.

2.6 EXAMPLES BASED ON THE SCHAEFER MODEL

For the Schaefer model (Schaefer, 1957), we have

$$F(x) = rx\left(1 - \frac{x}{K}\right), \quad \text{and} \quad c(x) = \frac{c}{x}. \tag{2.22}$$

Substituting these expressions into our basic formula [Eq. (2.19)], we obtain

$$\frac{d}{dx}\left[\left(p - \frac{c}{x}\right)rx\left(1 - \frac{x}{K}\right)\right] = \delta\left(p - \frac{c}{x}\right).$$

After differentiation this becomes a quadratic equation in x, the positive solution of which is given by

$$x^* = \frac{K}{4}\left[\left(\frac{c}{pK} + 1 - \frac{\delta}{r}\right) + \sqrt{\left(\frac{c}{pK} + 1 - \frac{\delta}{r}\right)^2 + \frac{8c\delta}{pKr}}\right]. \tag{2.23}$$

To simplify this expression we introduce the following dimensionless quantities:

$$z^* = \frac{x^*}{K}, \tag{2.24}$$

$$z_\infty = \frac{x_\infty}{K} = \frac{c}{pK}, \tag{2.25}$$

$$\gamma = \frac{\delta}{r}. \tag{2.26}$$

Thus z^* represents the biomass as a proportion of the environmental capacity K, and z_∞ is the corresponding open-access, rent-dissipating biomass level. (Note that z_∞ is determined by the *ratio* c/p of fishing costs to price.) Finally γ is the ratio of the discount rate to the intrinsic growth rate of the population; it seems appropriate to refer to γ as the *bionomic growth ratio*.

With the above substitutions, Eq. (2.23) now becomes

$$z^* = \tfrac{1}{4}[1 + z_\infty - \gamma + \sqrt{(1 + z_\infty - \gamma)^2 + 8z_\infty\gamma}]. \tag{2.27}$$

Values of z^* as a function of the parameters z_∞ and γ are shown in Table 2.2 and are displayed in Figure 2.9. These values should be examined in light of the general theory presented in Section 2.5. Zero discounting ($\gamma = 0$), for example, produces an optimal population level

**TABLE 2.2. OPTIMAL POPULATION LEVELS (NOR-
MALIZED) z^* AS A FUNCTION OF THE PARAMETERS
z_∞ (RENT-DISSIPATION LEVEL) AND γ (BIONOMIC
GROWTH RATIO). [See Eq. (2.27)].**

γ \ z_∞	0	0.1	0.3	0.5	0.7	0.9
0	0.50	0.55	0.65	0.75	0.85	0.95
0.10	0.45	0.51	0.62	0.73	0.84	0.95
0.25	0.38	0.45	0.59	0.71	0.83	0.94
0.50	0.25	0.37	0.54	0.68	0.81	0.94
1.0	0	0.25	0.47	0.64	0.79	0.93
2.0	0	0.16	0.40	0.59	0.77	0.92
3.0	0	0.14	0.37	0.57	0.75	0.92
5.0	0	0.12	0.34	0.54	0.73	0.91

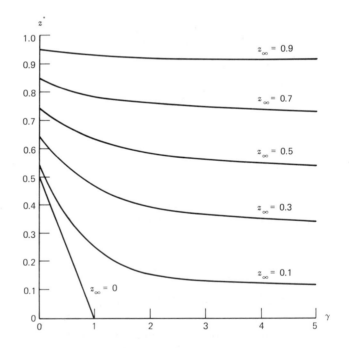

Figure 2.9 Optimal population levels (normalized) z^* as a function of the bionomic growth ratio γ for various values of z_∞.

$z^* = \frac{1}{2} + \frac{1}{2}z_\infty$ that is always greater than the MSY level $z_{MSY} = 0.50$. Positive discount rates lead to progressively decreasing optimal population levels z^* that approach z_∞ as $\gamma = \delta/r \to +\infty$.

An extreme case arises when $z_\infty = 0$; that is, when the costs of fishing are zero. [More precisely, this case arises whenever the costs of fishing are independent of the biomass level (see Section 2.8).] In this case Eq. (2.27) reduces to

$$z^* = \frac{1}{2} - \frac{1}{2}\gamma.$$

Thus z^* in this case is always less than z_{MSY}, and whenever the bionomic growth ratio $\gamma > 1$, the optimal equilibrium population level $z^* = 0$. An optimal harvest policy in such a situation leads to the most rapid possible extinction of the resource population.

This outcome can be easily explained on intuitive grounds. When $\gamma > 1$, the discount rate δ exceeds the intrinsic growth rate r of the resource. If it is assumed that revenues obtained from fishing can be transferred to alternative investment opportunities yielding a continuous rate of interest $\geq \delta$, then it clearly pays the sole owner to "cash out" his fishery investment and make the alternative investment. In other words, the "capital" that the sole owner has invested in the fish stock itself possesses an opportunity cost in terms of the revenue foregone by not transferring its value to the most profitable alternative investment opportunity.

The capital-theoretic aspects of resource management are discussed more fully in Chapter 3.

Next we apply the Schaefer model to two specific populations, the Pacific halibut (*Hippoglossus hippoglossus*) and the Antarctic fin whale (*Balaenoptera physalus*). (Schaefer has also applied his model to the Pacific yellowfin tuna fishery (Schaefer, 1967) (see Exercises 7 and 8).) It must be emphasized that the Schaefer model is a severe oversimplification of the biological and economic characteristics of these resources. The following results are presented mainly to illustrate the theory; they have limited operational significance. The halibut and whale populations are chosen because of their contrasting growth rates: high for halibut; low for whales.

The Pacific Halibut Fishery

The Pacific halibut is a large demersal (bottom-feeding) fish that ranges over an extensive area of the North Pacific. Halibut is a highly regarded table fish, prized for its flavor and texture, and the halibut fishery is second only to the Pacific salmon fisheries in commercial importance. Brought under the control of the Pacific Halibut Commission in 1924, the overfished halibut stocks were steadily rebuilt to levels of approximate maximum sustainable yield by the 1950s. The recent development of Russian and Japanese trawl fisheries in the North Pacific, however, appears to have seriously depleted the halibut stocks, and current landings are again far below MSY levels.

The problems of multispecies trawl fisheries are described in Chapter 9. For the present we apply the Schaefer model to a brief study of the halibut population occupying "area 2" in the classification of the Commission. A recent study by H. S. Mohring (1973) suggests the following parameters for this population:

$$r = 0.71, \quad K = 80.5 \times 10^6 \text{ kg.}$$

As a value for the bionomic parameter x_∞ we take the estimated biomass in 1930, the year for which Mohring's analysis begins. Prior to this time only minimal control of the fishery had been achieved, so we may assume that bionomic, open-access equilibrium had been reached by 1930. Mohring's figure is:

$$x_\infty = 17.5 \times 10^6 \text{ kg.}$$

The optimal biomass levels x^* and the corresponding annual sustained yields $Q^* = F(x^*)$ are given in Table 2.3.

TABLE 2.3. PACIFIC HALIBUT ("AREA 2"): OPTIMAL BIOMASS LEVELS (x^*) AND OPTIMAL ANNUAL SUSTAINED YIELDS (Q^*).

Discount Rate δ (%)	Optimal Population Level x^* ($\times 10^6$ kg)	Optimal Annual Yield Q^* ($\times 10^6$ kg)
0	49.0	13.6
5	47.2	13.9
10	45.5	14.1
15	43.9	14.2
20	42.3	14.25
25	40.9	14.3
30	39.6	14.3
40	37.0	14.2
50	34.9	14.0
100	27.9	12.9
$+\infty$	17.5	9.7

The Antarctic Fin Whale

Until the 1950s, the Antarctic whaling industry concentrated primarily on the population of blue whales (*Balaenoptera musculus*). As this species became depleted, however, exploitation extended to the smaller fin whales (*B. physalus*) and eventually to even smaller species. (The capture of blue whales has been prohibited since 1965 by the International Whaling Commission. By 1965 probably fewer than 5000 blue whales remained out of an original population estimated at approximately 150,000). Antarctic fin whales belong to the family of baleen whales; they are biologically distinct from the toothed whales, such as, for example, the sperm whale. The prime commercial value of baleen whales stems from the large amounts of edible oil they contain, although whale meat is also of some importance, particularly in the Japanese market. An average blue-whale carcass is worth about $10,000 to whalers; an average fin-whale carcass is worth about $5000.

We apply the Schaefer model to estimate optimal exploitation policies for the Antarctic fin-whale population. The dynamics of this population have been analyzed in some detail (Allen, 1973). It appears that the fin-whale growth curve (to the extent to which a simple model is applicable) is nonsymmetric, being strongly skewed to the right. For purposes of illustration, however, we continue to assume a symmetric (logistic) function $F(x)$ (see Exercise 1 at the end of this chapter). We use the following parameters:

$$r = 0.08, \quad K = 400,000 \text{ whales.}$$

(Note that the variables x and K refer to number of whales, rather than whale population biomass.)

The present fin-whale population level (1976) is approximately 70,000, with a sustainable yield $F(70,000) = 4620$. In an attempt to rebuild the fin-whale stocks, the International Whaling Commission has recently adopted progressively smaller annual quotas. It is difficult to estimate how far exploitation might proceed in the absence of quotas, but we somewhat arbitrarily assume that

$$x_\infty = 40,000 \text{ whales}$$

is the level at which open-access bionomic equilibrium would be reached. The results are given in Table 2.4. [It is unrealistic to treat the fin-whale fishery independently of other fisheries that have been and continue to be exploited concurrently. See Chapter 9 for a further discussion of this point.]

TABLE 2.4. ANTARCTIC FIN WHALE:
OPTIMAL   POPULATION   LEVELS
(x^*) AND OPTIMAL ANNUAL SUS-
TAINED YIELDS (Q^*).

Discount Rate δ (%)	Optimal Population Level x^*	Optimal Annual Yield Q^*
0	220,000	7920
1	200,000	8000
3	163,000	7726
5	133,000	7094
10	86,000	5406
15	67,000	4485
20	59,000	4024
$+\infty$	40,000	2880

Discussion

Tables 2.3 and 2.4 indicate how the intrinsic growth rate of the population affects the sensitivity of the optimal solution to the discount rate. In the case of the halibut species, biological overfishing does not become optimal unless δ exceeds 27% (i.e., unless the annual rate $i = e^{\delta} - 1$ exceeds 31%). The optimal sustained yield Q^* is particularly insensitive to the discount rate. [It should be noted that the Schaefer model disregards the age structure of the population. This may be a serious omission for slow-growing species such as halibut, and our conclusion that discounting is relatively unimportant for the halibut fishery is correspondingly suspect (see Chapter 8).] Clearly an increase in the price–cost ratio (i.e., a decrease in x_∞) would increase the sensitivity of both x^* and Q^* to δ.

The situation is quite different for the fin-whale population. A discount rate of 20% leads to a large reduction in the stock level and reduces sustained yield to just over one-half of MSY. This discounting effect may be at least partially responsible for the whalers' reluctance to agree to more stringent conservation measures. If we assume that the current population of 70,000 fin whales represents the whalers' estimate of the optimal population level, then we can conclude that the effective discount rate is 14% (annual equivalent $i = 15\%$). This is probably a low figure for a discount rate in such a risky resource industry as pelagic whaling.

Before concluding this section it should be repeated that this analysis and its applications are highly oversimplified. Many arguments—indeed, many different *kinds* of arguments—seem to suggest that optimal exploitation policies may be considerably more conservative than our present model indicates. To mention just one example, the effects of increasing the level of exploitation are almost always increasingly uncertain. For example the Schaefer model does not consider the possibility of depensation, which can result in unexpected collapses of exploited populations. Problems of this kind are examined throughout the book.

2.7 LINEAR VARIATIONAL PROBLEMS

The solution given in Section 2.5 to the optimal harvest problem depends on a formal (and actually invalid) application of the Euler condition from the classical calculus of variations. We now provide a mathematically rigorous justification of this solution, based on completely elementary methods that do not require the use of the calculus of variations. This fortunate circumstance is a consequence of the linearity of our optimization problem with respect to the control variable $h(t)$. Such elementary methods cannot be used to solve nonlinear problems (to be discussed in Chapter 5).

However, our method does cover various important extensions of the original problem in which the underlying parameters (r, δ, K, p, c) are replaced by given functions of time. These nonautonomous linear models are discussed in Chapter 3.

A Linear Control Problem

We now consider a general, one-dimensional, linear *control problem* of the form

$$\text{maximize} \left\{ \int_{t_0}^{t_1} [f_0(t, x(t)) + g_0(t, x(t))u(t)] \, dt \right\}, \qquad (2.28)$$

where $x(t)$ satisfies the *state equation*

$$\frac{dx}{dt} = f_1(t, x(t)) + g_1(t, x(t))u(t). \qquad (2.29)$$

(The terminology introduced here coincides with contemporary usage in optimal control theory, as discussed in Chapter 4.) The class U of *admissible* controls $u = u(t)$ consists of all piecewise-continuous functions $u(t)$, $t_0 \le t \le t_1$, satisfying the following given constraints [i.e., $u(t)$ is continuous with the possible exception of finitely many values of t where $u(t)$ may have simple jump discontinuities; more generally, we may allow arbitrarily bounded measurable controls $u(t)$, but this degree of generality is not necessary here]:

$$u_m \leq u(t) \leq u_M, \tag{2.30}$$

where u_m and u_M are given constants or, more generally, given functions of x and t.

The initial and terminal times t_0, t_1 are fixed and finite (although this requirement is relaxed later), and the *state variable* $x(t)$ assumes given initial and terminal values:

$$x(t_0) = x_0, \quad x(t_1) = x_1. \tag{2.31}$$

The given coefficient functions f_0, g_0 in the *objective functional* given by Eq. (2.28) and f_1, g_1 in the state equation [Eq. (2.29)] are assumed to be continuously differentiable. Thus each given control $u(t)$, $t_0 \leq t \leq t_1$, gives rise to a unique *response* $x(t)$ satisfying Eq. (2.29) in addition to the given initial condition $x(t_0) = x_0$. The control $u(t)$ is called *feasible* if it satisfies Eq. (2.30) and if the response also satisfies $x(t_1) = x_1$. We assume that the class of feasible controls is nonempty, for otherwise the problem is trivial.

Now assume that $g_1(t, x) \neq 0$. We can then eliminate $u(t)$ from our problem by substituting Eq. (2.29) into Eq. (2.28). The resulting problem has the form

$$\operatorname*{maximize}_{x \in X} \left\{ \int_{t_0}^{t_1} [G(t, x) + H(t, x)\dot{x}] \, dt \right\}, \tag{2.32}$$

where X is the class of piecewise-smooth functions $x(t)$ satisfying the conditions in Eqs. (2.31) and the constraints derived from the control constraints in Eq. (2.30), which take the form

$$A(x, t) \leq \dot{x}(t) \leq B(x, t). \tag{2.33}$$

Except for these constraints, this is a special case of the "simplest" problem of the calculus of variations: $\max \int f(t, x, \dot{x}) \, dt$. However, because $f_{\dot{x}\dot{x}} \equiv 0$, the classical theory cannot be applied. Indeed, it can be easily verified that in the absence of the constraints given by Eq. (2.33), the variational problem [Eq. (2.32)] cannot be solved as stated. For this reason, linear variational problems are seldom discussed in classical treatments of the subject.

Singular Solutions

If we naively attempt to apply classical methods to the variational problem [Eq. (2.32)]; that is, if we consider the Euler equation (here we are only using the Euler equation as a heuristic or mnemonic device; the following proof makes no use of the Euler equation)

$$\frac{\partial f}{\partial x} = \frac{d}{dt}\frac{\partial f}{\partial \dot{x}},$$

then, since

$$\frac{\partial f}{\partial x} = \frac{\partial G}{\partial x} + \frac{\partial H}{\partial x}\dot{x}$$

$$\frac{d}{dt}\frac{\partial f}{\partial \dot{x}} = \frac{d}{dt}(H) = \frac{\partial H}{\partial t} + \frac{\partial H}{\partial x}\dot{x},$$

we obtain the equation

$$\frac{\partial G}{\partial x} = \frac{\partial H}{\partial t}. \tag{2.34}$$

This equation establishes an implicit relationship between x and t. Any piecewise-smooth function $x^*(t)$ satisfying Eq. (2.34) is called a *singular solution* to our problem.

Now assume until further notice that Eq. (2.34) determines a unique singular path $x = x^*(t)$, $t_0 \le t \le t_1$. Obviously $x^*(t)$ cannot be the desired optimal solution unless by mere chance $x^*(t_0) = x_0$ and $x^*(t_1) = x_1$. However, $x^*(t)$ is the next best thing. More precisely, the optimal solution $x(t)$ is as close to $x^*(t)$ as possible, subject to the given boundary conditions in Eqs. (2.31) and to the constraints on $\dot{x}$ in Eq. (2.33). Stating this exactly is somewhat awkward.

THEOREM

Suppose that

$$\frac{\partial G}{\partial x} \gtreqless \frac{\partial H}{\partial t} \quad \text{whenever} \quad x \lesseqgtr x^*(t), \tag{2.35}$$

where $x^*(t)$ is the unique solution of Eq. (2.34). Then the optimal solution to the problem presented in Eqs. (2.32) and (2.33) is the *closest possible trajectory* $x(t)$, defined as follows [see Figure 2.10a]:

1. If $x_0 > x^*(t_0)$, then the optimal path $x(t)$ uses the fastest possible rate of decrease; that is:

$$\dot{x} = A(x, t), \quad x(t_0) = x_0$$

until the singular path $x^*(t)$ is reached. Conversely if $x_0 < x^*(t_0)$, then $x(t)$ uses the fastest possible rate of increase

$$\dot{x} = B(x, t), \quad x(t_0) = x_0$$

until $x^*(t)$ is reached. Let t_a denote the time at which $x(t)$ reaches $x^*(t)$.

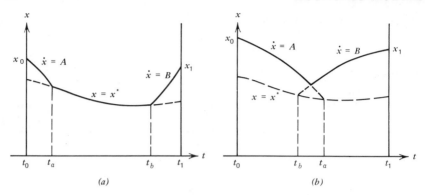

Figure 2.10 Optimal trajectories for the linear control problem.

2. Similarly if $x_1 \neq x^*(t_1)$, then $x(t)$ is the solution of the appropriate constraint equation for $t_b \leq t \leq t_1$.

3. Finally $x(t) = x^*(t)$ for $t_a < t < t_b$. An obvious modification of this description is required when $t_b < t_a$ [see Figure 2.10b].

If the inequalities given by Eqs. (2.35) are reversed, then the solution $x(t)$ described in the theorem is a minimizing rather than a maximizing solution.

Proof. The proof is based on a simple application of Green's theorem in the plane, which depends on the possibility of writing the objective functional given by Eq. (2.32) as a line integral:

$$\int_{t_0}^{t_1} [G(t, x) + H(t, x)\dot{x}] \, dt = \int_C [G \, dt + H \, dx]$$

where C is the curve $x = x(t)$, $t_0 \leq t \leq t_1$.

Suppose, for example, that the situation is as shown in Figure 2.11, so that the solid curve $x(t)$ denoted by *PSRTV* is the asserted optimal trajectory. Consider an arbitrary alternative admissible curve $x_1(t)$ from (t_0, x_0) to (t_1, x_1), shown as the dashed curve denoted by *PQRUV* in Figure 2.11. Because the curve $x(t)$ satisfies the constraint equality $\dot{x} = A(x, t)$ for $t < t_a$, it is clear that

$$x_1(t) \leq x(t) \quad \text{for} \quad t_0 \leq t \leq t_a.$$

Suppose, as shown, that

$$x_1(t) < x(t) \quad \text{for} \quad t_0 < t < t_R,$$

whereas $x_1(t_R) = x(t_R)$. Then we have

$$\int_{t_0}^{t_R} [G(t, x) + H(t, x)\dot{x}] \, dt - \int_{t_0}^{t_R} [G(t, x_1) + H(t, x_1)\dot{x}_1] \, dt$$

$$= \int_{PSR} \{G \, dt + H \, dx\} - \int_{PQR} \{G \, dt + H \, dx\}$$

$$= -\oint_{PQRSP} \{G \, dt + H \, dx\}$$

$$= \int\int \left\{ \frac{\partial G}{\partial x} - \frac{\partial H}{\partial t} \right\} dx \, dt$$

by Green's theorem, where the double integral extends over the region bounded by the curve $PQRSP$.

Since this region lies below the singular path $x^*(t)$, the hypothesis given by Eqs. (2.35) implies that

$$\frac{\partial G}{\partial x} - \frac{\partial H}{\partial t} > 0 \qquad (2.36)$$

on the region in question. Hence the above integral is positive.

A similar calculation applies for the interval $t_R \le t \le t_1$, where $x_1(t) > x(t)$. For in this range the inequality given by Eq. (2.36) is reversed, and the orientation of the corresponding boundary curve $RTVUR$ is also reversed. Hence we obtain

$$\int_{t_0}^{t_1} [G(t, x) + H(t, x)\dot{x}] \, dt > \int_{t_0}^{t_1} [G(t, x_1) + H(t, x_1)\dot{x}_1] \, dt.$$

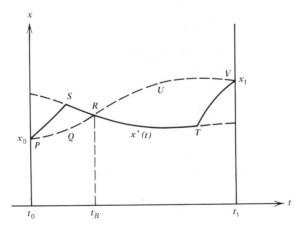

Figure 2.11 Proof of the most-rapid approach theorem.

This argument obviously extends to all possible geometric configurations generated by the curve $x(t)$ and another admissible curve $x_1(t)$. Thus the theorem is proved. (The proof of the foregoing theorem illustrates the well-known and important *principle of optimality:* the assertion that states that an optimal path $x(t)$ for $t_0 \le t \le t_1$ must also be optimal over any given subinterval such as $t_0 \le t \le t_R$.)

Summary

To summarize the procedure obtained above for solving linear control problems, first we express the problem in the form

$$\text{maximize} \int \phi(t, x, \dot{x}) \, dt,$$

observing that the integrand $\phi(t, x, \dot{x})$ depends linearly on $\dot{x}$. We then formally apply the Euler condition

$$\frac{\partial \phi}{\partial x} = \frac{d}{dt} \frac{\partial \phi}{\partial \dot{x}},$$

which reduces to an implicit equation

$$Q(t, x) \equiv 0$$

relating x and t. This equation is then solved for x to obtain the singular solution

$$x = x^*(t).$$

If the singular solution is unique and if the inequalities given by Eqs. (2.35), rather than their reverse, hold, then $x^*(t)$ is the desired optimum optimorum, and the optimal path $x(t)$ is obtained by the most-rapid approach method.

Several variations of this basic problem are now considered.

Blocked Intervals

K. J. Arrow (1964, 1968) has introduced the term *blocked interval* to refer to any time interval $[t_a, t_b]$ during which the constraints of a linear control problem prevent the path $x(t)$ from following the singular path $x^*(t)$. Initial and terminal adjustment phases are examples of blocked intervals, but they are not the only possible examples. Consider a singular path $x^*(t)$, as shown in Figure 2.12, but suppose that there is a constraint

$$\dot{x} \ge 0$$

[this precise case is considered by K. J. Arrow (1964) who is concerned

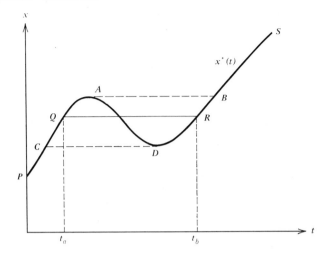

Figure 2.12 A blocked interval (t_a, t_b).

with the problem of irreversible investments]. Then $x(t)$ cannot follow $x^*(t)$ on its downward arc, but must follow some curve such as $PQRS$ with a blocked interval $[t_a, t_b]$ from Q to R.

We observe that our most-rapid approach theory does not suffice to determine the optimal blocked interval. However, the theory does imply that the optimal path must always lie as close as possible to the singular path, so that in Figure 2.12 the blocked interval cannot start to the left of C or to the right of A. Furthermore the theory readily implies that *whenever $x(t)$ does not coincide with the singular path $x^*(t)$, then one of the control constraints $u = u_m$ or u_M must be effective.* In other words, a blocked interval is literally blocked by the constraints.

Several instances of blocked intervals in resource-harvesting models are to be encountered later. Generally the optimal blocked interval can be determined by optimizing with respect to the time t_a at which the blocked interval begins. Overlapping blocked intervals, however, produce additional complications.

A Simple Example

As a simple application we consider the following optimal harvest problem, in which the price $p = p(t)$ is assumed to depend in a known manner on the time t:

$$\text{maximize} \int_0^\infty e^{-\delta t} p(t) h(t) \, dt$$

subject to

$$\frac{dx}{dt} = F(x) - h(t)$$

$$0 \le h(t) \le h_{max}.$$

It is left as an exercise for the reader to show that the singular path $x^* = x^*(t)$ is given by the equation

$$F'(x^*) = \delta - \frac{\dot{p}(t)}{p(t)}.$$

(See also Exercise 5 at the end of this chapter.)

The possibility of blocked intervals in this example will be discussed in Section 3.3.

Impulse Controls

Linear optimization problems are sometimes encountered in which one or both of the control constraints u_m and u_M can be assumed to be infinite. This simply means that discontinuous jumps in the state variable $x(t)$ are considered feasible. Under such conditions the most-rapid approach solution obviously utilizes such a discontinuous jump to transfer x instantly to the singular path. The control that effects such a discontinuous jump is referred to as an *impulse control.*

Infinite Time Horizon

From the mathematical point of view, the passage from a finite time interval $[t_0, t_1]$ to an infinite time interval can cause various difficulties. As a simple example suppose that the problem is to maximize

$$\int_0^\infty e^{-\delta t} c(t) \, dt$$

such that

$$\frac{dx}{dt} = rx - c(t); \quad x(0) = x_0; \quad x(t) \ge 0.$$

$$0 \le c(t) \le +\infty.$$

If $r > \delta$ this problem has no solution, because the integral can be made arbitrarily large by postponing *consumption* $c(t)$ to the distant future. Although the problem does not possess a singular solution, any finite time horizon $0 \le t \le T$ leads to an optimal solution (impulse control at $t = T$).

Such technicalities seldom arise in practical problems, however, although they may be encountered as the result of inappropriate models. In the optimal fishery-harvest problem discussed in Sections 2.4 and 2.5, for example, the resource stock is necessarily bounded between 0 and K. It is easy to verify that the optimal policy for this problem over an infinite time horizon is the most-rapid approach from $x(0)$ to the singular solution x^*, which is then maintained to $t = \infty$.

Nonuniqueness of Singular Paths

As an example of nonuniqueness, consider a depensation growth curve $F(x)$, as in Figure 2.13. For simplicity let $c(x) \equiv 0$. Then the basic equation

$$F'(x^*) = \delta$$

will have two solutions $x_1^* < x_2^*$ for certain values of δ. Which solution if either, is the optimal equilibrium?

This question cannot be fully answered without additional information. We can be sure, however, that x_1^* is *not* the optimum. Indeed, x_1^* is a minimizing rather than a maximizing solution, as should be clear from our theory, because the necessary inequality given by Eq. (2.35) is reversed in the vicinity of x_1^*.

Thus the optimal equilibrium population level may be either x_2^* or 0. Moreover, which solution is optimal may depend on the initial population level $x(0)$. It can be shown that there exists a critical value x_c $(0 \le x_c \le K)$ such that if $x(0) > x_c$ then the optimal harvest policy is the most-rapid approach to x_2^*, whereas if $x(0) < x_c$ then the most-rapid approach to zero (extinction) becomes optimal. We do not prove this in detail here, however.

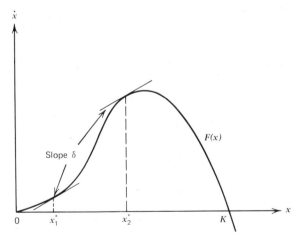

Figure 2.13 A case of nonuniqueness.

2.8 THE POSSIBILITY OF EXTINCTION

As mentioned in Section 2.3, the standard yield–effort relation $Q = kEx$ and the assumption that cost is proportional to E imply that the rent-dissipating, bionomic equilibrium population level x_∞ is positive, so that extinction of the fish population can only occur in the case of critical depensation. In other words, according to these conditions the cost of deliberately driving a fishery to extinction is infinite.

While this assumption may be considered realistic for marine fish populations, it is clear that extinction (in the strict biological sense) is a definite possibility for many populations of endangered species. In fact our general optimization model already allows for this possibility, since we may assume that

$$c(0) < +\infty. \tag{2.37}$$

(There is a slight interpretive difficulty at this stage, because $c(0)$ formally denotes the cost of a unit harvest when the population level is zero. The intended interpretation is simply that the cost of harvesting the last surviving member of the population is finite.)

Now suppose that

$$p > c(0); \tag{2.38}$$

that is, that the unit value of the resource exceeds the cost of harvesting the last unit. Then clearly open-access harvesting can be expected to lead to extinction, so that we have

$$x_\infty = 0. \tag{2.39}$$

We know in general that the optimal population level x^* approaches x_∞ as the discount rate $\delta \to \infty$, so it seems clear that $x^* = 0$ for sufficiently large but finite δ.

As a special case, suppose that unit harvesting costs do not depend on x:

$$c(x) = c = \text{constant}.$$

Then Eq. (2.16) becomes simply

$$F'(x^*) = \delta. \tag{2.40}$$

For large values of δ this equation may have no solution. In such cases it follows that extinction is the optimal exploitation policy, provided $p > c$.

For example, in the case of the logistic function $F(x) = rx(1 - x/K)$, Eq. (2.40) has no solution whenever $\delta > r$ (i.e., whenever the bionomic growth ratio $\gamma = \delta/r$ is greater than 1; see column 1 of Table 2.2). More generally, the following can easily be established (see Clark, 1973a).

THEOREM

Let $F(x)$ be a purely compensatory growth function (i.e., $F''(x)<0$) and assume that $c''(x)>0$. Then

1. $x^*>0$ (extinction is nonoptimal) if either $p<c(0)$ or $\delta<F'(0)$.
2. $x^*=0$ (extinction is optimal) if both $p\geq c(0)$ and $\delta>2F'(0)$.

Proof. Let

$$H(x) = -\frac{c'(x)F(x)}{p-c(x)}. \tag{2.41}$$

Then Eq. (2.16) can be written

$$\delta - F'(x) = H(x). \tag{2.42}$$

Assume that $p\geq c(0)$ and that $\delta>2F'(0)$. We show that in this case Eq. (2.42) has no solution $x\geq 0$, which clearly implies that extinction is optimal.

Since $H(x)$ is a decreasing function of p, it is sufficient to consider the limiting case $p=c(0)$. Then by the generalized mean value theorem we have

$$H(x) = -c'(x)\frac{F(x)}{c(0)-c(x)}$$

$$= -c'(x)\frac{F'(\xi)}{c'(\xi)} \quad \text{(for some } \xi, 0<\xi<x)$$

$$< F'(\xi)<F'(0).$$

Hence

$$\delta - F'(x) > 2F'(0) - F'(x) > F'(0) > H(x),$$

so that Eq. (2.42) has no solution, as claimed.

The proof in the converse direction is straightforward.

The technical sense in which the term "optimal" is used here should be emphasized. Under condition (2) of the theorem extinction is the *optimal* harvest policy only because it leads to the largest present value of economic revenues. We are certainly not suggesting that the deliberate extinction of a species is socially or aesthetically desirable just because extinction appears to be the most profitable course of action. Aesthetic or moral questions aside, the decision to exterminate a species is an irreversible decision that can only be justified in economic terms if we are *certain* that present conditions will persist into the distant future.

Even with such reservations it may appear surprising that under suitable circumstances extinction could constitute an economically optimal policy. [Much of the published literature on fishery economics explicitly asserts that regardless of time discounting, extinction can never be optimal. For example, several authors claim that no form of biological overfishing is practiced deliberately by the profit-maximizing sole owner. However, see Plourde (1970) or Brown (1974).] The explanation is simple enough however. The sole owner considers a resource asset to be a form of capital. If the asset fails to provide a suitable return on its capital value (in comparison with alternative forms of investment), it is profitable to liquidate the asset. If whaling companies expect to earn 10% on their investments and if the blue-whale population only increases 5% per annum, the whalers could quite rationally plan to liquidate the blue-whale stocks (rationally, that is, from their point of view).

The capital-theoretic aspects of resource-management policy are so fundamental that Chapter 3 in its entirety will be devoted to their analysis.

2.9 SUMMARY AND CRITIQUE

In this chapter we introduced certain economic aspects of renewable resource exploitation, concentrating primarily on a model of the commercial fishery. We presented the basic theory of the dissipation of economic rent in the unregulated open-access fishery, and we compared that theoretical situation with the harvest policy of a profit-maximizing sole owner.

The principal conclusion of our analysis is that as a consequence of time discounting, the sole owner also "dissipates" a certain fraction of the potential maximum economic rent by achieving an optimal level of sustained rent that is less than the maximum.

The extent to which a resource owner elects to "dissipate" sustainable rent (in favor of immediate revenues) depends on the rate of discount employed in the calculation of present values. More precisely, it depends on the *bionomic growth ratio*, or the ratio between the discount rate and the intrinsic growth rate of the stock. It must be emphasized that there is nothing particularly surprising about these results (although they have not always been clearly recognized in the literature). It is a well-known, general result in capital theory that the optimal level of investment in capital assets critically depends on the discount rate, with high rates implying low levels of capital accumulation.

What may be surprising to some readers, however, is the observation that the sole owner may in some cases actually prefer to deplete a resource stock to below the level of maximum sustainable yield. This is particularly likely in situations where biological resources exhibit small growth rates; in such cases the profit-maximizing policy (in terms of present values) may be extremely sensitive to the discount rate. In subsequent chapters this sensitivity is seen to be by no means unusual. Most temperate-zone forests, for example, cannot sustain a growth rate of much more than 5% per annum. When discount rates greater than 5% are employed, forest management may become unprofitable; virgin forests may simply be chopped down and the land abandoned. Whether this is a desirable outcome from a national viewpoint depends on various other considerations, some of which will be discussed in later chapters. The complex question of what determines the rate of discount is discussed in Chapter 3.

Other economic parameters that affect resource exploitation include the price of the product and the cost of harvesting the product. Both the bionomic equilibrium of the open-access fishery and the optimal equilibrium of the controlled fishery depend on the relationship between price and cost. In general the level of exploitation increases as the cost–price ratio decreases. We have already observed how regulation agencies may take advantage of this fact; the control measures of some agencies often increase the costs of fishing, thus controlling exploitation by a more or less deliberate introduction of inefficiency.

Another important aspect of our theory is the most-rapid approach characteristic of optimal harvest policies. According to this result, the profit-maximizing owner of a previously unexploited stock will attempt to harvest the excess population as rapidly as possible to realize an early profit. Observation of developing resource industries (forestry as well as fisheries) agrees well with this prediction. On the other hand, the reverse prediction that an overexploited resource will not be harvested until it has recovered to an optimal level may seem less realistic. Various explanations for this discrepancy can be suggested, particularly the imperfection of loan markets in dealing with uncertain investments of this kind. However, these questions are beyond the scope of the present analysis.

It might be argued that the results of this chapter have little practical significance because of the large number of unrealistic assumptions that have been adopted. Judgment on this issue should of course be reserved until we have had the chance to study these assumptions more fully. It will indeed turn out that many interesting and important aspects of resource exploitation have been "swept under the rug" by virtue of our simplifying assumptions. Nevertheless the basic bioeconomic interrelations between costs, prices, and discount rates on one hand and biological growth rates and carrying capacity on the other hand will remain valid throughout our studies.

We outlined the biological limitations of the Schaefer model at the end of Chapter 1. On the economic side, the main aspects that require a more sophisticated analysis include:

1. Variation in economic parameters over time (Chapter 3).
2. Questions of supply and demand (Chapter 5).
3. Fixed costs and irreversibility of investment (Chapter 4).
4. Uncertain and insufficient data.
5. Institutional and political constraints.

EXERCISES

1. Let

$$F(x) = \begin{cases} rx (x < K) \\ -\infty (x > K). \end{cases}$$

   If $c(x) = c/x$, show that

$$x^* = \begin{cases} K & \text{if} \quad \gamma < \dfrac{K}{K - x_\infty} \\[2ex] \dfrac{\gamma x_\infty}{\gamma - 1} & \text{otherwise,} \end{cases}$$

   where $x_\infty = c/p$ and $\gamma = \delta/r$. Apply this model to the data given in the text for the Antarctic fin whale ($r = 0.08$; $K = 400{,}000$; $x_\infty = 40{,}000$) and compare your results with the results from the Schaefer model.

2. In the case of the Gompertz model $F(x) = rx \ln (K/x)$, show that $x^* > 0$ even if $c(0)$ is finite, for all $\delta < \infty$.

3. Let $c(x)$ denote the unit harvest cost when the population level is x, as in this chapter. However, suppose that no natural increase or decrease in the population level occurs during the harvest process. Show (a) that the total net revenue derived from a total harvest $h \geq 0$, given an initial population level x_0, equals approximately

$$TR(h) = \int_{x_0 - h}^{x_0} [p - c(x)] \, dx,$$

   and (b) that $TR(h)$ is maximized when the "escapement" $x_0 - h$ equals x_∞. Explain your answer (cf. Scott, 1955).

4. With $c(x) \equiv 0$, consider the finite time-horizon problem

$$\text{maximize} \int_0^T e^{-\delta t} h(t) \, dt,$$

where

$$\dot{x} = F(x) - h(t)$$

and

$$0 \le h(t) \le h_{max}.$$

Suppose that the terminal stock level $x(T)$ is not specified (except, of course, as $x(t) \ge 0$). Determine the optimal harvest policy. [*Hint:* For simplicity let $x(0) = x^*$. The optimal policy clearly has the form

$$h^*(t) = \begin{cases} F(x^*) & \text{for} \quad 0 < t < \tau \\ h_{max} & \text{for} \quad \tau < t < T, \end{cases}$$

where τ is an undetermined switching time. Find the optimal τ.]

5. If price is a given function of time $p = p(t)$, show that the singular path $x = x^*(t)$ satisfies

$$F'(x) - \frac{c'(x)F(x)}{p(t) - c(x)} = \delta - \frac{\dot{p}(t)}{p(t) - c(x)}.$$

(The significance of the additional term on the right side of the equation is discussed in Chapter 3.)

6. Show that if the growth function $F(x)$ exhibits critical depensation, then extinction may be optimal even if $c(0)$ is infinite.

7. The logistic model has been applied to the Eastern Pacific yellowfin tuna (*Thunnus albacares*) population by M. B. Schaefer (1967), who estimates the parameters as

$$r = 2.61, \quad K = 1.34 \times 10^8 \text{ kg}.$$

Assuming $x_\infty = 0.3K$, calculate x^* for values of δ between 0.0 and 1.0. What is the MSY? (*Note:* The value $x_\infty = 0.3K$ is chosen here for illustrative purposes. Schaefer's data suggests that x_∞ is actually closer to $0.5K$.)

8. (*Continuation of Problem 7.*) Schaefer (1967) also estimates the catchability coefficient $q = 3.8 \times 10^{-5}$, so that

$$h = qEx,$$

where E is the fishing effort in number of standard fishing days, x is the stock level in kg, and h is the annual harvest in kg. Use the data in problem 7 to calculate the effort levels corresponding to maximum sustained economic rent, MSY, and open-access bionomic equilibrium. (*Answers:* 24,000; 34,300; 48,100, respectively.)

9. Assume that the sole owner attaches a "preservation value" $V(x)$ to the resource stock x itself, so that the objective functional takes the

form

$$J = \int_0^\infty e^{-\delta t}\{[p - c(x)]h(t) + V(x)\}\, dt.$$

How does this affect the optimal harvest policy? Assume that $V'(x) >$ 0 (see Vousden, 1973).

10. Suppose that two fishing companies only exploit a given fishery, but that for some reason negotiation between the companies is impossible. What bionomic equilibrium will be reached? (*Note:* Assuming the Gordon–Schaefer model, Cheung (1970) attempts to show that altogether the two companies will utilize two-thirds of the open-access, rent-dissipating effort level E_∞, and in general that n companies will utilize $n/(n+1)\, E_\infty$. In the author's view, however, this proof is fallacious, because it is based solely on the equilibrium yield-effort model.)

BIBLIOGRAPHICAL NOTES

Although usually associated with the names of Gordon (1954) and Schaefer (1957), several earlier authors discuss the fishery model based on logistic growth. In particular, see Graham (1952), who gives an extensive review of the earlier literature. A recent reiteration of this theory appears in Christy (1973). The work of Gulland (1974), though based primarily on static models, is well worth reading.

There is considerable disagreement regarding the meaning of the terms "open-access" resources and "common-property" resources. Here we follow the suggestion of Ciriacy-Wantrup (1971), using the term "open-access" to refer to cases in which access to the resource stock is completely uncontrolled, and the term "common-property resources" to refer to public resource stocks, regulated or otherwise. Examples of common-property resources include public grazing land subject to permit and game-animal herds managed to produce sustained yields.

A famous expository article on common-property resources (or open-access resources in our present terminology) is Hardin (1968).

The concept of a sole owner is introduced by Scott (1955), who shows that the neglect of future revenues would lead the sole owner to reach the same bionomic equilibrium that open-access exploitation would produce. Scott also discusses the maximization of present values, but falls short of an analytic solution.

The mathematical analysis of Scott's version of the Gordon–Schaefer model is attempted by Crutchfield and Zellner (1962). Although these authors obtain the correct optimal equilibrium x^*, they fail to observe the most-rapid approach characteristic of optimal harvest policies. But more seriously they also fail to recognize that discounting can lead to biological overfishing; indeed they assert that this is impossible (Crutchfield and Zellner, 1962, p. 19). The complete analytic solution of the model of Gordon, Schaefer, Scott, and Crutchfield and Zellner is secured by Clark (1973a), and is further analyzed by Clark and Munro (1975). More esoteric models that include time-discounting are discussed by Quirk and Smith (1970) and by Plourde (1971).

Linear variational problems (in one dimension) are discussed by Clark and de Pree (1975); the most-rapid approach solution described earlier by Spence (1973) and Clark (1974b) does not observe the straightforward application of Green's theorem. The possibility of using Green's theorem in linear variational problems is noted by Miele (1958), La Salle (1959), and others. Linear variational problems in n dimensions have been studied extensively in recent years (Lee and Markus, 1968). We shall see in Chapter 4 that these problems can be treated in a unified manner by means of the Pontrjagin maximum principle. When available, however, the method of Green's theorem is vastly simpler than the general theory. Further results concerning most-rapid approach paths are given by Spence and Starrett (1975).

The question of extinction is discussed by Bachmura (1971), Gould (1972), Clark (1973a,b), Beddington, Taylor and Watts (1975), and others.

3

CAPITAL-THEORETIC ASPECTS OF RESOURCE MANAGEMENT

By definition the value of a capital asset is equal to the present value of the net future revenues that it is expected to yield. This applies not only to traditional assets such as property and buildings, but also to natural resource stocks such as minerals and forest stands.

Similarly a common-property resource stock is valuable to society because of its productive potential. Unregulated open-access exploitation of such a resource, however, disregards the future productivity of the resource. Overexploitation is basically a problem of the misallocation of resources over time in the sense that the excessive current use of a resource leads to the loss of future resource benefits.

In this chapter we consider explicitly these capital-theoretic aspects of resource management, beginning with a brief summary of the theory of interest. Recognizing the capital-theoretic nature of resource stocks is essential to a clear understanding of resource economics. From this viewpoint resource management simply becomes a special problem in capital theory, although it is an especially interesting and difficult problem. For example, the results derived in Chapter 2 turn out, when properly interpreted, to be special cases of well-known general principles of capital theory.

3.1 INTEREST AND DISCOUNT RATES

When a given cash payment P is invested at compound interest, its value increases exponentially according to the formula

$$\text{Future value} = P \cdot (1 + i)^n, \tag{3.1}$$

where i represents the *annual rate of interest* (compounded annually) and n denotes the number of years from the present. Introducing

$$\delta = \ln(1 + i) \tag{3.2}$$

as the *instantaneous* annual rate of interest, we obtain the slightly more general formula

$$\text{Future value} = P \cdot e^{\delta t} \tag{3.3}$$

for the future value at an arbitrary time $t \geq 0$. Alternatively δ is sometimes referred to as the annual rate of interest (compounded continuously). (Here we assume that the interest rate δ remains constant over time; for the important case where $\delta = \delta(t)$ see Exercise 1 at the end of this chapter).

Discounting the value of future payments is simply the reverse process of compounding the interest on present payments. Thus by definition the *present value* of a payment P due t years from now is given by

$$\text{Present value} = Pe^{-\delta t}. \tag{3.4}$$

In this setting it is customary to refer to δ as the instantaneous annual *rate of discount.* "Discount rate" and "interest rate" are then synonomous terms normally applied when future payments are discounted or when present payments are compounded, respectively. Although interest rates are usually thought to be positive, zero and negative rates are also possible.

The *total present value PV* of a sequence of payments $P_0, P_1, \ldots, P_N$, due in years $0, 1, \ldots, N$, respectively, is given by

$$PV = \sum_{k=0}^{N} \frac{P_k}{(1 + i)^k}. \tag{3.5}$$

Similarly the present value of a continuous time-stream of revenues $P(t)$, $0 \leq t \leq T$, is given by

$$PV = \int_0^T P(t)e^{-\delta t}\, dt. \tag{3.6}$$

In these expressions the *time horizon* (N or T) may be either finite or infinite, a criterion of convergence being required in the infinite case. In Eq. (3.5) or Eq. (3.6) the revenue stream $P = P_N$ or $P(t)$ may assume both positive and negative values at various times. With $P = R - C$ (the difference between gross revenues R and costs C), we have $P < 0$ if and only if $C > R$.

Real Versus Inflationary Rates Of Interest

A bewildering array of interest rates faces the potential investor in any economy. Much of this diversity results from the various degrees of uncertainty that pertain to different types of investment. But even in the case of highly secure investments such as savings-bank deposits and government bonds, interest rates vary considerably over time. This variation is highly correlated with the general rate of inflation that currently affects the economy.

When the value of money is measured in terms of purchasing power, it is clear that the real rate of return on invested funds is generally not equal to the nominal (monetary) rate. Instead it equals the excess of the nominal rate over the rate of inflation:

$$\begin{pmatrix} \text{Real rate} \\ \text{of interest} \end{pmatrix} = \begin{pmatrix} \text{nominal rate} \\ \text{of interest} \end{pmatrix} - \begin{pmatrix} \text{rate} \\ \text{of inflation} \end{pmatrix}.$$

In this book the terms "interest rate" and "discount rate" always refer to real rates unless explicitly stated to the contrary. Because many resource-management policies are sensitive to the discount rate, it is essential to keep this distinction in mind.

Whereas monetary interest rates can vary over an immense range (for example, rates up to 200% per month occurred during Germany's inflation in the 1930s) in recent times real interest rates have remained within a relatively narrow range (usually from about 2 to 4% per annum) as reflected by rates of return on government bonds.

Time Preference and Marginal Productivity of Capital

What economic forces determine interest rates is a question that has intrigued many generations of economists. Here we summarize the neo-classical theory of interest in brief form.

First it is assumed that the investment decisions of individuals in the economy result in a supply curve for investment funds (see Figure 3.1a) (the theory of supply and demand is discussed in Chapter 5). This curve determines the total amount of funds that investors wish to make available at each rate of interest i. There exists a minimum rate i_0 below which no investment funds are forthcoming. For $i > i_0$ the supply is assumed to be an increasing function of the interest rate. The interest rate i_0 is considered to reflect society's basic time-preference rate.

By definition investment is an increment to capital assets K. Let $F(K)$ denote the net productivity of capital; that is, the rate of return from a capital stock of size K. The curve in Figure 3.1b represents the marginal productivity of capital $F'(K)$, which is assumed to be a decreasing function of K. [The term "marginal" as used in economics is equivalent to "derivative"; it applies to any function of one or several independent variables. For convenience, however, the marginal value of a function is

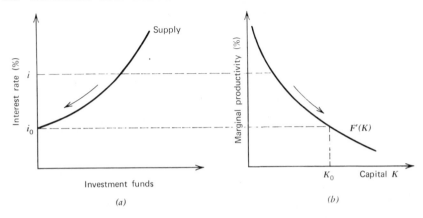

Figure 3.1 The neoclassical theory of interest.

frequently interpreted as the incremental value corresponding to a unit change in the independent variable. This amounts to the approximation $f'(x) \approx f(x+1) - f(x)$.] Note that $F'(K)$ is the maximum rate of interest that the owners of capital can afford to pay on additional (marginal) invested funds. Thus short-run equality of supply and demand in the market for investment funds requires that

$$F'(K) = i. \tag{3.7}$$

But now whenever the interest rate i exceeds i_0 the supply of investment funds is positive. The level of capital K therefore increases and marginal productivity consequently decreases. The process continues until $K = K_0$, where

$$F'(K_0) = i_0.$$

At this stage investment ceases and capital stops growing.

Thus in equilibrium we obtain the equalities

Interest rate = marginal productivity of capital

= social time-preference rate.

The equilibrium referred to here is actually a double equilibrium: market equilibrium in investment funds plus equilibrium in the growth of the economy. We know that in a growing economy the interest rate exceeds the social time-preference rate, the difference being related to the economic growth rate. [An appreciation of the complexity of these various growth-rate relationships can be gained from Arrow and Kurz (1970, Chapters 1–3).]

The Opportunity Cost of Capital

The relationship between the interest rate and the marginal productivity of capital [Eq. (3.7)] is of basic importance in capital theory. As derived above, the formula in Eq. (3.7) applies to the marginal productivity of capital in the economy as a whole. The formula is also valid, however, for the investment decisions of individual firms within the economy.

Thus suppose that a firm is contemplating making an addition to its capital assets. What criterion should be used to judge the desirability of such an action? First, management should be sure that no alternative investment can provide a greater rate of return. Thus the investment ΔK is not to be undertaken unless $F'(K)$ equals the marginal *opportunity cost* of capital; that is, the rate of return on the most profitable alternative investment opportunity. Second, the firm should insist that the return on investment be "adequate." In particular this obviously requires that $F'(K) \geq i$, where i is the rate of interest in the economy as a whole. If indeed $F'(K) > i$ then the investment should probably be undertaken, because the firm can borrow funds at rate i thereby making a net profit. Consequently, equilibrium at the firm's level also requires $F'(K) = i$.

To summarize, in the ideal world depicted by the foregoing model under global equilibrium conditions, the rate of interest is equal to the marginal productivity of capital, to the marginal opportunity cost of capital, and to the social time-preference rate.

In actuality numerous influences tend to upset this simple equilibrium solution. Equilibrium conditions seldom prevail. Capital markets are never perfect. Taxation may have distorting effects. Risk and uncertainty affect investment decisions, and insurance markets are frequently inadequate. Most contemporary economists seem to believe that market-determined interest rates and private opportunity-cost rates exceed the socially optimal time-preference rate (see the references quoted in the Notes at the end of this chapter.) The present work makes no attempt to contribute to these questions, except to point out that in problems of renewable resource management the rate of interest often plays a critical role. Perhaps this observation can lend some sense of urgency to theoretical debates in the literature.

3.2 CAPITAL THEORY AND RENEWABLE RESOURCES

We now return to the fishery model in Chapter 2. Recall the basic formula [Eq. (2.16)] for the optimal stock level x^*:

$$F'(x^*) - \frac{c'(x^*)F(x^*)}{p - c(x^*)} = \delta. \tag{3.8}$$

We observe the similarity between this formula and the marginal-productivity rule given by Eq. (3.7). Indeed when harvesting costs are independent of the population level (i.e., when $c'(x) \equiv 0$), the two formulas become identical. And of course the interpretation is also the same: in the fishery model the natural growth rate $F(x)$ represents the productivity of the fish population; in Eq. (3.7) $F(K)$ represents the productivity of capital. Maximization of the present value of future revenues with respect to a given discount rate δ therefore leads to the standard marginal-productivity rule $F'(x^*) = \delta$.

When the initial stock level x_0 differs from x^*, the optimal policy is either to "invest" (when $x_0 < x^*$) or to "disinvest" (when $x_0 > x^*$) as rapidly as possible, where investment now implies building up the capital (fish) stock. For example, when $x_0 < x^*$, the value of the asset (i.e., the fish population) is growing at a rate greater than the opportunity-cost rate δ. Thus it is a "superior asset" that should clearly be retained and expanded. Conversely when $x_0 > x^*$, the asset should be disposed of (not the entire asset, of course, but the excess $x_0 - x^*$). Moreover, the desired investment or disinvestment should be conducted as rapidly as possible; in our model this means simply that the optimal harvest rate h is zero for $x < x^*$ and h_{max} for $x > x^*$. [An exception to the most-rapid approach rule (to be discussed in Chapter 5) arises if a large supply of fish to the market leads to a decrease in the price level p. This clearly reduces the optimal harvest rate.]

The Stock Effect

How can the second term in Eq. (3.8) be interpreted in the present framework? When the costs of fishing increase as the fish population is reduced, it is natural to expect a modification in favor of increased population levels. Because

$$\frac{c'(x^*)F(x^*)}{p - c(x^*)} < 0 \tag{3.9}$$

we see by Eq. (3.8) that $F'(x^*) < \delta$, which indeed implies that x^* is greater than the level corresponding to the simple marginal-productivity rule $F'(x) = \delta$. (Here we assume decreasing marginal productivity $F''(x) < 0$.) We use the phrase *marginal stock effect* to describe the increase in x^* brought about by the sensitivity of harvesting costs to the level of the stock. In practice this stock effect is often quite significant, as the examples discussed in Chapter 2 indicate.

That the basic rule $F'(x^*) = \delta$ must be modified when $c(x)$ depends on x can also be understood by noting that in this case the variable x cannot be taken as a measure of the *value* of the fish population. Clearly the marginal value of the population (i.e., the net value of a unit of harvested fish) is $p - c(x)$, which is an increasing function of x. When everything is expressed in terms of net values, rather than in terms of biomass x, Eq. (3.8) can be seen to reduce to the basic marginal-productivity rule (see Exercise 3 at the end of this chapter).

Many further instances of Eq. (3.8) with various modifications are examined throughout the remainder of this book.

3.3 NONAUTONOMOUS MODELS

An unnecessary limitation of our optimal fisheries-management model is the assumption that the economic parameters (price, cost, and discount rate) remain constant over time. Suppose for example that the price $p = p(t)$ varies with time, but that the cost and the discount rate remain constant. (Variations in δ are dealt with in Exercises 1 and 2; and cost variations are discussed in Section 3.4). By a straightforward modification of the derivation of Eq. (2.16), it is easy to verify that the singular solution $x^* = x^*(t)$ now satisfies

$$F'(x^*) - \frac{c'(x^*)F(x^*)}{p(t) - c(x^*)} = \delta - \frac{\dot{p}(t)}{p(t) - c(x^*)}. \qquad (3.10)$$

The interpretation of the additional term on the right side of this equation is simple. Price changes imply that the value of the asset x is changing independently of its natural growth. The term $\dot{p}(t)/[p(t) - c(x)]$ is just the relative rate of growth of the marginal value of the asset.

When $c(x) \equiv 0$, Eq. (3.10) reduces to

$$F'(x^*) = \delta - \frac{\dot{p}(t)}{p(t)}, \qquad (3.11)$$

a well-known formula in capital theory, which simply indicates that relative price increases (for example) have the same effect as a reduction in the discount rate. (Since δ denotes the real discount rate, we only need to consider real price increases in this analysis.)

How do price changes affect the optimal harvest policy? Since price increases reduce the effective value of δ, they clearly imply increased stock levels x^*. In other words, it is desirable to exploit the stock less heavily when prices are increasing, because it will be worthwhile to exploit the stock more heavily at a later time.

Myopic Policies

A significant and perhaps surprising result of this analysis is the fact that the optimal population level $x^*(t)$ depends only on the current rate of change of price $\dot{p}(t)$ and not on price changes predicted in the future. Such "investment" policies depend only on short-term price fluctuations and have been termed *myopic* (Arrow, 1964). The validity of myopic policies depends upon the possibility of adjusting the stock level $x(t)$ continually to follow the singular path $x^*(t)$. If this is always possible, future price changes can be taken care of when they arise.

An important exception to the myopic rule occurs if constraints exist that prevent $x(t)$ from following the singular path $x^*(t)$; that is, if blocked intervals arise. Under these conditions the optimal policy may no longer be myopic, and foreknowledge of future price changes may be required.

A Sudden Drop in Price

As an example consider a case in which the price drops suddenly at a certain time T:

$$p(t) = \begin{cases} p_1 & t < T \\ p_2 & t > T, \end{cases} \tag{3.12}$$

where p_1 and p_2 are assumed constant with $p_1 > p_2$. Intuitively we would expect the resource owner who foresees a price drop to increase the rate of harvesting for $t < T$ to take advantage of the currently favorable price. How does our theory prescribe the optimal policy?

Because $\dot{p}(T)$ is unspecified, we cannot use Eq. (3.10) directly. Let us approximate $p(t)$ by a continuous function, as shown in Figure 3.2a. The corresponding singular path has two constant sections $x_1^* < x_2^*$ and an intervening segment resulting from the $\dot{p}$ term in Eq. (3.10). Passing to the limiting case we obtain the situation that appears in Figure 3.2b. The singular path $x^*(t)$ consists of the segments x_1^* and x_2^*, plus a vertical segment at $t = T$ that may be seen to arise from the fact that $\dot{p}(T) = -\infty$.

Obviously $x(t)$ cannot track this discontinuous singular path. Thus there will be a blocked interval $t_1 \leq t \leq t_2$ surrounding the point $t = T$. During the interval $t_1 \leq t < T$, the maximum harvest rate is used to reduce the stock to some level below x_1^*. At time $t = T$ harvesting ceases entirely to allow the stock to recover to the new optimal level x_2^*. These properties of the optimal solution can be easily verified by applying the most-rapid approach theory in Chapter 2.

The only remaining problem is to determine the *switching time* t_1 optimally. Treating t_1 as a parameter we may calculate the total present

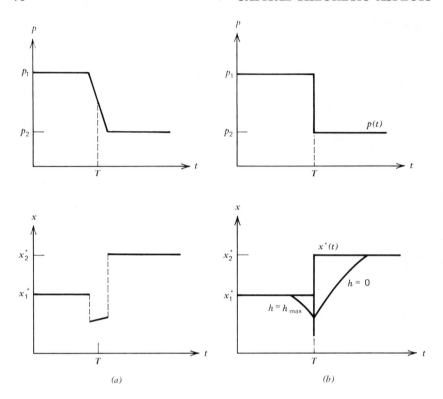

Figure 3.2 The effect of a sudden price drop: (a) approximate functions $p(t)$, $x^*(t)$; (b) the limiting case.

value of the entire harvest policy as a function of t_1. Hence the optimization of t_1 reduces to a calculus problem not pursued here. We observe that the formal solution agrees with our intuitive guess: The owner, foreseeing a price decline at $t = T$, increases the harvest rate to the maximum level during some period prior to T and is then faced with an overexploited stock which he allows to recover.

The simple model above suggests that private resource owners may react in rather extreme ways to changing economic conditions—a prediction supported by the observed behavior of mine owners to name one example from exhaustible resource economics.

Finally we observe that the reaction to price changes depends on whether such changes are anticipated. Resource owners increase their current harvest rates when price decreases are anticipated, but decrease their harvest rates once price decreases occur unexpectedly.

3.4 APPLICATIONS TO POLICY PROBLEMS: AN EXAMPLE*

Although the linear renewable-resource model developed in Chapters 2 and 3 appears to be free of major complications, it can actually be used as a powerful analytic tool. We attempt to illustrate this fact by applying the model to a complex problem encountered in fisheries.

One source of difficulty encountered in many fishing industries throughout the world arises from the fact that segments of the labor force engaged in fishing are isolated from the rest of the economy, both in physical terms and in terms of marketable skills. [See Crutchfield (1975, p. 13ff).] The social opportunity cost of such isolated labor is very low, if not zero. If we are to maximize the social benefit to be derived from fishing, we must value the labor at its social cost. However, private fishery operators are faced with a positive cost for labor; thus the private cost of labor exceeds the social cost. This in turn raises the possibility that unregulated private fishing cound actually lead to an underexploitation of the fish stock (from an economic point of view). In other words it is possible that the optimal level of fishing effort would exceed the level associated with bionomic equilibrium as perceived by the private sector.

If the aforementioned situation actually prevailed, the private sector could be persuaded to expand beyond the point of bionomic equilibrium through a government subsidy program. Policies of this type have actually been followed in various parts of the world. [See Copes (1972).] Politicians and other government officials usually describe such policies as being designed to alleviate unemployment in fishing communities.

While the social cost of labor in fishing may currently be below the private cost of labor due to the immobility of part or all of the labor force in the fishing sector, we cannot assume that this situation is permanent. It is possible that with appropriate government efforts in retraining and resettlement programs, the problem of labor mobility can be overcome over time. [See Copes (1972) and Crutchfield (1975).] If indeed such governmental efforts are feasible, we can think of the true social cost of labor—the potential output of labor in other parts of the economy—rising over time.

The optimal strategy would seem to be obvious. Authorities should be prepared to subsidize the fishing industry while preparations are under way to overcome the barriers to labor mobility. Once such programs are put into effect, the subsidy should be reduced gradually and in time should become negative (i.e., be transformed into a tax.) The one problem with such a strategy is that severe political constraints may be

*This section was prepared by Professor Gordon R. Munro, Department of Economics, University of British Columbia.

encountered. Economic history shows us that once a subsidy, open or disguised, is introduced it is often politically difficult to remove the subsidy when conditions change and the reason for the subsidy no longer exists. Thus in the example of the fishing industry, once the subsidy is introduced it can remain indefinitely. The programs to overcome barriers to labor mobility either are not put into effect because the unemployment problem has been disguised or if they are put into effect, fail to attract fishermen out of the industry to other parts of the economy. Thus the subsidy program, which could well produce substantial benefits for society in the short run, will confront society with a permanent burden in the long run.

In actuality we can be confronted with a situation in which (1) subsidies are no longer justified and can be easily reversed, or (2) subsidies prove irreversible, or (3) subsidies can be reversed, but only with a time lag. We analyze this problem by considering the two polar extremes: in the first case the subsidy is totally reversible; in the second case the subsidy is totally irreversible once granted. To do this, we use nonautonomous linear models and exploit the concepts of "myopic" decision rules and blocked intervals.

In both cases we assume the following:

1. Full employment prevails in the nonfishing sector of the economy.

2. The fishery commences in bionomic equilibrium; given any change in conditions, the industry will expand or contract until bionomic equilibrium is restored. The industry is considered in bionomic equilibrium when total revenue − [total private cost − subsidy] = 0. The initial level of subsidization is assumed to equal zero.

3. While labor in the fishing industry is completely immobile initially, capital is fully mobile.

4. The initial socially optimal level of fishing exceeds the level achieved by the private fishing industry in bionomic equilibrium.

In the first case the authorities introduce a subsidy that causes an expansion in fishing and a consequent reduction in the biomass. In time the policies designed to eliminate the barriers to labor mobility are implemented; consequently the social cost of fishing begins to increase and eventually coincides with the private cost of fishing. We assume that the private cost of fishing remains stationary throughout.

As the social cost of fishing rises, the optimal biomass rises and the optimal subsidy rate falls. This is illustrated in Figure 3.3. The total social cost of harvesting is denoted by TC_s and the total private cost is TC_p. The initial optimal stock level is $x^\dagger$ where $x^\dagger < x_0$, the stock level associated with initial bionomic equilibrium. As TC_s shifts upwards over time the

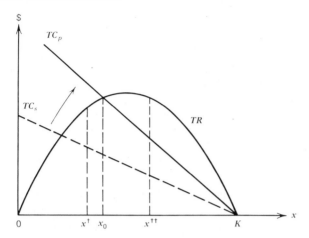

Figure 3.3 Social cost TC_s versus private cost TC_p in fishery exploitation.

optimal stock level $x^*(t)$ rises and approaches $x^{\ddagger}$ as $t \to \infty$. [In fact $x^{\dagger}(t)$ may temporarily fall before rising toward $x^{\ddagger}$, as is explained below.]

The model we require to determine the optimal stock level over time is clearly nonautonomous, because one of its parameters—the cost function—is a function of time. Our objective functional can now be expressed as

$$PV = \int_0^{\infty} e^{-\delta t}[p - \phi(t)c(x)]h(t)\, dt, \qquad (3.13)$$

where $\phi(t)$ is a variable coefficient. The coefficient $\phi(t) = 1$ for all time periods prior to the upward shift in social harvesting costs. When TC_s begins to shift upwards, $\phi(t)$ increases and approaches the upper bound $\phi(t) = \bar{\phi}$ as $TC_s \to TC_p$; that is,

$$\lim_{t \to \infty} \phi(t) = \bar{\phi}.$$

By routine calculation it can be shown that the singular solution $x^* = x^*(t)$ satisfies

$$F'(x^*) - \frac{\phi(t)c'(x^*)F(x^*)}{p - \phi(t)c(x^*)} = \delta + \frac{\dot{\phi}(t)c(x^*)}{p - \phi(t)c(x^*)}. \qquad (3.14)$$

The similarity between Eq. (3.14) and Eq. (3.10) is obvious. The relative rate of growth of the marginal value of the resource asset is equal to $-\dot{\phi}(t)c(x)/[p - \phi(t)c(x)]$. In this instance the growth is affected by shifting harvesting costs, rather than by price changes through time.

It is worth reemphasizing here that the investment policy is "myopic" in the sense that it is necessary to know only the immediate change in harvesting costs to determine $x^*(t)$. It is not necessary to predict the whole future course of harvesting costs over time. It should also be emphasized that while $x^*(t)$ can be expected to increase over time as costs rise, the term $\dot{\phi}c(x)/\{p - \phi c(x)\}$ actually decreases x^*. In other words the anticipation of higher costs tomorrow—in, of, and by itself—can cause intensified fishing effort today. Thus the time path of $x^*(t)$ can look like the path shown in Figure 3.4. Here T_U is the point in time when social costs begin to shift upwards.

Next let us turn to the question of the subsidy rate. It is difficult to specify the optimal subsidy precisely unless we know the industry's reaction function; that is, the function that describes the collective response of the individual firms to perceived profits and losses. This we do not presume to know. (However, see Section 4.6.) We can, however, at least specify the subsidy rate that will drive the industry to operate at the optimal stock level. The subsidy rule we propose is simply to set the subsidy rate at each point in time in such a way that the perceived rent of the industry at the optimal stock level is zero. If the actual stock level is in excess of x^*, the subsidy policy generates perceived rents and causes an expansion in fishing effort. Conversely if the actual stock level is below $x^*(t)$, the subsidy policy generates a perceived loss for the industry and therefore induces a contraction in fishing effort. The subsidy rate for the industry as a whole can be expressed as

$$S(t) = [\bar{\phi}c(x^*) - p]F(x^*). \tag{3.15}$$

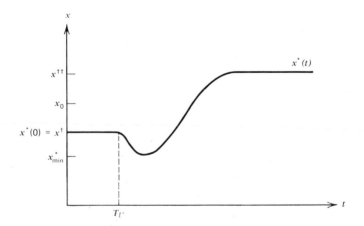

Figure 3.4   The singular path $x^*(t)$.

[The costs appearing in Eq. (3.15) obviously refer to the costs perceived by the *private* fishing industry. We have already assumed that these private costs are the same as the ultimate social costs viz., $\bar{\phi}c(x)$.] The optimal subsidy per unit of harvest is expressed simply as

$$\frac{S(t)}{h(t)} = [\bar{\phi}c(x^*) - p]. \qquad (3.16)$$

The desired subsidy can also be based on the level of effort employed in the fishery. Assuming as in Eq. (2.9) that

$$h = G(x)E, \quad \text{where} \quad G(x) = \frac{c}{c(x)}, \quad c = \text{cost per unit effort,}$$

we see that the subsidy per unit of effort is given by

$$\frac{S(t)}{E(t)} = c\left\{\bar{\phi} - \frac{p}{c(x^*)}\right\}. \qquad (3.17)$$

We turn now to the second case, one that is more difficult—but certainly more interesting—to analyze. We assume that subsidies can be increased, but not decreased; consequently the stock level $x(t)$ can only be reduced. We also assume that the true social costs of fishing will increase at some time in the future. The authorities are assumed to have the power to remove the barriers to labor mobility. If the authorities do not remove the barriers to labor mobility because they are complacent about the apparent success of the subsidy program in alleviating unemployment, future opportunity costs will still be considered to have increased; that is, the true measure of alternative output forgone by devoting resources to fishing must account for what these inputs could have produced elsewhere if the authorities had taken the necessary steps to eliminate the barriers to labor mobility.

Having said all of this we can refer to Figure 3.4 and express the problem in the following way. At any point in time t, we want to be as close as possible to the optimum optimorum $x^*(t)$. However, if we move close to $x^*(t)$ in the near future when unemployment is high, we will be far from $x^*(t)$ in the distant future. An expansionary policy today will impose heavy burdens on society in the future. One must select a constrained optimal stock level, hereafter denoted as x_z, balancing off present gains against future losses. Once the stock level x_z is reached, no upward change in x can be made. The decision to move to x_z is therefore irreversible.

At an earlier point in this chapter (Section 3.3) we discussed the concept of blocked intervals. Recall that a blocked interval is defined as a situation in which $x(t)$ is prevented from following the singular path by the presence of binding constraints on the control variable. Here we are confronted with a blocked interval of infinite length. The control variable $h(t)$ is constrained by the limitations on the subsidy policy. To underline the length of the blocked interval we need only note that even if no subsidy is introduced so that $x_z = x_0$, a point in time will be reached at which $x_z < x^*(t)$. The stock level x_z will remain below $x^*(t)$ thereafter.

Let us suppose that the singular path $x^*(t)$ is as shown in Figure 3.4. Note that at $t = T_U$ the path initially dips downward before rising to its ultimate equilibrium at $x = x^{\ddagger}$. One of two possibilities may then arise concerning the optimal approach to the constrained equilibrium x_z. First, the single *bang-bang approach* (see Figure 3.5a), may use $h = h_{max}$ until x is reduced to the level $x_z > x^{\dagger}$. Second a two-step approach may be required in which x is first reduced to $x^{\dagger}$ then subsequently to x_z. [If the slope of $x^*(t)$ near $t = T_U$ is excessive, an additional blocked interval may arise in the neighborhood of this point.] Here we restrict our attention to the first case; the second case is slightly more complex but involves no new principles of analysis.

Let us now proceed to derive the decision rule for obtaining the constrained optimal stock level x_z. The nature of the constraint we impose upon the rule is that once x_z is chosen, society must live with it forever. In our equation we will not attempt to constrain $x_z \leq x_0$. However, if on inspection we find that x_z generated by our decision rule exceeds x_0, we will consider the solution inadmissible and reject it. The policy recommendation in that case will simply be to remain at the initial level x_0 and impose a subsidy rate of zero.

Now let the time at which the constrained optimal stock level is reached be denoted at T_z. The adjustment phase is then from $t = 0$ to $t = T_z$. Once the adjustment phase is completed, the control variable $h(t)$

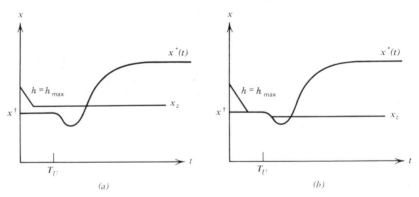

Figure 3.5   Optimal solutions to the constrained subsidy problem: (a) $x_z > x^{\dagger}$; (b) $x_z < x^{\dagger}$.

becomes permanently fixed (or locked into position, as it were) at $h(t) = F(x_z)$. The objective functional then can best be expressed as

$$PV = \int_0^{T_z} e^{-\delta t}[p - \phi(t)c(x)]h(t)\,dt + \int_{T_z}^{\infty} e^{-\delta t}[p - \phi(t)c(x_z)]F(x_z)\,dt.$$

(3.18)

Differentiating PV with respect to x_z and setting $dPV/dx_z = 0$ we obtain

$$e^{-\delta T_z}[p - \phi(T_z)c(x_z)] - pF'(x_z)\frac{e^{-\delta T_z}}{\delta}$$

$$+ [F(x_z)c(x_z)]' \cdot \int_{T_z}^{\infty} e^{-\delta t}\phi(t)\,dt = 0 \quad (3.19)$$

[here we use the relation $dx_z/dT_z = F(x_z) - h(T_z)$].

Equation (3.19) is cumbersome, but it can be more manageable if a linear transformation is performed on the integral so that it becomes

$$e^{-\delta T_z}\int_0^{\infty} e^{-\delta t}\phi(t + T_z)\,dt.$$

Upon so doing Eq. (3.19) can be reexpressed as

$$\alpha\left\{F'(x_z) - \frac{\beta_\delta(T_z)c'(x_z)F(x_z)}{p - \beta_\delta(T_z)c(x_z)}\right\} = \delta, \qquad (3.20)$$

where

$$\beta_\delta(T_z) = \delta\int_0^{\infty} e^{-\delta t}\phi(t + T_z)\,dt \qquad (3.21)$$

and

$$\alpha = \frac{p - \beta_\delta(T_z)c(x_z)}{p - \phi(T_z)c(x_z)}. \qquad (3.22)$$

It is instructive to compare the decision rules represented by Eq. (3.20) and Eq. (3.14). Equation (3.20) contains no corrective comparable to the term $\dot{\phi}c(x^*)/[p - \phi c(x^*)]$ appearing in Eq. (3.14), because it is no longer adequate to consider the change in the social cost of fishing only in the immediate future. The fact that an irreversible decision is being made means that we must consider the time path of costs far into the future—indeed in theory as far as $t = \infty$. The "myopic" policy or rule we discussed earlier must be replaced by an "omniscient" policy.

This all-seeing requirement is captured in the multiplier $\beta_\delta(T_z)$, applied in the first instance to the marginal stock effect. The multiplier $\beta_\delta(T_z)$ can best be understood if we first assume that $T_z = 0$ (that the adjustment from $x(0)$ to x_z is instantaneous). Then $\beta_\delta(T_z)$ can be seen as a weighted average of present and future harvesting costs.

Note that the larger the multiplier the larger will be the marginal stock effect, which in turn serves to raise the optimal stock level x_z. All of this implies that reducing the stock today will have adverse and permanent cost consequences in the future. Thus the higher the future costs, the sooner they will be realized, the less they are discounted, the larger the optimal stock will be (i.e., the less incentive there will be to expand fishing effort in the present).

However, the impact of $\beta_\delta(T_z)$ upon the marginal stock effect is mitigated by the corrective α applied to the marginal yield of the biomass

$$F'(x_z) - \frac{\beta_\delta(T_z)c'(x_z)F(x_z)}{p - \beta_\delta(T_z)c(x_z)} .$$

This corrective accounts for the fact that the return to any marginal increment to the biomass will be less in the future than it is today. Since $\alpha < 1$, the corrective in, of, and by itself serves to reduce x_z. If the decision rule is applied according to the strategy outlined earlier, however, the impact of $\beta_\delta(T_z)$ on x_z via α can never exceed the impact of $\beta_\delta(T_z)$ on x_z via the marginal stock effect. [In other words we can be certain that $x_z \geq \min x^*(t)$. This is easily proved by reductio ad absurdum. If in fact $x_z < \min x^*(t)$, let x' be a biomass level such that $x_z < x' < \min x^*(t)$. Then the harvest policy that reduces $x(t)$ to x' rather than to x_z is easily shown to provide a larger present-value integral, so that x_z is inoptimal. The latter assertion can be proved by applying Green's theorem to the difference between the two present-value integrals in question.]

Still assuming that $T_z = 0$, we can review some of the factors that influence the level of x_z. The deeper the initial level of unemployment and hence the lower the initial level of social costs, the smaller x_z will be. On the other hand, the more rapid the adjustment to full employment, the larger $\beta_\delta(T_z)$ will be and therefore the higher the optimal stock level will be. The reason for this is obvious. The more rapid the adjustment to full employment, the greater the role of future costs will be in the decision-making process. Finally the discount rate plays a critical role. At one extreme, if $\delta = 0$ then the future and higher costs dominate and no justification for an expansion of fishing effort exists. It can be noted that

$$\lim_{\delta \to 0} \beta_\delta(T_z) = \bar{\phi}, \tag{3.23}$$

which in turn implies that Eq. (3.20) reduces to

$$[p - \bar{\phi}c(x_z)]F'(x) - \bar{\phi}c'(x_z)F(x_z) = 0. \tag{3.24}$$

This reduction implies that x_z is the stock level that maximizes sustainable (social) rent from the fishery. This in turn implies $x_z > x_0$. Because x cannot be increased above $x_0 = x(0)$, the optimal subsidy is equal to zero. It can be noted in passing that caeteris paribus there will exist a social rate of discount $\delta^\dagger > 0$ such that $x_z = x_0$. Thus a subsidy of zero will be called for if the actual discount rate δ^* is such that $0 \leq \delta^* \leq \delta^\dagger$.

At the other extreme, if $\delta = \infty$ then the future is irrelevant and expanding fishing effort à outrance is desirable. Returning to Eq. (3.20) we can see that if $\delta = \infty$ the equation reduces to $p = \beta_\delta(T_z)c(x_z)$. Fishing should proceed to the point where sustainable rent is eliminated even in the immediate future.

The discussion of the discount rate provides us with some insight into the penalty that society incurs by neglecting to account for the increase in future social costs. If we simply neglect the future increase in costs, at some point we will arrive at $x^*_{min} \leq x^*(0)$ by acting as if there are no binding constraints on $h(t)$. Once x^*_{min} is reached, the system will be "locked in" at this biomass level. The true optimal stock level, taking future cost shifts into consideration, is $x_z > x^*_{min}$ (unless $\delta = \infty$). The stock levels x_z and x^*_{min} are determined assuming the existence of a given social discount rate δ^*. Now note that the optimal stock level x_z could have been equal to the existing x^*_{min} if a higher discount rate had been in existence. We can express this higher discount rate $\delta^{**} = \theta\delta^*$, where $\theta > 1$. Thus neglecting future cost increases and ultimately opting for x^*_{min} is equivalent to employing a discount rate that exceeds the true social rate of discount by a factor of θ. We refer to the coefficient θ as the *coefficient of implicit overdiscounting*.

Thus far we have avoided the problem of the noninstantaneous adjustment of x from x_0 to x_z (assuming that $x_z < x_0$). Equation (3.21) reveals that $\beta_\delta(T_z)$ is an increasing function of T_z. In other words the longer the adjustment phase to x_z, the higher x_z will be. The reason is obvious. The benefits from expansion will be felt in the near future; the costs will be felt in the more distant future. If a long period of time elapses before the fishing industry can be expanded to the optimal level, then this will diminish the temporary benefits. While the problem can be stated without difficulty in intuitive terms, the determination of x_z is not easy, for T_z and x_z are interdependent. Nevertheless a solution is possible, as revealed by Eq. (3.20). The multiplier $\beta_\delta(T_z)$ translates the cost of long-term adjustments to x_z into future cost equivalents.

The preceding discussion has made it obvious that the problems raised by irreversible policy decisions are extraordinarily complex. Placing these problems in a dynamic context aggravates the complexity, but unfortunately it is nonsensical to consider the problems by any other method.

The discussion here shows that we can progress a long way forward analyzing these complexities with the aid of our relatively simple linear model.

EXERCISES

1. Let the instantaneous discount (interest) rate $\delta = \delta(t)$ be a function of time. Show that the present value of a future payment P due at time t is given by

    $$PV = P \exp \left(-\int_0^t \delta(s) \, ds \right).$$

    If δ is constant, this equation reduces to the usual expression. (*Hint:* Divide the time interval $[0, t]$ into n subintervals Δt_k and approximate $\delta(t)$ by $\delta(t_k)$ on each subinterval. Compute the compounded value at time t of an initial payment Q; then let $n \to \infty$.]

2. Show that the basic optimality rule, [Eq. (3.8)] remains valid when $\delta = \delta(t)$. (See Exercise 1.) To what extent must future changes in the interest rate be taken into account?

3. This exercise establishes the fact that the optimality rule [Eq. (3.8)] is *identical* to the marginal productivity rule [Eq. (3.7)], provided that all values are measured in terms of cash values.
    (a) Show that the "net cash value" of the resource stock x is given by

    $$V = V(x) = \int_{x_\infty}^x \{p - c(z)\} \, dz.$$

    [*Hint:* Obviously $dV/dx = p - c(x)$.]
    (b) Let

    $$\rho = \rho(x) = \{p - c(x)\}F(x),$$

    which represents the cash value of the production $F(x)$. Show that Eq. (3.8) is equivalent to

    $$\frac{dV}{d\rho} = \delta.$$

4. *Optimal deer hunting:* Logged-over forest land often supports a large deer population during the early stages of regrowth. Let this effect be modeled by assuming that the deer population satisfies

    $$\frac{dx}{dt} = rx \left(1 - \frac{x}{K(t)} \right),$$

where the carrying capacity $K(t)$ now depends on the age t of the logged-over area measured from the time of logging. Determine the harvest policy that maximizes the total deer harvest for $0 \le t \le T$. (This problem was suggested by Professor C. J. Walters.)

5. Let $x = x(t)$ denote the inventory of some salable asset and let $q = q(t) \ge 0$ denote the rate of sales, making $q(t)$ the control variable. Given a variable price $p = p(t)$, determine the sales policy that maximizes the total present value of sales (neglecting all costs). (A generalized version of this model is used in the study of exhaustible resource exploitation in Chapter 5.)

BIBLIOGRAPHICAL NOTES

The classical reference works on the theory of interest include I. Fisher (1930) and F. A. Lutz (1968); there are many others. The reader will probably find the recent book by Hirshleifer (1970) most approachable. The relevance of discounted present value in management decisions is discussed in Hirshleifer's book and in various works on cost–benefit analysis, including Mishan (1971) and Layard (1972).

Socially optimal discount rates have been discussed by Feldstein (1964), Baumol (1968), and others. See also Solow (1974, 1975) for related questions on the intergenerational distribution of welfare.

The concept of resource stocks as capital assets is as old as capital theory itself; indeed, one of the standard textbook examples in capital theory is the forest-rotation problem (see Chapter 8). Scott (1955) describes the capital-theoretic aspects of fisheries and shows that a private owner who neglects the future productivity of the stock reaches the same stock level x_∞ that the open-access fishery reaches. Further aspects of fisheries from the capital-theoretic viewpoint are described by Clark and Munro (1975), where further references can be found.

Nonautonomous models in resource management resulting from time-changes in costs, prices, and discount rates are seldom studied, and the simplicity of *linear* models has yet to be fully exploited. Myopic investment policies are frequently encountered in the management-science, operations-research literature. We learn in a later chapter that myopic harvesting policies can only be optimal under several rather special assumptions, particularly the assumption of linearity.

For an application of the theory of myopic decision rules and blocked intervals to a problem of dam construction versus environmental preservation, see Fisher, Krutilla and Cicchetti (1972).

4
OPTIMAL CONTROL THEORY

In Chapters 2 and 3 we have succeeded in solving certain dynamic optimization problems by using a simple method based on Green's theorem (see Section 2.7). Now we turn to various multidimensional and nonlinear problems. To solve these we require the more powerful optimization techniques provided by the calculus of variations. In this chapter we describe these techniques in a contemporary setting called "optimal control theory."

The techniques of optimal control theory—in particular L. S. Pontryagin's famous maximum principle (1962)—are outgrowths and extensions of the classical variational techniques of Euler, Lagrange, Legendre, Weierstrass, Hamilton, and Jacobi. Indeed the maximum principle, which is a necessary condition (more precisely a collection of necessary conditions) for optimality, encompasses most of the classical necessary conditions of these early mathematicians in a unified way that simplifies applications. But optimal control theory is more than a mere simplification of the classical theory; optimal control theory is a significantly improved theory in that it covers both nonlinear and linear optimization problems and inequality constraints (problems that could not be handled using the classical techniques). The maximum principle also possesses a clear economic interpretation not evident in the classical formulation (although implicit in the Hamilton–Jacobi theory).

We make no attempt in this book to record the difficult proof of the maximum principle; the reader is referred to standard works such as Pontryagin et al. (1962) and Lee and Markus (1968). For a unified discussion of continuous- and discrete-time optimal control, see Canon, Cullum, and Polak (1970); see also Sec. 7.8 of the present work.

4.1 ONE-DIMENSIONAL CONTROL PROBLEMS

We begin by introducing the terminology of optimal control theory in a simple, one-dimensional setting. Consider the differential equation

$$\frac{dx}{dt} = f[x, t, u(t)], \quad 0 \le t \le T \tag{4.1}$$

with initial condition

$$x(0) = x_0, \tag{4.2}$$

where $f(x, t, u)$ is a given, continuously differentiable function of three real variables x, t, and u. The variable $x = x(t)$, called the *state variable*, in practice normally describes the "state" of some given system (in this case a one-dimensional system) at time t. Equation (4.1), called the *state equation*, describes the evolution of the system from its initial state x_0, resulting from the application of a given *control* $u = u(t)$. The terminal time T is called the *time horizon*; here we suppose that T is finite.

Given a particular control $u(t)$, $0 \le t \le T$, the corresponding solution $x(t)$ to Eq. (4.1) is referred to as the *response*. From a rigorous mathematical point of view, various technical problems concerning the existence and the uniqueness of such responses ought to be considered. Inasmuch as these questions do not cause any difficulties in our intended applications, we bypass most of the technicalities here. It is of some importance, however, to delineate the class of functions $u(t)$ that we wish to allow as control functions.

The class U of *admissible controls* is by definition the class of all piecewise-continuous real functions $u(t)$ defined for $0 \le t \le T$ and satisfying

$$u(t) \in U_t \tag{4.3}$$

where U_t is a given interval (which may be infinite) called the *control set*. [The function $u(t)$ is said to be piecewise continuous if it is continuous except at a finite number of points of discontinuity, at each of which the function possesses finite right- and left-hand limits. This is the class of controls usually considered in variational calculus, although more general measurable controls are sometimes introduced.]

In many problems we also impose a *terminal condition*

$$x(T) = x_T. \tag{4.4}$$

We then define a *feasible control* to be any admissible control such that the response satisfies this terminal condition as well as the assumed initial condition [Eq. (4.2)].

No doubt the reader recognizes that the fishery model in Chapter 2 is an example of this situation. The equation

$$\frac{dx}{dt} = F(x) - h(t)$$

is the state equation, with control variable $h(t)$ subject to the constraints

$$0 \le h(t) \le h_{max},$$

which specify the control set $U_t = [0, h_{max}]$. As in Section 2.7, here we consider only the case of a finite time horizon T.

The Objective Functional

Next we introduce an *objective functional*

$$J\{u\} = \int_0^T g[x(t), t, u(t)] \, dt, \tag{4.5}$$

where $g(x, t, u)$ is a given, continuously differentiable function and $x(t)$ denotes the response to $u(t)$. The fundamental problem in optimal control theory is to determine a feasible control $u(t)$ that maximizes $J\{u\}$. Such a control, if it exists, is called an *optimal control.* The *maximum principle* (see below) gives certain necessary conditions that must be satisfied by an optimal control.

The Classical Variational Problem

Before stating the maximum principle, we note that the above formulation encompasses the so-called "simplest" problem of the calculus of variations: to maximize an integral expression of the form

$$\tilde{J}(x) = \int_0^T g(x, t, \dot{x}) \, dt \tag{4.6}$$

over the class of piecewise-smooth functions $x(t)$, satisfying given initial and terminal conditions

$$x(0) = x_0, \quad x(T) = x_T.$$

By introducing the simple state equation

$$\dot{x} = u,$$

we immediately transform this problem into the control-theoretic setting, without control constraints; that is, $U_t = (-\infty, \infty)$. The maximum principle implies the standard Euler equation, as well as certain other results from the classical calculus of variations (see Exercise 1 at the end of this chapter).

The Maximum Principle

The maximum principle is most conveniently formulated in terms of the following expression, called the *Hamiltonian:*

$$\mathcal{H}[x(t), t, u(t); \lambda(t)] = g[x(t), t, u(t)] + \lambda(t)f[x(t), t, u(t)]. \quad (4.7)$$

Here $\lambda(t)$ is an additional unknown function called the *adjoint variable.* If $u(t)$ is an optimal control and $x(t)$ is the corresponding response, the maximum principle asserts the existence of an adjoint variable $\lambda(t)$ such that the following equations are satisfied, for all t, $0 \le t \le T$:

$$\frac{d\lambda}{dt} = -\frac{\partial \mathcal{H}}{\partial x} = -\frac{\partial g}{\partial x} - \lambda(t)\frac{\partial f}{\partial x}, \quad (4.8)$$

$$\mathcal{H}[x(t), t, u(t); \lambda(t)] \equiv \max_{u \in U_t} \mathcal{H}[x(t), t, u; \lambda(t)]. \quad (4.9)$$

It should be emphasized that these equations refer to the optimal control $u(t)$, its associated response $x(t)$, and the corresponding adjoint variable $\lambda(t)$. They are *necessary* conditions that must be satisfied by an optimal control and response.

The differential equation, Eq. (4.8), called the *adjoint equation*, is easily understood, but Eq. (4.9), sometimes referred to by itself as the maximum principle, requires explanation. Equation (4.9) asserts that at every given time t, the value $u(t)$ of the optimal control must *maximize the value of the Hamiltonian expression* over all "admissible" values $u(t)$ satisfying the control constraints [Eq. (4.3)]. It is useful to observe that if the optimal control $u(t)$ happens to lie in the *interior* of the control interval U_t (i.e., if the control constraints are not binding), then Eq. (4.9) implies that

$$\frac{\partial \mathcal{H}}{\partial u} = 0. \quad (4.10)$$

It is easy to verify that Eq. (4.8) and (4.10) imply the classical Euler equation for solutions of the variational problem [Eq. (4.6)]. (See Exercise 1.)

How is the maximum principle applied? Note that we now have three unknown functions to determine: $x(t)$, $u(t)$, and $\lambda(t)$. For these three functions we have three equations: the state equation [Eq. (4.1)], the adjoint equation [Eq. (4.8)], and the maximum principle [Eq. (4.9)]. The state and the adjoint equations are first-order, ordinary differential equations; their solutions require the specification of initial or terminal conditions. Two such conditions are included in the problem: the initial condition Eq. (4.2), and the terminal condition Eq. (4.4). Thus in principle we have just the right number of conditions to determine the unknown functions $x(t)$, $u(t)$, and $\lambda(t)$.

In practice the actual solution of these equations can be distinctly nontrivial. Analytic solutions can be found in various special cases, but in general it is necessary to revert to numerical techniques. These may be quite difficult in themselves, due to the fact that Eqs. (4.1) and (4.8) and the conditions given in Eqs. (4.2) and (4.4) comprise a *two-point boundary-value problem*. However, these technical difficulties do not concern us here.

An additional difficulty of considerable importance is the *existence* of optimal controls, particularly for problems that arise from economic models. Later (especially in Section 5.5), we will see that the lack of a solution sometimes indicates a failure of the model to reflect a real-world situation accurately. Two not entirely independent methods exist for dealing with such contingencies: (1) the model can be revised, or (2) the concept of admissible control can be altered.

Linear Variational Problems

As a first application of the maximum principle, we again consider the linear variational problem in Section 2.7:

$$J = \int_0^T [G(x, t) + H(x, t)\dot{x}] \, dt \tag{4.11}$$

$$A(x, t) \leq \dot{x}(t) \leq B(x, t). \tag{4.12}$$

We introduce the state equation

$$\frac{dx}{dt} = u. \tag{4.13}$$

The Hamiltonian of this problem is

$$\mathcal{H} = [G + Hu] + \lambda u = G + (H + \lambda)u. \tag{4.14}$$

According to the maximum principle [Eq. (4.9)], the optimal control $u(t)$ must maximize this expression. Define

$$\sigma(t) = H(x, t) + \lambda(t). \tag{4.15}$$

Then clearly $u(t)$ must satisfy

$$u(t) = \begin{cases} B(x, t) & \text{whenever} \quad \sigma(t) > 0 \\ A(x, t) & \text{whenever} \quad \sigma(t) < 0. \end{cases} \tag{4.16}$$

A control that takes on these extreme values is called a *bang-bang* control, and for obvious reasons the function $\sigma(t)$ is referred to as the *switching function*.

However, note that when the switching function $\sigma(t)$ vanishes, the

Hamiltonian becomes independent of u, so that the maximum principle does not specify the value of the optimal control. The most important case (called the *singular* case) arises when $\sigma(t)$ vanishes identically over some time interval of positive length:

$$\sigma(t) = H(x, t) + \lambda(t) \equiv 0. \tag{4.17}$$

The corresponding *singular control* $u(t)$ can now be determined as follows. First differentiate Eq. (4.17):

$$\frac{d\sigma}{dt} = \frac{\partial H}{\partial x}\frac{dx}{dt} + \frac{\partial H}{\partial t} + \frac{d\lambda}{dt}$$

$$= \frac{\partial H}{\partial x} u + \frac{\partial H}{\partial t} - \frac{\partial \mathcal{H}}{\partial x}$$

$$= \frac{\partial H}{\partial x} u + \frac{\partial H}{\partial t} - \frac{\partial G}{\partial x} - \frac{\partial H}{\partial x} u$$

$$= \frac{\partial H}{\partial t} - \frac{\partial G}{\partial x} = 0.$$

Note that this is the equation of the *singular path* $x = x^*(t)$ in Section 2.7. Thus the singular case $\sigma(t) \equiv 0$ exactly corresponds to the singular solution $x = x^*(t)$ described there. The corresponding singular control is simply $u^*(t) = \dot{x}^*(t)$.

Thus the maximum principle implies that the optimal control $u = \dot{x}$ for our linear problem must be a combination of bang-bang and singular controls. Further ad hoc arguments can be adduced to determine the correct juxtaposition of these elements, but our most-rapid approach theory (Section 2.7) already suffices and the topic is not pursued further here.

The reader should not be discouraged to find that the maximum principle provides a rather awkward approach to the simple linear control problem described above. It is only to be expected that a general theory is less than optimally efficient when applied to trivial problems!

Transversality Conditions

In certain problems still to be encountered in this book, no terminal value $x(T)$ is specified for the state variable. In such problems, referred to as *free terminal-value problems*, the maximum principle is extended by adding the *transversality condition*

$$\lambda(T) = 0. \tag{4.18}$$

Thus we still have a two-point boundary-value problem for the differential Eqs. (4.1) and (4.8).

For example, suppose that the terminal stock level $x(T)$ is not specified in our fishery model:

$$\frac{dx}{dt} = F(x) - h(t), \quad 0 \le t \le T$$

$$x(0) = x_0$$

$$J\{h\} = \int_0^T e^{-\delta t} \{p - c(x)\} h(t)\, dt.$$

The Hamiltonian for this problem is

$$\mathcal{H} = e^{-\delta t} \{p - c(x)\} h + \lambda(t) \{F(x) - h\}.$$

The switching function is therefore

$$\sigma(t) = e^{-\delta t} \{p - c(x)\} - \lambda(t).$$

Consequently, along the singular path $\sigma(t) = 0$ we have

$$\lambda(t) = e^{-\delta t} \{p - c(x)\}. \tag{4.19}$$

[The economic significance of Eq. (4.19) will be discussed in Section 4.3.] Because $p > c(x^*)$, the transversality condition [Eq. (4.18)] implies that we must leave the singular path $x = x^*$ before $t = T$. While off the singular path, we must use bang-bang control $h = 0$ or $h_{\max}$. Clearly $h = 0$ is not optimal, because $h = h_{\max}$ for t near T provides a positive contribution to present values. Thus the transversality condition supports our intuition: the optimal harvest policy follows the singular path for $t \le t_0$, and then uses the maximum harvest rate $h = h_{\max}$ for $t_0 < t < T$ (see Figure 4.1).

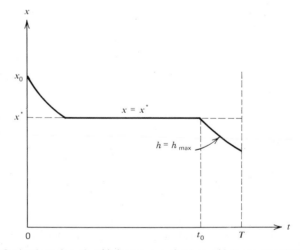

Figure 4.1 Optimal stock path $x(t)$ (heavy curve) when $x(t)$ is not specified.

The actual value $x(T)$ attained, which must obviously satisfy $x_\infty \leq x(T) < x^*$, can be determined by numerical computation.

Another type of problem to be encountered later (with regard to exhaustible resources) is the *free terminal-time problem*, in which T is not specified in advance (although the "target" $x(T)$ is specified). In this case the maximum principle includes the extra condition

$$\mathcal{H}_T = \mathcal{H}[x(T), T, u(T); \lambda(T)] = 0. \tag{4.20}$$

4.2 A NONLINEAR FISHERY MODEL

We now turn to an idealized nonlinear optimization problem based on our fishery model. Our state equation remains·

$$\frac{dx}{dt} = F(x) - h(t) \tag{4.21}$$

with given initial and terminal conditions

$$x(0) = x_0, \quad x(T) = x_T. \tag{4.22}$$

Now we suppose that the revenue obtained from harvesting is a nonlinear function $R(h)$, with $R(h)$ a smooth, convex, nonnegative function of $h \geq 0$. [If, for example, the price p of fish depends on the rate of output h, then we have $R(h) = p(h) \cdot h$. Such "demand curves" $p = p(h)$ are discussed in Chapter 5.] For simplicity we neglect the costs of harvesting here; these costs will be reintroduced in the next chapter. The objective functional is then

$$J\{h\} = \int_0^\infty e^{-\delta t} R(h) \, dt. \tag{4.23}$$

How does the nonlinearity in $R(h)$ affect the optimal harvest policy? [The preceding nonlinear harvesting model is mathematically identical to the Ramsey–Cass–Samuelson aggregated model of optimal economic growth. See, for example, Intriligator (1971, p. 399ff).]

The Hamiltonian for this problem is

$$\mathcal{H} = e^{-\delta t} R(h) + \lambda\{F(x) - h\}. \tag{4.24}$$

If the control constraint $h \geq 0$ is not binding, the maximum principle implies that

$$\frac{\partial \mathcal{H}}{\partial h} = e^{-\delta t} R'(h) - \lambda = 0$$

or

$$\lambda = e^{-\delta t} R'(h). \tag{4.25}$$

Hence

$$\frac{d\lambda}{dt} = e^{-\delta t} \left[-\delta R'(h) + R''(h) \frac{dh}{dt} \right].$$

On the other hand the adjoint equation is

$$\frac{d\lambda}{dt} = -\frac{\partial H}{\partial x} = -\lambda F'(x) = -e^{-\delta t} R'(h) F'(x).$$

Equating these two expressions we obtain

$$\frac{dh}{dt} = \frac{R'(h)}{R''(h)} [\delta - F'(x)]. \tag{4.26}$$

We also record the state Eq. (4.21) again:

$$\frac{dx}{dt} = F(x) - h. \tag{4.27}$$

Equations (4.26) and (4.27) are therefore necessary conditions for the optimal control $h(t)$ and the corresponding response $x(t)$.

Equations (4.26) and (4.27) constitute what mathematicians call a *plane autonomous system* of ordinary differential equations; the word "autonomous" refers to the fact that the right-hand sides of these equations are not explicit functions of the time t. Such systems have a well-developed theory, which we discuss in some detail in Chapter 6. Our present analysis of this system is therefore rather brief.

From the elementary theory of ordinary differential equations, the system represented by Eqs. (4.26) and (4.27) possesses a unique solution $(x(t), h(t))$ passing through any given initial point (x_0, h_0). These solutions form a family of *trajectories*, one through each point of the (x, h)- plane (see Figure 4.2). The geometrical analysis of these trajectories is facilitated by considering the *isoclines*; that is, the curves on which $\dot{x} = 0$ and $\dot{h} = 0$, respectively. From Eqs. (4.26) and (4.27), the x- isocline is the curve $h = F(x)$, whereas the h- isocline is the vertical line $x = x^*$, where

$$F'(x^*) = \delta. \tag{4.28}$$

[If $R'(h_1) = 0$ then the horizontal line $h = h_1$ is also an h isocline of the preceding system. We neglect this possibility here, but return to it in Chapter 5.] The intersection of these two isoclines is the point (x^*, h^*); this is the *equilibrium point* for the given system of differential equations.

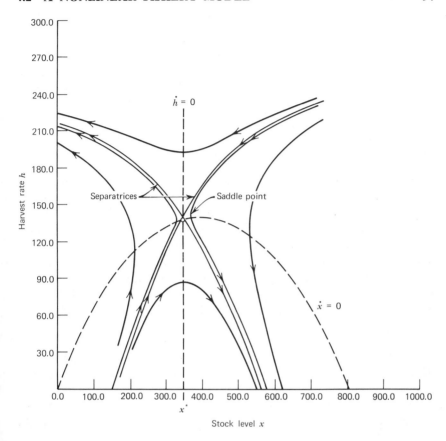

Figure 4.2 Phase-plane diagram of the nonlinear optimal harvest problem (computer-generated graph).

Thus we find that our nonlinear harvesting problem, like the linear harvesting problem in Chapter 2, possesses an optimal equilibrium solution $x = x^*$. Furthermore, in the present model with zero harvesting costs, the equilibrium solutions are the same, as can be seen by comparing Eq. (4.28) with the earlier result in Eq. (2.16). The optimal approach to equilibrium, however, is no longer a bang-bang approach, as we now demonstrate.

To determine the optimal solution $(x(t), h(t))$ corresponding to the given initial and terminal conditions represented by Eq. (4.19), we proceed as follows. First we observe that the equilibrium point (x^*, h^*) is a *saddle point* for the differential system in Eqs. (4.26) and (4.27). Associated with the saddle point are particular trajectories, called *separatrices* (these terms are described more fully in Chapter 6), that either converge toward the saddle point as $t \to \infty$ or diverge away from it as t increases from $-\infty$. All other trajectories first approach the saddle point, but then veer off and diverge from it (again see Figure 4.2). The entire collection of these trajectories is referred to as the *phase-plane diagram* of the system represented by Eqs. (4.26) and (4.27).

Two possible configurations for the initial and terminal stock levels x_0, x_T appear in Figure 4.3, where the dashed curves are the separatrices. In Figure 4.3*a* $x_T < x^* < x_0$. We must choose some trajectory $[x(t), h(t)]$ which starts at $t = 0$ on the line $x = x_0$ and terminates on $x = x_T$ at time T. Since (x^*, h^*) is an equilibrium of the system, the velocity of points on trajectories near (x^*, h^*) is small. Thus the length of time taken to traverse a given trajectory increases as the trajectory passes closer to the equilibrium point. If T is small then the required optimal trajectory remains far from (x^*, h^*), whereas if T is large then the trajectory lies near the separatrices and the point $(x(t), h(t))$ spends "most of its time" in the vicinity of the equilibrium point (x^*, h^*). [This observation constitutes the *catenary turnpike theorem* of Samuelson (1965).] Similar considerations apply to other initial and terminal conditions (e.g., see Figure 4.3*b*).

Now let the time horizon $T \to +\infty$ and consider the limiting position of the optimal trajectory shown in Figure 4.3*a*. This limiting position clearly coincides with the separatrices. When $T = \infty$, the part of the trajectory from (x^*, h^*) to the terminal point $x = x_T$ simply drops out; that is, the terminal condition does not apply at $T = \infty$. This is fortunate in that the

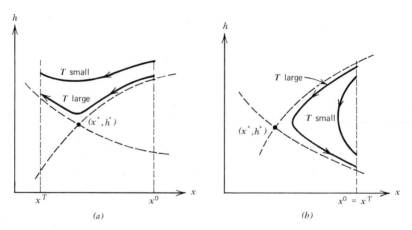

Figure 4.3 Various optimal trajectories.

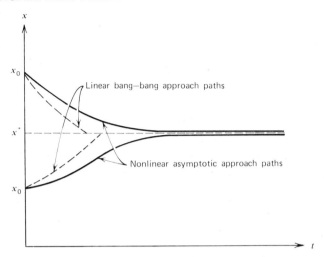

Figure 4.4   Optimal stock paths $x(t)$ for linear and nonlinear fishery models. Two cases are shown: $x_0 > x^*$ and $x_0 < x^*$.

terminal condition is an ad hoc device we adopted solely for mathematical convenience.

We conclude therefore that for the case of an infinite time horizon, the optimal harvest rate $h(t)$ is determined by following the separatrix that begins at $x = x_0$ and approaching the equilibrium point $x = x^*$ asymptotically. The resulting stock-level path $x = x(t)$ is indicated in Figure 4.4; the bang-bang approach obtained in the linear theory also appears in the figure.

What is the practical significance of this result? Recall that in the linear model where the revenues obtained from fishing were assumed to be directly proportional to the harvest, the optimal policy consisted of adjusting $x(t)$ as rapidly as possible to the equilibrium level x^*. The problems of "flooding the market" and driving down the price of fish by harvesting at a high rate were not assumed. Conversely, low harvests were not assumed to lead to increased prices. But these are precisely the phenomena introduced in the nonlinear model, and they have the expected effect shown in Figure 4.4. The optimal approach path is not the most-rapid approach; rather it is a more gradual approach engendered by market reactions to the harvest rate. [It is conceivable that the harvest rate $h(t)$ determined from Eqs. (4.23) and (4.24) would violate predetermined control limits $0 \leq h \leq h_{max}$, in which case the optimal policy would be a combination of bang-bang and asymptotic approaches.]

Feedback Control

Clearly nonlinear optimization problems are significantly more complex than linear problems—not only from the theoretical viewpoint, but also from the practical viewpoint of actually calculating the optimal solution. In the nonlinear case we no longer can apply any method that resembles the most-rapid approach harvest policy; instead we must solve the differential Eqs. (4.26) and (4.27) to deduce the optimal policy.

The most-rapid approach policy applied in linear theory is a special case of what engineers call a *feedback control law*. This term refers to control policies (either optimal or not optimal) with the property that the control $u(t)$ at time t is expressed directly as a function of the current state variable $x(t)$. Such control laws are simple to describe and to implement, and they are capable of responding to random fluctuations in the state variable and in the parameters of the problem. (Of course our deterministic models do not address the question of random fluctuations.)

A feedback control law for the above nonlinear problem with an infinite time horizon is easily obtained by numerically computing the separatrices shown in Figure 4.2. Of course this computation is numerically unstable in the vicinity of the saddle-point equilibrium, but this instability is of little practical significance, because we can realistically employ a linear model (viz the linearization of our nonlinear model) in the neighborhood of the equilibrium point (x^*, h^*).

The practical implication of all this theory is that a bang-bang control policy should probably be considered optimal, except in cases where the required harvest rate ($h = 0$ or h_{max}) will produce obviously deleterious effects.

4.3 ECONOMIC INTERPRETATION OF THE MAXIMUM PRINCIPLE

We now present an intuitive "proof" of the maximum principle, in which the economic significance of the adjoint variable and of the Hamiltonian expression become more transparent. Actually the proof given here is rigorously valid in cases where certain strong differentiability assumptions hold. Unfortunately these assumptions are rarely satisfied in practice, and a fully correct proof of the maximum principle is far more complex than the methods employed here suggest.

Let us consider the optimal control problem formulated in Section 4.1:

$$\frac{dx}{dt} = f(x, t, u), \quad t_0 \leq t \leq t_1$$

$$x(t_0) = x_0$$

$$x(t_1) = x_1$$

$$u(t) \in U$$

$$J\{u\} = \int_{t_0}^{t_1} g[x(s), s, u(s)] \, ds.$$

The terminal point (t_1, x_1) is considered fixed, but we treat the initial point (t_0, x_0) as a variable. For each initial point we define the function

$$w(x_0, t_0) = \max J\{u\} = J\{u^*\}, \tag{4.29}$$

where u^*, the optimal control for the above problem, starts at $x(t_0) = x_0$. Such an optimal control is assumed to exist for all initial points (t_0, x_0) under consideration.

The function $w(x_0, t_0)$ is sometimes referred to as the *payoff function*, or the *optimal-return function*. In economic problems where $J\{u\}$ is a present-value integral, $w(x_0, t_0)$ represents simply the present value of the stock level x_0 at time t_0, *presuming that an optimal exploitation policy is to be utilized for $t_0 \le t \le t_1$.* As we know from Chapter 3, $w(x_0, t_0)$ therefore equals the value of the capital asset x_0 at time t_0 (given a time horizon t_1).

First we observe that for any t between t_0 and t_1 we have

$$w(x_0, t_0) = \int_{t_0}^{t} g[x^*(s), s, u^*(s)] \, ds + w[x^*(t), t], \tag{4.30}$$

where $x^*(t)$ is the optimal trajectory and $u^*(t)$ is the optimal control given $x(t_0) = x_0$. This formula, often referred to as the *principle of optimality*, simply says that the optimal policy for the entire interval $t_0 \le s \le t_1$ must be separately optimal for $t_0 \le s \le t$ and for $t \le s \le t_1$. Differentiating Eq. (4.30) with respect to t we obtain

$$\frac{d}{dt} w[x^*(t), t] = -g[x^*(t), t, u^*(t)],$$

or, carrying out the derivative on the left,

$$\frac{\partial w}{\partial x}[x^*(t), t] f(x^*(t), t, u^*(t)) + \frac{\partial w}{\partial t}[x^*(t), t] = -g[x^*(t), t, u^*(t)]. \tag{4.31}$$

Next consider an arbitrary number $v \in U$ and define the "perturbed" control

$$u(t) = \begin{cases} v & \text{for} \quad t_0 \le t \le t_0 + \Delta t \\ u_v^*(t) & \text{for} \quad t_0 + \Delta t \le t \le t_1, \end{cases}$$

where $u_v^*(t)$ denotes the optimal control corresponding to the initial point $(t_0 + \Delta t, x_v(t_0 + \Delta t))$ with x_v representing the response to v. Because $u(t)$ is a suboptimal control we have

$$w(x_0, t_0) = J\{u^*\} \geq J\{u\}$$

$$= \int_{t_0}^{t_0+\Delta t} g(x_v, s, v) \, ds + w[x_v(t_0+\Delta t), t_0+\Delta t].$$

We rewrite this in the form

$$w[x_v(t_0+\Delta t), t_0+\Delta t] - w[x_v(t_0), t_0] \leq - \int_{t_0}^{t_0+\Delta t} g(x_v, s, v) \, ds.$$

Dividing by Δt and letting $\Delta t \to 0$ we obtain

$$\frac{d}{dt} w[x_v(t), t]\bigg|_{t=t_0} \leq - g(x_0, t_0, v).$$

Carrying out the differentiation as above and then replacing the variable point (x_0, t_0) with an arbitrary point (x, t), we have

$$\frac{\partial w}{\partial x}(x, t)f(x, t, v) + \frac{\partial w}{\partial t}(x, t) \leq - g(x, t, v), \qquad (4.32)$$

which is valid for all x, for all $t \in [t_0, t_1]$, and for all $v \in U$.

Thus if we define the function

$$G(x, t, v) = \frac{\partial w}{\partial x}(x, t)f(x, t, v) + \frac{\partial w}{\partial t}(x, t) + g(x, t, v),$$

we have established that

$$\text{(a)} \quad G(x, t, v) \leq 0 \quad \text{for all} \quad x, t, v.$$

$$\text{(b)} \quad G[x^*(t), t, u^*(t)] \equiv 0.$$

We now can see that both the maximum principle [Eq. (4.9)] and the adjoint equation [Eq. (4.8)] follow from these two conditions. Fixing $x = x^*(t)$, we see immediately from conditions (a) and (b) that

$$\max_{v \in U} G[x^*(t), t, v] = G[x^*(t), t, u^*(t)]. \qquad (4.33)$$

We now *define*

$$\lambda(t) = \frac{\partial w}{\partial x}[x^*(t), t] \qquad (4.34)$$

and

$$\mathcal{H}(x, t, u, \lambda) = g(x, t, u) + \lambda f(x, t, u). \qquad (4.35)$$

Because the term $\partial w / \partial t$ can be canceled out, Eq. (4.33) reduces to

$$\max_{v \in U} \mathcal{H}[x^*(t), t, v, \lambda(t)] = \mathcal{H}[x^*(t), t, u^*(t), \lambda(t)],$$

which is precisely the maximum principle given by Eq. (4.9).

Now fixing $u = u^*(t)$, we see that conditions (a) and (b) imply that $G[x, t, u^*(t)]$ attains its maximum value when $x = x^*(t)$. Hence assuming appropriate differentiability,

$$0 = \frac{\partial G}{\partial x}[x^*(t), t, u^*(t)]$$

$$= \frac{\partial^2 w}{\partial x^2}[x^*(t), t]f[x^*(t), t, u^*(t)] + \frac{\partial w}{\partial x}\frac{\partial f}{\partial x} + \frac{\partial^2 w}{\partial x \partial t} + \frac{\partial g}{\partial x}$$

where all functions on the right have arguments $x^*(t), t, u^*(t)$. Noting from Eq. (4.34) that

$$\frac{d\lambda}{dt} = \frac{\partial^2 w}{\partial x^2}[x^*(t), t]f[x^*(t), t, u^*(t)] + \frac{\partial^2 w}{\partial x \partial t},$$

we finally conclude that

$$\frac{d\lambda}{dt} = -\lambda \frac{\partial f}{\partial x} - \frac{\partial g}{\partial x} = -\frac{\partial \mathcal{H}}{\partial x},$$

which is the adjoint Eq. (4.8). This completes the proof of the maximum principle. (The same proof can be easily extended to include the multidimensional case, to be discussed in Section 4.4)

Economic Interpretation

The principal economic fact arising from the foregoing derivation is embodied in definition [Eq. (4.34)] of the adjoint variable

$$\lambda(t) = \frac{\partial w}{\partial x}[x^*(t), t].$$

Because $w(x, t)$ represents the *value* of the capital (or resource) stock x at time t, this definition identifies $\lambda(t)$ as the *marginal value* of the capital stock x at time t. For example, if the stock level is reduced one unit (e.g., by harvesting), its value at time t will be reduced by $\lambda(t)$. J. M. Keynes coined the term *marginal user cost* for the loss in value when a capital asset is reduced by one marginal unit. In this terminology the adjoint variable is, therefore, the marginal user cost along the optimal trajectory $x^*(t)$.

Modern usage leans toward the term *shadow price of capital* for the adjoint variable $\lambda(t)$; the term "shadow price" refers to the fact that the asset's value is not its direct sale value but the value imputed from its future productivity. In any case it should be remembered that $\lambda(t)$ represents the marginal value of the asset at time t.

Now how is the Hamiltonian expression

$$\mathcal{H}(x, t, u, \lambda) = g(x, t, u) + \lambda f(x, t, u)$$

interpreted? The two terms on the right side of the expression can be recognized as *value flows:* $g(x, t, u)$ is the flow of accumulated "dividends" to the objective functional

$$J\{u\} = \int_{t_0}^{t_1} g(x, t, u) \, dt,$$

whereas $f(x, t, u)$ is the flow of "investment" in capital

$$\frac{dx}{dt} = f(x, t, u);$$

to express this investment flow in value terms, it must be multiplied by the shadow price of capital $\lambda(t)$.

Consequently the Hamiltonian $\mathcal{H}(x, t, u, \lambda)$ represents the total rate of increase of total assets (accumulated dividends + capital assets). The maximum principle then asserts that an optimal control $u(t)$ must maximize the rate of increase of total assets. Before the optimal choice of $u(t)$ can be made, however, we must know the shadow price $\lambda(t)$. In a sense the maximum principle reduces the optimal control problem to the problem of determining $\lambda(t)$. (It should be noted that this reduction was actually achieved in the nineteenth century by means of the Hamilton–Jacobi formulation of the calculus of variations. The main contribution of the modern maximum principle lies in its generalization of the classical condition $\partial \mathcal{H}/\partial u = 0$ by the statement that u must maximize $\mathcal{H}$.) Although this leads to an improved understanding of the nature of control problems, it does not generally result in any technical simplification. The remainder of the book makes clear that the determination of $\lambda(t)$ by analytic methods is often extremely difficult and in many cases impossible.

Finally we attempt to interpret the adjoint equation

$$\frac{d\lambda}{dt} = -\frac{\partial g}{\partial x} - \lambda(t) \frac{\partial f}{\partial x}$$

in economic terms. According to R. Dorfman (1969, p. 821), $-\dot{\lambda}$ can be understood as the rate of depreciation of capital, and the adjoint equation then asserts that along the optimal path, this rate of depreciation must equal the sum of the marginal flows to accumulated dividends and to capital assets. Unlike the maximum principle itself, this condition hardly seems self-evident. [In the economic literature, attention is usually focused on the *current shadow price* $\mu(t) = e^{\delta t}\lambda(t)$, rather than on the *discounted shadow price* $\lambda(t)$. Similarly $\tilde{\mathcal{H}} = e^{\delta t}\mathcal{H}(x, t, u, \lambda) = g(x, t, u) + \mu(t)f(x, t, u)$ is called the *current-value Hamiltonian*. With these notational changes the adjoint equation becomes

$$\dot{\mu} = \delta\mu - \frac{\partial\tilde{\mathcal{H}}}{\partial x},$$

which is the form usually encountered (see Exercise 4).]

4.4 MULTIDIMENSIONAL OPTIMAL CONTROL PROBLEMS

In this section we present a general formulation of the Pontryagin maximum principle (Pontryagin et al., 1962), extending the result quoted in Section 4.1 in several directions: (1) the state and control variables become multidimensional vector quantities; (2) the terminal conditions are generalized; (3) a *terminal payoff* is included in the objective functional. Although we do not utilize all aspects of this general formulation for any individual application, it is convenient to prepare a unified statement from which special cases can readily be derived as required.

We now consider the system of n state equations

$$\frac{dx^i}{dt} = f^i(x^1, \ldots, x^n, u^1, \ldots, u^m, t), \quad 0 \le t \le T, \quad i = 1, 2, \ldots, n.$$

$$(4.36)$$

The vector $x = (x^1, \ldots, x^n)$ is called the *state vector*, and the individual components x^i are called the *state variables*. Similarly $u = (u^1, \ldots, u^m)$ is the *control vector*, and the components u^i are the *control variables*; each u^i is assumed to be a piecewise-continuous function of time t. The state equation [Eq. (4.36)] can be written in the simplified vector form

$$\frac{dx}{dt} = f(x, u, t), \quad 0 \le t \le T. \tag{4.37}$$

The initial values of the state variables are assumed known:

$$x(0) = x_0. \tag{4.38}$$

Controls that lie in a given *control set* $u(t) \in U_t$ for all t are called *admissible controls*. The set U_t is a closed, convex set.

We also impose terminal conditions on *some* of the state variables, which by appropriate renumbering may be assumed $x^1, \ldots, x^k$ $(0 \le k \le n)$:

$$x^i(T) = x_T^i, \quad i = 1, 2, \ldots, k. \tag{4.39}$$

Finally we introduce an objective functional of the form

$$J\{u\} = \int_0^T f^0(x, u, t)\, dt + G[x(T)]. \tag{4.40}$$

The term $G[x(T)]$ is called the *terminal payoff*; here we assume that the terminal time T is fixed. The problem is then to determine an *optimal control*; that is, a feasible control [an admissible control such that the response $x(t)$ satisfies the terminal conditions given in Eqs. (4.39)] that maximizes $J\{u\}$ with respect to the class of all feasible controls.

This control problem has various aspects, the first of which is the question of *controllability*; that is, the question of the existence of at least one feasible control. The controllability of multidimensional nonlinear systems in general is a difficult problem that has engendered a large amount of literature. [See, for example, the *Journal of Optimization Theory and Control*, the *SIAM Journal for Control*, and many other specialized publications. An introduction to controllability is given by Lee and Markus (1968, Chapter 2).] In the problems in this book we usually assume controllability; in most cases it is clear that this assumption is valid.

Given that the problem under consideration is controllable (i.e., that feasible controls exist), the next question concerns the existence of an optimal control. Even when $J\{u\}$ is bounded, *the existence of an optimal control is by no means automatically ensured.* Later several feasible control problems are to be encountered that fail to possess an optimal solution, at least under certain conditions. Frequently, however, it can be seen that the given problem merely needs to be reformulated somewhat. Such reformulations are typical of the modern treatment of variational problems and often consist of altering the class of mathematically acceptable "solutions." In other cases, however, the problem itself needs reformulation.

Indeed we are already familiar with one instance of this reformulation process: the linear variational problem in Section 2.7:

$$\text{maximize} \int_{t_0}^{t_1} f(t, x, \dot{x})\, dt, \quad x(t_i) = x_i$$

with $f(t, x, \dot{x})$ linear in $\dot{x}$. This problem does not possess a solution in the usual class of piecewise-smooth curves $x(t)$ joining the given end points.

The problem can be rendered solvable either by *restricting* the class of admissible solutions (for example, by means of inequalities on $\dot{x}$) or by *extending* it (for example, by allowing impulse controls). Which process is preferable depends on the original model that gave rise to the optimization problem.

The preceding remarks are intended to serve as a mild warning to potential users of the maximum principle: the problem you have formulated may not possess an optimal solution. The author's own experience is that blind application of the maximum principle in such cases soon leads to ridiculous results and quickly necessitates a reconsideration of the original problem. In most economics problems the inadequacies of the model usually become transparent, and a suitable modification suggests itself. [A technicality of considerable importance in questions of the existence of optimal controls concerns the *convexity* of the given control problem in a definite sense (Warga, 1962). The reformulation that is appropriate when this convexity condition fails is described briefly in Section 5.4.]

The Maximum Principle

Provided that the control problem defined by Eqs. (4.37)–(4.40) possesses an optimal solution $u(t)$ with response $x(t)$, the maximum principle asserts a number of necessary conditions that must be satisfied. First we introduce the *Hamiltonian*

$$\mathcal{H}(x, u, t; \lambda) = \lambda^0 f^0(x, u, t) + \sum_{i=1}^{n} \lambda^i(t) f^i(x, u, t), \qquad (4.41)$$

where λ^0 is constant and $\lambda^i(t)$, $i = 1, 2, \ldots, n$, are additional unknowns called the adjoint variables. These adjoint variables must satisfy the adjoint equations

$$\frac{d\lambda^i}{dt} = -\frac{\partial \mathcal{H}}{\partial x^i} = -\sum_{j=0}^{n} \lambda^j \frac{\partial f^j}{\partial x^i} \quad (i = 1, 2, \ldots, n), \qquad (4.42)$$

and we must have $(\lambda^0)^2 + \sum_{1}^{n} \lambda^i(t)^2 \not\equiv 0$ (i.e., all multipliers cannot vanish simultaneously) and

$$\lambda^0 = 0 \quad \text{or} \quad \lambda^0 = 1. \qquad (4.43)$$

In addition we must have for all t, $0 \le t \le T$,

$$\mathcal{H}[x(t), u(t), t; \lambda(t)] = \max_{u \in U_t} \mathcal{H}[x(t), u, t; \lambda(t)]. \qquad (4.44)$$

Finally, for each of the state variables whose terminal values are not specified by Eqs. (4.39), we have a corresponding *transversality condition*

$$\lambda^i(T) = \frac{\partial G}{\partial x^i}, \quad i = k+1, \ldots, n. \tag{4.45}$$

[A generalized form of this transversality condition is given in Exercise 8.] If the terminal time T is not specified, then we have the additional condition

$$\mathcal{H}(T) = 0. \tag{4.46}$$

To repeat, if $u(t)$ is an optimal control for a given problem and if $x(t)$ is the corresponding response, then there must exist an adjoint vector $\lambda(t)$ such that all the conditions in Eqs. (4.42)–(4.46) are satisfied. With regard to the condition in Eqs. (4.43), the case $\lambda_0 = 0$ is somewhat pathological and is referred to as the *abnormal case*. This case cannot arise when $n = 1$. With this proviso the reader can easily verify that the one-dimensional maximum principle (see Section 4.1) is a special case of the above principle.

The following sections describe some simple applications of the maximum principle to linear multidimensional control problems. For linear problems the maximum principle implies that an optimal control must be a combination of singular and bang-bang controls.

4.5 OPTIMAL INVESTMENT POLICIES

The unifying theme of this book is that resource stocks should be considered as capital assets. The exploitation of a resource stock, however, normally requires a certain amount of capital assistance in the traditional sense. A fishery requires boats, gear, docking facilities, and processing plants; forestry requires logging equipment, roads, and sawmills; and so on. In this section we discuss two models of renewable-resource exploitation in which the role of traditional capital is explicit.

Let us reintroduce the basic Gordon–Schaefer model of commercial fishing:

$$\frac{dx}{dt} = F(x) - Ex \quad x(0) = x^0, \tag{4.47}$$

$$0 \le E(t) \le E_{max}. \tag{4.48}$$

In previous investigations of this model we treated the maximum effort E_{max} as an exogenously determined constant. Whereas the lower constraint $E_{min} = 0$ has a perfectly rational conceptual basis (renewable resources cannot usually be created at will), the upper constraint E_{max} seems quite artificial even if it is allowed to vary over time. We now adopt the point of view that the maximum effort rate E_{max} is determined by the level of capital invested in the resource industry under study. Of course this is somewhat of an abstraction; in practice capital investment usually takes several different forms. In a fishery, for example, maximum effort can be determined by the number and size of vessels, whereas the maximum rate of harvest can be limited by the capacity of processing plants. In some circumstances fleet size can be the limiting variable; under other circumstances plant capacity can limit the harvest rate.

Irreversibility

Let $K = K(t)$ denote total capital assets in the fishery at time t. We assume that

$$E_{max} = \alpha K(t), \tag{4.49}$$

where α is a positive constant. Thus $1/\alpha$ equals the *marginal cost of increasing capacity*. Our problem now is to model $K(t)$ in a realistic fashion. One property often considered characteristic of investment in resource-harvesting equipment is *irreversibility*. Fishing vessels have little resale or scrap value unless sold to other fishermen. Norwegian whaling vessels made obsolete by the depletion of whale populations were simply drydocked, except for a few that were refitted for use as factory ships in other fisheries. Many other examples readily come to mind.

Here we model irreversibility by supposing that

$$\frac{dK}{dt} = I(t) \geq 0, \tag{4.50}$$

where $I(t)$ denotes the rate of investment. Admittedly this is an extreme form of the assumption of irreversibility; we comment later on the effect of adopting less extreme hypotheses. Note that we do *not* assume that because the level of capital cannot be reduced, all of the harvesting capacity must be fully utilized. Our assumption is only that

$$E(t) \leq E_{max} = \alpha K(t) \quad \text{for all } t.$$

Concerning the possibility of increasing $K(t)$, we study two cases:

1. *External financing*: Invested capital $K(t)$ can be increased at a given maximum rate:

$$0 \leq I(t) \leq I_{max}. \tag{4.51}$$

The limiting case $I_{max} = \infty$ allows for jump increases in $K(t)$.

2. *Internal financing*: Initial capital K_0 is positive, but $K(t)$ can only be increased by a process of reinvesting the net proceeds from harvesting:

$$0 \le I(t) \le [px(t) - c]E(t). \tag{4.52}$$

In either case the net present value can be written as

$$PV = \int_0^\infty e^{-\delta t}\{[px(t) - c]E(t) - I(t)\}\, dt. \tag{4.53}$$

We wish to maximize this expression, subject to Eqs. (4.47)–(4.50) and either Eq. (4.51) or Eq. (4.52).

The first result to be established is that the optimal rate of investment $I^*(t)$ is always a bang-bang control: $I^*(t) = 0$ or I_{max} in case (1); $I^*(t) = 0$ or $\{px(t) - c\}E^*(t)$ in case (2). To verify this we use the maximum principle, the Hamiltonian being given by

$$\mathscr{H} = e^{-\delta t}\{[px - c]E - I\} + \lambda_1\{F(x) - Ex\} + \lambda_2 I. \tag{4.54}$$

[We consider only the normal case, in which $\lambda_0 = 1$ (see Exercise 9).] Because of the linearity of $\mathscr{H}$ in both control variables E and I, only singular and bang-bang controls can arise. But I cannot be singular; if it is we have

$$\frac{\partial \mathscr{H}}{\partial I} = \lambda_2 - e^{-\delta t} \equiv 0.$$

Hence $d\lambda_2/dt = -\delta e^{-\delta t} < 0$, whereas we must have

$$\frac{d\lambda_2}{dt} = -\frac{\partial \mathscr{H}}{\partial K} = 0,$$

a contradiction. Thus $I^*(t)$ must be a bang-bang control.

The singular case for the effort variable E is readily seen to be identical with the elementary model. Singular control occurs only when $x = x^*$ satisfies the basic rule

$$F'(x^*) - \frac{c'(x^*)F(x^*)}{p - c(x^*)} = \delta, \tag{4.55}$$

where in the present model $c(x) = cx^{-1}$.

It remains to synthesize the overall optimal investment-and-harvest policy. Consider case (1) and suppose first that

$$x(0) > x^*.$$

The optimal investment policy is then simply to increase K to some level

$K^* \geq K(0)$ as rapidly as possible:

$$I^*(t) = \begin{cases} I_{max} & \text{for} \quad 0 \leq t \leq T_1 \\ 0 & \text{for} \quad t > T_1. \end{cases} \tag{4.56}$$

The problem is therefore reduced to determining the optimal level K^*. For simplicity assume that $I_{max} = +\infty$, so that an impulse control is used to increase $K(0)$ to some level $K = K^* \geq K(0)$ at $t = 0$. Now let $PV(E_{max})$ denote the present value of the optimal program, given that $E(t) \leq E_{max}$ for all t. Thus

$$PV(E_{max}) = \int_0^\infty e^{-\delta t}[px(t) - c]E(t) \, dt, \tag{4.57}$$

where $E(t) = E_{max}$ until $x(t)$ is reduced to x^*, after which $E(t) = E^* = F(x^*)/x^*$.

Because $x(0) > x^*$ it follows that $PV(E_{max})$ is a strictly increasing function of E_{max} (see Figure 4.5). This is true because an increase in E_{max} causes the initial bang-bang adjustment to occur more rapidly and the equilibrium to be reached more quickly. Thus the present value of both

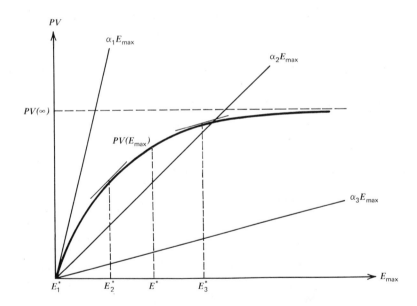

Figure 4.5   Determination of the optimal maximum effort level E_{max}^*: $\alpha_1 < \alpha_2 < \alpha_3$. The corresponding optimal levels are E_1^*, E_2^*, E_3^*. The optimal equilibrium effort level is at $E = E^*$.

harvesting phases is increased, so that $PV(E_{max})$ also increases. Furthermore, $PV(E_{max})$ is bounded by $PV(\infty)$, which is the present value when an initial-impulse harvest reduces the stock instantaneously to x^*.
For any given value of K with $E_{max} = \alpha K$, we have

$$\text{Net present value} = PV(E_{max}) - K = PV(E_{max}) - \frac{1}{\alpha} E_{max}. \qquad (4.58)$$

If $E_{max} = E^*_{max}$ maximizes Eq. (4.58) so that

$$\frac{dPV}{dE_{max}} = \frac{1}{\alpha}, \qquad (4.59)$$

then the optimal level of invested capital is $K^* = (1/\alpha)E^*_{max}$. From Figure 4.5 we can identify three possibilities:

1. High marginal cost of increasing capacity $\alpha = \alpha_1$. In this case $E^*_{max} = \alpha_1 K^* = 0$. Although the fishery can be profitably exploited once the initial investment is made, the cost of this initial investment is too great to allow the industry to develop.

2. Intermediate marginal cost of increasing capacity $\alpha = \alpha_2$. We now have $0 < E^*_{max} < E^* = F(x^*)/x^*$. Investment, while profitable, is too expensive to allow the industry to expand sufficiently to harvest at the otherwise optimal sustained yield level E^*.

3. Low marginal cost of increasing capacity $\alpha = \alpha_3$. In this case $E^*_{max} > E^*$. Initial investment in the fishery expands to a level *greater* than the level required to achieve the ultimate optimal sustained yield. The excess effort level E^*_{max} is utilized until the optimum population x^* is reached, at which time effort must be sharply reduced to E^*.

Excess Capacity

Like most resource industries, marine fisheries have experienced many decades of increasing real prices and increasing efficiency of equipment. The present model shows how a particular fishery can be expected to react to increasing efficiency when it leads to increased capital effectiveness. [Increasing prices and decreasing unit harvesting costs exaggerate the effect by increasing the function $PV(E_{max})$.] At first the fishery is unexploited; eventually exploitation begins, and a temporary equilibrium may be established at an effort level $E < E^*$ and population level $x > x^*$. The critical stage appears when the effort level E^*_{max} exceeds the level E^* corresponding to optimal sustainable yield: once x is reduced to x^*, the fishery experiences an excess capacity that *must not be utilized* under an optimal management policy.

The phenomenon of excess capacity in fisheries is extremely common; it has been described (and decried) by numerous authors. The received theory attributes this phenomenon entirely to the effect of common-property exploitation. Our analysis suggests, however, that excess capacity can even occur under optimal management of the fishery.

Regulatory agencies are frequently presented with the argument that existing excess capacity should be utilized rather than allowed to "go to waste." (This argument has been particularly evident at recent meetings of the International Whaling Commission.) Yet our analysis strongly suggests the opposite conclusion. Even if the excessive equipment has no alternative use and no scrap value, it is better not to use the equipment at all rather than to deplete the fish population to keep the equipment in use. The case is even stronger if the equipment can be sold or refitted for other profitable uses.

The argument to continue to utilize excess equipment "until it wears out" may involve alternative motives. The fishing industry can be expected to perceive a much larger discount rate than the rate employed (perhaps unconsciously) by the regulating agency. The industry frequently wishes to harvest more fish than agency biologists recommend. Making full use of existing equipment is only one of many fallacious economic arguments that the industry may present. (One argument in favor of temporary "overfishing" that does seem to be justifiable in certain cases regards scale effects to be discussed in Section 5.4.)

Finite Investment Rate

In the case in which $I_{max} < \infty$ the marginal decision rule [Eq. (4.59)] must be modified to account for the variation in the population level $x = x(t)$ over time. If $PV(E_{max}, x)$ denotes the present value of future revenues given a present population x and constant E_{max}, then the rule becomes

$$\frac{\partial PV}{\partial E_{max}} = \frac{1}{\alpha}. \tag{4.60}$$

The Initially Overexploited Fishery

Now suppose that $x(0) < x^*$. If the initial level of invested capital $K(0)$ is excessive (i.e., if $K(0) > K^*$), the optimal investment policy is obvious: refrain from harvesting until the fish population has recovered to the optimal level x^*, and then harvest at less than full capacity. If $K(0) < K^*$, clearly K should be increased to its optimal level once the population has recovered. Whether the ultimate sustained yield is at $E = E^*$ or at some lower effort level depends on the marginal cost $1/\alpha$ of increasing capacity as before.

Internal Financing

We now consider case (2), in which the only source of funds for increasing the level of capital investment is the revenue derived from harvesting:

$$0 \leq I(t) \leq [px(t) - c]E(t).$$

We know that the optimal investment policy is a bang-bang policy, so that (assuming $x(0) \geq x^*$)

$$I^*(t) = \begin{cases} [px(t) - c]E^*(t) & \text{for} \quad 0 \leq t \leq T_1, \\ 0 & \text{for} \quad t > T_1. \end{cases}$$

Moreover, $E^*(t)$ is a singular-bang-bang policy:

$$E^*(t) = \begin{cases} E_{\max}(t) & \text{for} \quad 0 \leq t \leq T_2 \\ E^* = \dfrac{F(x^*)}{x^*} & \text{for} \quad t > T_2. \end{cases}$$

(If equilibrium at x^* is never reached, then we have $T_2 = +\infty$.)

It is clear that additional investment should not be undertaken once the optimal equilibrium is reached; that is,

$$T_1 \leq T_2.$$

The optimal investment-and-harvest policy can therefore be described as follows. At first there is a stage of maximum growth of capacity, during which all net revenue is reinvested in additional harvesting equipment. Investment ceases abruptly at $t = T_1$, but harvesting continues at full capacity, thereby reducing the standing fish population. Finally there may be a third stage, beginning at $t = T_2$, in which harvesting continues in equilibrium with less than full utilization of the harvesting capacity. This process is the same as that in case (1).

Therefore, just as in case (1), the problem reduces to the determination of the optimal final level of investment $K^* = (1/\alpha)E^*_{\max}$. As before, it is easy to see that K^* depends on the value of the coefficient α. For consider the problem of deciding at time $t = t_1$ whether to reinvest marginal current revenues to increase capacity $E_{\max}$. A marginal increase $\Delta E_{\max} = 1$ requires an increase $\Delta K = 1/\alpha$. The corresponding marginal gain in the present value of future revenue is $\partial PV/\partial E_{\max}$. Thus the marginal decision rule is

$$\frac{\partial PV}{\partial E_{\max}} = \frac{1}{\alpha},$$

as in case (1). Not surprisingly, the method of determining the optimal level of invested captial K^* turns out to be independent of the method of financing. But K^* itself may depend on the method of financing, because the decision rule also involves the stock level x at the time of the decision, which in turn depends on the previous rate of harvesting and thus the previous rate of increase of E_{max}.

The Internal Rate of Return

In the literature on cost-benefit analysis, a lengthy controversy is associated with the choice between the criteria of maximizing discounted present value and maximizing the rate of return on investment (i.e., maximizing the "internal" rate of return). [This controversy has been discussed with respect to Canadian forest-management by Goundry (1960).] Contemporary economic opinion seems to favor the present-value criterion strongly (Hirschleifer, 1970; Massé, 1962; Mishan, 1971). Hirschleifer points out that the internal-rate criterion involves the tacit assumption that revenues can be reinvested at the same rate ad infinitum and therefore that the investment project can expand indefinitely. No such assumption pertains to the present-value criterion.

Case (2) of the model described in this section suggests a reconciliation of these two investment criteria. When sufficient external financing is not available, exploitation of a finite (but renewable) resource stock undergoes two distinct phases: an initial-growth phase is followed ultimately by a sustained-yield phase. During the growth phase, capital is the limiting factor: the resource is harvested as rapidly as possible, and all revenues are reinvested in the industry. During the sustained-yield phase, the resource stock becomes the limiting factor: further investment is unwarranted, and revenues are diverted to the most attractive alternative investment that yields returns at the opportunity discount rate δ. During the growth phase, maximization of the internal rate of return is the optimal policy, whereas maximization of present value is optimal during the sustained-yield phase.

We also identify an intermediate phase (if the coefficient α is sufficiently large), during which neither capital nor resource stock is strictly limiting, but rather the *anticipated* resource limitation inhibits further capital investment. This phase is clearly a blocked interval, as discussed in Sections 2.7 and 3.3.

Depreciation, Amortization, and Replacement Policy

Problems associated with the depreciation and replacement of equipment are important components of management policy. Fortunately we do not need to concern ourselves with these problems, except to remark that they do *not* affect the conclusions drawn in this section. Even if excess capacity is developed, during the expansion phase of a resource industry—whether deliberately or through lack of proper management—there is no valid reason to overexploit the resource stock. The fact that the excess equipment will ultimately depreciate and wear out is irrelevant. The fact that amortization of the excess equipment may not have been completed is a sign of erroneous amortization and has no effect on the optimal harvest policy whatsoever. As long as the correct discount rate is utilized, the best policy is always to maintain the optimal population level and to refrain from utilizing any excess capacity that may exist. In practice, of course, various subsequent developments, such as an unexpected increase in the resource population or an increase in price, may necessitate either a temporary or a permanent increase in the level of invested capital.

4.6 REGULATION OF FISHERIES: TAXATION

The standard methods state regulatory agencies use to derive public funds from the exploitation of natural resources by private industries include taxes, license fees, and the sale of property rights. Economists are particularly attracted to taxation, partly because of its flexibility and partly because many of the advantages of a competitive economic system can be better maintained under taxation than under other regulatory methods. But more important to the present discussion, taxation can be used—at least in principle—to force private industries to behave in socially desirable ways.

Returning to our fishery model, let us now suppose that the regulating authority imposes a tax of τ per unit of landed fish. We assume that this is the only form of control applied to the fishery, which otherwise operates under open-access conditions. Then a new bionomic equilibrium is established at the population level x for which

$$p' = p - \tau = c(x). \tag{4.61}$$

If $c(x)$ is a strictly decreasing function of x, the tax τ can be chosen to achieve any desired equilibrium level x, such as x_{MSY} or perhaps x^*. [Note that if $p < c(x_{\text{MSY}})$, then a negative tax τ would be required to achieve equilibrium at x_{MSY}. Negative taxes, usually called *subsidies*, are far from unusual in resource industries.] The corresponding sustained economic rent

$$[p - c(x)]F(x) = \tau F(x) \tag{4.62}$$

now accrues to the regulating agency, rather than being dissipated in an open-access fishery.

In terms of the Gordon–Schaefer yield-effort diagram (Figure 2.1), the introduction of a fixed tax τ simply has the effect of multiplying the total revenue curve TR by a constant $\lambda < 1$. The same result may be achieved by multiplying the total cost curve TC by $\lambda^{-1} > 1$. The latter is equivalent to a tax on effort, which can therefore be considered an alternative to the tax τ on catch. In practice, however, taxing the catch appears to be simpler and more effective than taxing effort, particularly since the catch can be easily measured whereas effort is notoriously difficult to quantify.

Of course Eqs. (4.61) and (4.62) are only equilibrium conditions. To discuss a dynamically optimal taxation policy, we must hypothesize the reaction of the fishery to disequilibrium conditions resulting from the imposition of a tax. As a simple but extreme case, suppose that the reaction is instantaneous; that is, the harvest rate is adjusted in such a way that net revenues to the fishermen are continuously dissipated. If $\tau = \tau(t)$ denotes the tax at time t, we then have

$$p - c[x(t)] \equiv \tau(t). \tag{4.63}$$

For a given tax policy $\tau(t)$, we therefore obtain

$$\frac{dx}{dt} = -\frac{\dot{\tau}}{c'(x)}, \quad x(0) = x_0$$

and hence

$$h(t) = F(x) - \dot{x} = F(x) + \frac{\dot{\tau}}{c'(x)}. \tag{4.64}$$

Thus the harvest rate $h(t)$ is determined by the tax policy, and vice versa.

Now suppose that the objective of the regulatory agency is to maximize the total rent (i.e., the present value) produced by the fishery. Because by assumption no rent accrues to the fishermen, the only rent obtained is the tax revenue:

$$\text{Rent flow} = \tau(t)h(t) = [p - c(x)]h(t).$$

Thus the agency's objective is to maximize the total present value of taxes:

$$PV = \int_0^\infty e^{-\delta t}\{p - c(x)\}h(t)\,dt. \tag{4.65}$$

In other words the agency faces the optimization problem of a sole owner: it simply adopts a taxation policy $\tau(t)$, given by Eq. (4.64), that results in the optimal harvest policy $h(t)$.

For example, suppose that $x(0) = x_\infty$ and $\tau(0) = 0$; that is, suppose that the agency begins with an open-access fishery in bionomic equilibrium.

Then the optimal policy uses $h(t) = 0$ until x increases to the optimal level x^*. According to Eq. (4.64), the appropriate tax is given by

$$\tau(t) = \begin{cases} \displaystyle\int_0^t - c'[x(t)]F[x(t)]\, dt & \text{when} \quad x(t) < x^* \\ \tau^* = p - c(x^*) & \text{when} \quad x(t) = x^*. \end{cases}$$

Because of the constraint $h(t) \geq 0$, however, a much simpler tax policy than this one serves the same purpose. If the tax is set at the fixed rate τ^*;

$$\tau(t) \equiv \tau^* = p - c(x^*) \quad (t > 0),$$

the same result obviously occurs, because the fishery cannot operate when rent is negative.

If the complete closure of the fishery is unacceptable, Eq. (4.64) specifies the tax policy corresponding to any desired harvest rate $h(t)$, still under the assumption [Eq. (4.63)] of identically zero revenues (net of opportunity costs) to fishermen.

A Dynamic Reaction Model

Equation (4.63) assumes (quite unrealistically) that the fishery will instantaneously adjust itself to dissipate perceived economic rent at all times. A more realistic model assumes that the level of effort utilized in the fishery expands or contracts when the perceived rent is positive or negative, respectively. Let us consider the following model of this dynamic interaction between the fishery and the effort expended by fishermen:

$$\left. \begin{aligned} \frac{dx}{dt} &= F(x) - Ex \\ \frac{dE}{dt} &= kE[(p - \tau)x - c]. \end{aligned} \right\} \tag{4.66}$$

Equations (4.66) imply that effort increases or decreases in proportion to the perceived rent flow $\{(p - \tau)x - c\}E$. The constant k is a *stiffness parameter* measuring the strength of the reaction of effort to perceived rent. As $k \to \infty$ in the limiting case, the reaction becomes instantaneous, and we obtain $(p - \tau)x - c \equiv 0$.

The initial values of x and E are assumed to be

$$x(0) = x_0, \quad E(0) = E_0, \tag{4.67}$$

and the objective is to determine a tax policy $\tau = \tau(t)$ so as to maximize

$$J\{\tau\} = \int_0^\infty e^{-\delta t}\{px(t) - c\}E(t)\, dt. \tag{4.68}$$

Because $px - c = (p - \tau)x - c + \tau x$, we see that the rent flow is divided between the fishermen, who receive the portion $[(p - \tau)x - c]E$, and the taxing authority, which receives the remainder $\tau x E$.

This problem is linear in the control variable τ, so that we require constraints of the form

$$\tau_{min} \leq \tau \leq \tau_{max}. \tag{4.69}$$

The case in which $\tau_{min} < 0$ allows us to consider subsidies, which in this case would have the effect of increasing the rate of expansion of the fishery.

The Hamiltonian is

$$\mathcal{H} = e^{-\delta t}[px - c]E + \lambda_1(t)[F(x) - Ex] + \lambda_2(t)kE[(p - \tau)x - c]. \tag{4.70}$$

(Again we consider only the normal case.) Singular control occurs when the coefficient of τ vanishes:

$$\lambda_2(t)kE(t) = 0.$$

Later we note that $E(t) > 0$ for all t; hence we may suppose that $\lambda_2(t) \equiv 0$. From the adjoint equation for λ_2

$$\frac{d\lambda_2}{dt} = -\frac{\partial \mathcal{H}}{\partial E} = -e^{-\delta t}[px - c] + \lambda_1 x - \lambda_2 k[(p - \tau)x - c)], \tag{4.71}$$

we obtain

$$\lambda_1(t) = e^{-\delta t}\left\{p - \frac{c}{x}\right\}. \tag{4.72}$$

Thus along the singular path, $\mu(t) = e^{\delta t}\lambda_1(t)$ is the usual shadow price (see Eq. 4.19). Furthermore because $\lambda_2 = 0$, we have

$$\frac{d\lambda_1}{dt} = -\frac{\partial \mathcal{H}}{\partial x} = -e^{-\delta t}pE - \lambda_1[F'(x) - E]. \tag{4.73}$$

By differentiation of Eq. (4.72) we obtain, upon using Eqs. (4.73) and (4.66)

$$F'(x) + \frac{cF(x)}{x^2(p - c/x)} = \delta,$$

which is once again our basic optimal-equilibrium formula [Eq. (2.16)]. We conclude that for the present problem, singular control occurs only at the optimal equilibrium

$$x = x^*, \quad E = E^* = \frac{F(x^*)}{x^*}, \quad \tau = \tau^* = p - \frac{c}{x^*}.$$

Next we consider the optimal approach path. Since $\tau(t)$ must maximize the linear Hamiltonian expression in Eq. (4.70), a bang-bang approach $\tau = \tau_{min}$ or τ_{max} must be used. The problem here is somewhat more complicated than it was in the original one-dimensional fishery model, however. [The present model is actually a modification of the famous "linear regulator" problem with which modern optimal control theory began. This linear, time-optimal control problem is discussed in detail in most optimal control theory texts (cf. Lee and Markus, 1968; Hermes and La Salle, 1969; Pontryagin et al., 1962).] First we solve the system given by Eqs. (4.66) with the control τ set equal in turn to the two extreme values $\tau = \tau_{min}$ and $\tau = \tau_{max}$. This gives rise to two families of trajectories in the (x, E) plane, the *min-control* trajectories (see Figure 4.6a), and the *max-control* trajectories (see Figure 4.6b). [The nature of the equilibrium point in Figure 4.6(a) is discussed in Chapter 6, where it is also established that these trajectories do not produce a limit cycle unless $F(x)$ is depensatory.] The desired optimal equilibrium point (x^*, E^*) is approached by using some combination of these trajectories.

First we construct the min-control and the max-control trajectories through the point (x^*, E^*), as shown in Figure 4.7. If the initial point (x_0, E_0) happens to lie on one of these curves, as at point A or point D, then the optimal approach simply uses the appropriate tax τ_{min} or τ_{max} to drive the system to (x^*, E^*), at which point the tax is switched to τ^*. The optimal tax policy thus drives the system to equilibrium as rapidly as possible.

If the initial point (x_0, E_0) does not lie on one of the above curves, then it is impossible to drive the system to (x^*, E^*) by using a single tax rate.

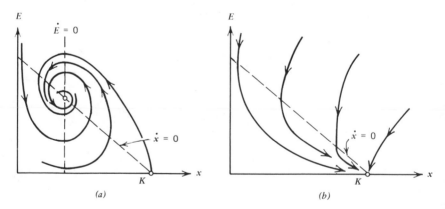

Figure 4.6 Trajectories of the system represented by Eq. (4.66): (a) min-control $\tau = \tau_{min}$; (b) max-control $\tau = \tau_{max}$.

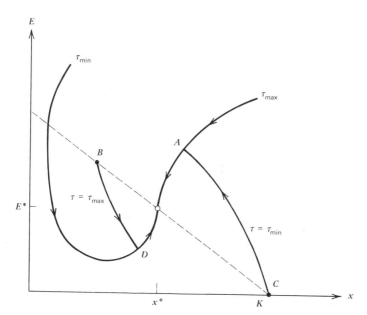

Figure 4.7 Solution of the optimal taxation problem.

For example, suppose that the system begins at the completely unexploited state (point C in Figure 4.7). The min-control path $\tau = \tau_{min}$ should then be used until the system evolves to a point A on the previous trajectory through (x^*, E^*). At this point, the tax is switched to τ_{max} and the system is driven to (x^*, E^*) as before.

The implications of such a policy should be clear. At first the fishery is allowed to expand at its normal rate (if $\tau_{min} = 0$) or the expansion can be subsidized if desired. (Our model suggests no method of determining the optimal values of τ_{min} and τ_{max}.) But unless the rate of expansion is controlled *before* the population level is reduced to x^*, the inertia of the expansion process inevitably results in economic overfishing. The optimal solution therefore calls for the imposition of the maximum tax τ_{max} at some early stage in the developmental process.

Similarly, if the fishery is initially overexploited as it is at point B in Figure 4.7, the tax τ_{max} should be imposed to reduce effort and to build up the fish population. But, as before, this tax should be reduced to τ_{min} at some state (point D in Figure 4.7) before x^* is reached. Finally the optimal tax τ^* is imposed to maintain optimal equilibrium.

In general then the optimal tax program is a rather complicated, three-stage policy consisting of a combination of maximum and minimum taxation, followed eventually by a fixed equilibrium tax rate τ^*. [A rigorous mathematical proof that the optimal policy is actually as described here appears formidable. We therefore allow our intuition to guide us as to the correctness of the given solution. It is easy to see that the problem is controllable in the sense that any initial point (x_0, E_0) with $x_0 > 0$ can be driven to (x^*, E^*) by a tax policy of the type described. Of course this assumes that $\tau_{max} > \tau^*$; in fact Figures 4.6 and 4.7 are drawn under the assumption that $\tau_{max} > p - c/K$, so that any desired population level $x \leq K$ can be achieved through taxation.] Of course it would be a mistake to take the implications of this model too seriously. Nevertheless the model does seem to point out several useful observations, which appear to be valid with somewhat greater generality. First, the control of a common-property resource should be initiated before resource development gets "out of hand"; otherwise control may become extremely difficult. [This preparatory phase may be especially important if, as Smith (1969) suggests, the natural expansion rate of effort exceeds its natural contraction rate; that is, if the coefficient k in Eq. (4.66) takes on different values $k_1 > k_2$ when rent is positive or negative, respectively.] On the other hand the model also suggests why in practice it is often so difficult to achieve control at a sufficiently early developmental stage. Taxation, for example, would transfer some of the initial benefits from exploiters to the taxing authority. While such a result is obviously highly desirable from a social viewpoint, clearly it will be vigorously resisted unless the exploiters can be convinced that resource control is in their own long-run interests.

EXERCISES

1. Consider the calculus-of-variations problem

$$\underset{x \in X}{\text{maximize}} \int_{t_0}^{t_1} g(x, t, \dot{x}) \, dt,$$

where X is the class of all piecewise-smooth functions $x(t)$ [i.e., $\dot{x}(t)$ is piecewise continuous] satisfying given boundary conditions $x(t_i) = x_i$ $(i = 1, 2)$. Assuming that g is nonlinear in $\dot{x}$, derive the classical Euler equation

$$\frac{\partial g}{\partial x} = \frac{d}{dt} \frac{\partial g}{\partial \dot{x}}$$

from the maximum principle. (*Hint*: The solution is simple. Put $u = \dot{x}$, and combine the condition $\partial \mathcal{H}/\partial u = 0$ with the adjoint equation.)

2. Continuing with Exercise 1, show from the maximum principle that if

$x(t)$ is the maximizing function then

$$g(x, t, \dot{x}) - g(x, t, \xi) \geq (\dot{x} - \xi)g_{\dot{x}}(x, t, \dot{x}) \quad \text{for all } \xi \in R.$$

This is the *Weierstrass excess condition* of the classical theory.

3. Consider the *autonomous* control problem

$$\frac{dx}{dt} = f(x, u)$$

$$J\{u\} = \int_0^T g(x, u) \, dt.$$

Show that the Hamilitonian $\mathcal{H}(x, u, \lambda)$ is constant for the optimal control $u(t)$ and its response $x(t)$.

4. Let $\mu(t) = e^{\delta t}\lambda(t)$ be the shadow price associated with the problem:

$$\text{maximize} \int_{t_0}^{t_1} e^{-\delta t} g(x, t, u) \, dt$$

$$\text{subject to } \dot{x} = f(x, t, u).$$

Show that the adjoint equation can be written in the form

$$\frac{d\mu}{dt} = \delta\mu - \frac{\partial \tilde{\mathcal{H}}}{\partial x},$$

$\tilde{\mathcal{H}} = e^{\delta t}\mathcal{H} = g + \mu f$ where $\tilde{\mathcal{H}}$ is the current-value Hamiltonian. This is the form in which the adjoint equation is usually quoted in the economic literature.

5. Discuss the modifications to the maximum principle required when the *minimum* of $J\{u\}$ is desired. (*Hint*: Let $J_1\{u\} = -J\{u\}$, so that the maximization of $J_1\{u\}$ corresponds to the minimization of $J\{u\}$.) NOTE: this exercise explains why the condition in Eq. (4.43) is replaced by $\lambda^0 = 0$ or -1 in many texts.

6. Show that the following problem does not possess an optimal control:

$$\frac{dx}{dt} = u(t), \quad t \geq 0$$

$$x(0) = 0, \quad x(T) = 1, \quad T \text{ unspecified}$$

$$0 \leq u(t) \leq 1$$

$$\text{minimize } J_\phi\{u\} = \int_0^T \phi(t)u(t) \, dt,$$

where $\phi(t)$ is a given, positive, decreasing, continuous function of $t \geq 0$. (*Hint*: Simply observe that inf $J_\phi\{u\} = 0$ but $J_\phi\{u\} > 0$ for any feasible control.) On the other hand if T is specified, then an optimal control exists. What is it? (This example should indicate that a suitable degree of caution should be exercised in applying the maximum principle. *Existence* of optimal controls is a subtle matter not to be considered lightly.)

7. *Exhaustible resources*: Let $x(t)$ denote the reserves of some exhaustible resource at time t. If $q(t)$ is the rate of exhaustion we have

$$\frac{dx}{dt} = -q(t), \quad x(0) = K = \text{original reserves.}$$

Let $p = p(q)$ denote the price at which the resource can be sold if it is produced at rate q. (Thus $p(q)$ is the *demand function*; see Chapter 5.) Assume that $p(0)$ is finite, and that $p'(q) < 0$. Suppose that the resource is to be exhausted at some unspecified time T; that is, assume

$$x(T) = 0.$$

Determine the exhaustion rate that maximizes total discounted revenues

$$J\{q\} = \int_0^T e^{-\delta t} p(q) \cdot q \, dt.$$

Explicitly show (in the order given) that
(a) $\lambda = $ constant.
(b) $q(T) = 0$.
(c) Hence $\lambda = e^{-\delta t} p(0)$.
(d) The optimal extraction rate satisfies $R'[q(t)] = e^{\delta(T-t)} p(0)$, where $R'(q) = d/dq[qp(q)]$ is the marginal revenue.
(e) Hence the rate $q(t)$ decreases, and the corresponding price $p = p[q(t)]$ increases with time.

8. Let M be a given submanifold of R^n. If the terminal condition for the multidimensional control problem in Section 4.4 is

$$x(T) \in M,$$

then the transversality condition becomes

$$\lambda(T) - \nabla G[x(T)] \perp T_{x(T)}(M),$$

where ∇ denotes gradient, $\perp$ denotes normality, and $T_x(M)$ denotes the tangent space to M at x. Verify that this condition reduces to Eq. (4.43) for the case considered in Section 4.4.

9. Prove that the optimal control problem in Section 4.5 is normal, at least for a finite time horizon T, provided that $x(T)$ and $K(T)$ are unspecified. [*Hint*: If abnormal, then $\lambda_0 = 0$. By transversality $\lambda_i(T) = 0$ for $i = 1, 2$. Show that $\lambda_i(t) \equiv 0$, in contradiction to the condition in Eq. (4.43).]

BIBLIOGRAPHICAL NOTES

Modern optimal control theory is considered to originate with the thesis of D. Bushaw (1958), who conjectured a bang-bang principle for linear control problems. Bushaw's conjecture was proved by La Salle (1960). By 1958, the Russian mathematician L. S. Pontryagin had formulated the maximum principle, which unified the linear theory and the classical (nonlinear) calculus of variations. The Pontryagin method is closely related to the Hamilton–Jacobi theory in the calculus of variations.

In addition to the original work of Pontryagin et al. (1962) recommended references are Lee and Markus (1968), an excellent, readable account of the rigorous mathematical theory; and Bryson and Ho (1969), a completely intuitive approach with many engineering applications. There are also many specialized works: on stochastic control (Aström, 1970; Kushner, 1971); control of time-lag systems (Oguztörelli, 1966); control of distributed-parameter systems (Lions, 1971); and others. The books of Bellman (1957), and Bellman and Kalaba (1965) on dynamic programming have many points of contact with optimal control theory.

Of special interest to economists are the works of Dorfman (1969), Burmeister and Dobell (1970), and Intriligator (1971). The third of these works discusses the intuitive relationship between the adjoint variable $\lambda(t)$ and the classical method of Lagrange multipliers. This connection is described in Section 7.8, where a discrete-time version of the maximum principle is discussed.

5

SUPPLY AND DEMAND: NONLINEAR MODELS

The linear fishery models discussed in previous chapters are based on the assumption that the rate of harvesting does not affect the price received for the fish product. In other words the fishery faces an infinitely elastic demand curve (see Section 5.1). While this assumption may be reasonable for a small fishery that supplies a negligible fraction of the total fish product available to consumers, its validity is questionable for large and important fisheries that are likely to be regulated by government. The assumption is also clearly unjustifiable in many other renewable-resource industries, such as forestry.

This chapter is devoted to the study of supply and demand relationships and their implications for dynamic resource-harvesting models. These implications inevitably lead to *nonlinear* dynamic optimization problems and therefore to an increase in the level of difficulty of the theory. In the fishery model various new phenomena arise, both for the open-access fishery and for the managed fishery. Of particular interest is the possibility that severe dynamic instability may be introduced specifically as the result of demand inelasticity. (Of course many other sources of instability, both biological and economic, may also exist.)

Before studying the nonlinear fishery model, however, we digress (in Section 5.2) to examine briefly the economics of exhaustible resources, a subject in which there has been a considerable surge of interest lately perhaps in response to the energy crisis. The mathematical treatment of this subject goes back to H. Hotelling (1931), who emphasized the necessity of employing the methods of the calculus of variations in the theory of exhaustible resource economics. Although the theory possesses special features pertaining to the exhaustibility of the resource, many of

126

the results are also relevant to renewable resources. A case in point is the theorem that from among several pools of a given resource, the pools of highest quality (or of lowest extraction cost) tend to be utilized first. This result obviously pertains equally to forestry, fisheries, and agriculture, particularly in the initial "mining" stage when virgin stocks are reduced to their ultimate sustained-yield levels.

5.1 THE ELEMENTARY THEORY OF SUPPLY AND DEMAND

In this section we briefly outline the fundamental concepts of supply and demand curves and the market equilibrium established by the interplay of supply and demand. The treatment is unavoidably superficial, and the reader who is unfamiliar with this topic is advised to consult a good elementary economics text to obtain an appreciation for the qualifications underlying the theory.

Demand Curves

The *demand curve* or *demand schedule* for a given product defines the total amount Q that consumers will purchase (per unit time) at each price level p. It is a uniform custom in economics to display the demand curve with Q as the abscissa and p as the ordinate (see Figure 5.1). The demand curve is assumed to be nonincreasing; that is an increase in price can only lower the demand.

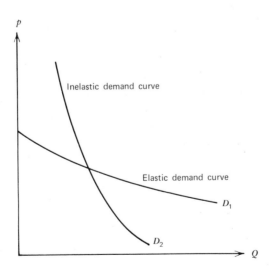

Figure 5.1   Demand curves: D_1 is elastic; D_2 is inelastic.

If $p = P(Q)$ is the equation of the demand curve, then the expression

$$e = -\frac{p}{Q}\frac{dQ}{dp} = -\frac{P(Q)}{QP'(Q)} \tag{5.1}$$

is called the *elasticity of demand.* Since for small Δp

$$e \approx \frac{|\Delta Q|/Q}{\Delta p/p},$$

we can see that the elasticity of demand represents the *percentage* decrease in the quantity demanded that is brought about by a 1% increase in price. For example, the statement that "the demand elasticity for medium-sized American automobiles is 2.5" means that a 1% increase (decrease) in price will lead to a 2.5% decrease (increase) in the sale of these automobiles.

If $e \gg 1$ (i.e., if a small price change results in a relatively large change in the level of demand), then the demand is said to be *highly elastic.* Commodities for which there are readily available substitutes typically have highly elastic demand curves, because an increase in price causes consumers to switch to a substitute product. An extreme case arises when p is constant, so that $e = +\infty$: constant price corresponds to infinitely elastic demand.

Conversely if $e \ll 1$ the demand is highly inelastic, and price changes have little effect on demand. Examples are essential commodities, especially those that constitute a negligible proportion of consumer expenditures, such as sugar.

An interesting property of inelastic demand schedules is the fact that *total revenue* $TR = p \cdot Q = Q \cdot P(Q)$ is a decreasing function of output Q. More precisely $TR = QP(Q)$ is a decreasing function of Q if and only if $e < 1$. The proof of this simple fact is left as an exercise for the reader. This property of demand curves can be invoked to explain the recurrent phenomenon of farmers destroying their crops in bumper years; too large an apple harvest, for example, may "flood the market" and result in a decrease in total revenues to apple growers.

EXAMPLE 1 (FIGURE 5.2)

Suppose demand is a linear function:

$$P(Q) = a - bQ.$$

Then $TR = aQ - bQ^2$, and marginal revenue dTR/dQ equals

$$MR = a - 2bQ.$$

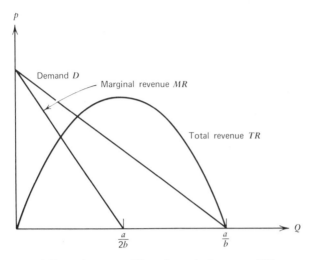

Figure 5.2   Demand D, total revenue TR, and marginal revenue MR curves.

Also elasticity is

$$e = \frac{a}{bQ} - 1.$$

For $Q < a/2b$, total revenue is increasing, marginal revenue is positive, and elasticity > 1; these three conditions are reversed for $Q > a/2b$. This example indicates the fact that in general elasticity depends on output level Q. ∎

EXAMPLE 2.

It can be easily verified that the demand function

$$P(Q) = aQ^{-\alpha} \quad (\alpha > 0)$$

has constant elasticity $e = 1/\alpha$. Conversely any *isoelastic* demand function ($e =$ constant) must have this form (see Exercise 1 at the end of this chapter). ∎

Utility of Consumption and Consumers' Surplus

The area under the demand curve from 0 to Q is called the *social utility of consumption*

$$U(Q) = \int_0^Q P(q)\, dq. \tag{5.2}$$

Thus price $P(Q)$ is equated to *marginal utility of consumption* $U'(Q)$. The difference between total utility $U(Q)$ and the total amount $P(Q) \cdot Q$, paid

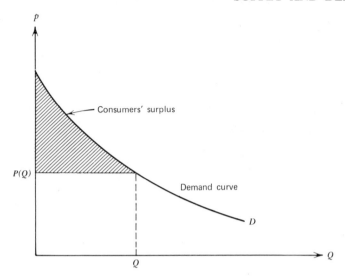

Figure 5.3   Consumers' surplus.

for output Q (see Figure 5.3) is called *consumers' surplus:*

$$\text{Consumers' surplus} = U(Q) - Q \cdot P(Q) = \int_0^Q [P(q) - P(Q)]\, dq. \quad (5.3)$$

Roughly speaking, the significance of these concepts lies in the observation that at price levels $p > P(Q)$, consumers would be "satisfied" to purchase some smaller (but positive) quantity Q' of the commodity. But consumers actually purchase *more than* this quantity at the lower price $P(Q)$. Consumers' surplus is a measure of the total excess "satisfaction" or utility derived from purchasing the amount Q at the price $P(Q)$.

Supply Curves

With the relationship between price and demand behind us, we now turn to the questions of cost and supply. The *supply curve* for a given product (see Figure 5.4) determines the quantity Q of the commodity that producers will supply at various levels of price p. If $p = S(Q)$ is the equation of the supply curve, we define the *elasticity of supply* as

$$e = \frac{S(Q)}{QS'(Q)}. \quad (5.4)$$

Note that e is positive when the supply curve is increasing, which is the normal case.

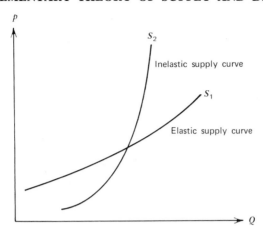

Figure 5.4   Supply curves: S_1 is elastic; S_2 is inelastic.

Analysis of the Firm: Pure Competition

A firm is said to face *pure competition* when the price that it can obtain for its product is independent of its output Q. Although a sloping demand curve may exist for the output of the industry as a whole, the competitive firm's output is too small to have any appreciable effect on the overall price level. Competitive firms are thus said to be "price takers."

Let $TC = C(Q)$ denote the total cost rate for the firm producing output at the rate Q. We suppose that the total cost curve for this firm is as shown in Figure 5.5a, with marginal cost decreasing for $0 < Q < Q_1$ but increasing for $Q > Q_1$. Production efficiency is thus assumed to increase up to a certain level, but to decrease at progressively higher levels of output. (This model refers to short-term costs, and neglects the possibility of long-term adjustments in the cost curve.) The associated *marginal cost curve* $MC = C'(Q)$ and the *average cost curve* $AC = C(Q)/Q$ are shown in Figure 5.5b. Note that the intersection of the marginal cost and average cost curves occurs at the point Q_2 where average cost is minimal (compare Figures 5.5a and b).

What level of production will be realized by the profit-maximizing competitive firm? Because p is fixed, profit $\Pi = pQ - C(Q)$ is maximized by setting *marginal cost equal to price*. There is, however, an additional condition to be met: price p must exceed $p_0 = C'(Q_2) = $ minimum average cost, for otherwise we have

$$\Pi = pQ - C(Q)$$

$$= Q\left[p - \frac{C(Q)}{Q}\right] < 0 \quad \text{for all } Q.$$

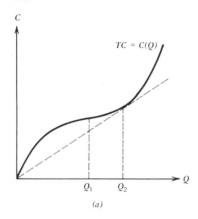
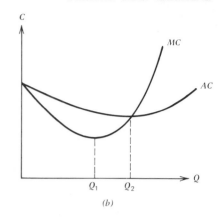

Figure 5.5   Cost curves: (a) total cost, (b) marginal cost and average cost.

Obviously the firm will elect to shut down, rather than to operate at a loss.

The supply curve for the single competitive firm is therefore given by the branch of the marginal cost curve that lies to the right of $Q = Q_2$:

$$S_{\text{firm}}(Q) = C'(Q), \quad Q \geq Q_2.$$

To obtain the total supply curve $S(Q)$ for an entire industry, we simply sum (horizontally) all the supply curves for the individual firms. This sum determines the total supply corresponding to each given price p.

Competitive Equilibrium

Equilibrium of production in a competitive industry occurs when supply price equals demand price; that is, at the point (Q^*, p^*) shown in Figure 5.6. For if $Q > Q^*$, the demand price $P(Q)$ is less than the supply price $S(Q)$ that producers must receive in order to produce at level Q. Output therefore declines toward Q^*, and price increases toward p^*. The process is reversed for $Q < Q^*$.

Competitive equilibrium is socially efficient in the sense that it maximizes the net difference between total social utility of consumption and total cost of production. To see this, let Q_i represent the amount produced by firm i, $i = 1, 2, \ldots, N$. The difference between total utility and total cost is then

$$U\left(\sum_1^N Q_i\right) - \sum_1^N C_i(Q_i),$$

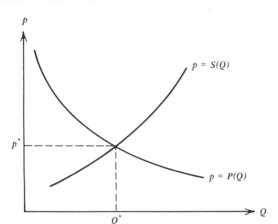

Figure 5.6 Equilibrium of supply and demand.

where $C_i(Q_i)$ is the ith firm's cost function. For a maximum we require

$$U'\left(\sum Q_i\right) - \frac{\partial C_i}{\partial Q_i} = 0 \quad i = 1, 2, \ldots N$$

or

$$\frac{\partial C_i}{\partial Q_i} = U'(Q),$$

where $Q = \sum Q_i$ is total production. Thus all marginal costs must equal a fixed value $p^* = U'(Q) = P(Q)$, which is just the condition of competitive equilibrium.

Monopoly

The polar extreme of pure competition is the case of *monopoly*, in which a single producer controls the entire production and can set the price of a commodity subject to the demand curve $p = P(Q)$. [Of course the monopolist may also manipulate the demand schedule (for example, through advertising). We do not consider such questions here, however.] At output level Q the monopolist's profit is given by

$$\Pi = Q \cdot P(Q) - C(Q). \tag{5.5}$$

To maximize this, the monopolist equates marginal cost $C'(Q)$ with marginal revenue $MR = (d/dQ)[Q \cdot P(Q)] = P(Q) + QP'(Q)$. Because by assumption $P'(Q) < 0$ and the supply curve $C'(Q)$ is rising, we see in Figure 5.7 that the monopolist produces less and charges more than the social optimum. (This assumes that the monopoly supply curve is the

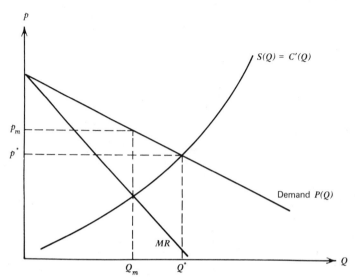

Figure 5.7 Competitive equilibrium (Q^*, p^*) versus monopoly solution (Q_m, p_m) with output $Q_m < Q^*$ and price $p_m > p^*$.

same as the competitive supply curve.) In Section 5.3, however, the supply curve for a renewable resource is seen to be typically backward bending, so that this description of monopolistic production does not necessarily remain valid for this case.

The social implications of the monopoly solution can be seen by examining Figure 5.8. At any level of production Q the monopoly profit [Eq. (5.5)] is the shaded area in Figure 5.8b, whereas consumers'

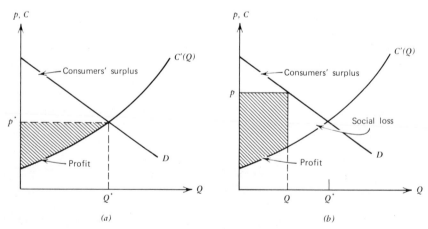

Figure 5.8 Social implications of monopoly: (a) social optimum; (b) monopoly solution.

surplus is the triangular area indicated in the figure. Total social benefits to producers and consumers can therefore be identified as the sum of these two areas (in other words the area between the supply and demand curves) for $0 \leq q \leq Q$. Obviously these total benefits are maximized at $Q = Q^*$. By producing at a lower level $Q < Q^*$, the monopolist increases profit Π at the expense of consumers' surplus. Moreover total social benefits are reduced in the process. Thus monopolistic production leads not only to a redistribution of benefits in favor of the monopolist, but also to a net overall loss of social benefits.

Of course the foregoing analysis is entirely static and therefore cannot be directly applied to the variable production policies that typify the exploitation of both renewable- and exhaustible-resource stocks. These dynamic aspects naturally introduce significant mathematical difficulties.

5.2 THE ECONOMICS OF EXHAUSTIBLE RESOURCES

We now digress to a study of the economics of exhaustible resources. This interesting and important field provides a dynamic setting in which to apply the principles of supply and demand, using variational methods. Some of the results are also applicable to renewable resources; this is particularly true of the initial harvesting stage, which in many ways resembles the mining of an exhaustible resource. Indeed it could well be argued that renewable-resource economics is distinguished from the exhaustible case by the existence of a critical transition from an initial mining stage to an ultimate sustained-yield phase. Many contemporary instances of overexploitation arise because this transition is not carried out in an optimal manner.

Linear Models

We begin with a simple linear model of mineral exploitation; the effects introduced by nonlinearity are examined later.

Let $x = x(t)$ denote the reserves of ore in a particular mine at time t. We assume throughout that the initial reserves $R = x(0)$ are known. If $q = q(t)$ denotes the rate of extraction, then our state equation is simply

$$\frac{dx}{dt} = -q(t), \quad x(0) = R. \tag{5.6}$$

The state variable must satisfy the constraint

$$x(t) \geq 0, \tag{5.7}$$

which is of course crucial to the problem. Also the control variable $q(t)$

satisfies

$$0 \le q(t) \le q_{max}, \qquad (5.8)$$

with the case of $q_{max} = +\infty$ corresponding to the possibility of impulse controls.

We begin by assuming that there are no mining costs and that the market is competitive, with price $p = p(t)$ a known function of time $t \ge 0$. (This particular model is perhaps best described as an "inventory" model, because the problem is merely to optimize the time of disposing of the mineral inventory.) The mine owner's objective is therefore to maximize

$$PV = \int_0^\infty e^{-\delta t} p(t) q(t) \, dt. \qquad (5.9)$$

In the case $q_{max} = +\infty$ the solution to this problem is intuitively clear. Consider the graph of the price function $p = p(t)$ in Figure 5.9, and superimpose on this graph the one-parameter family of "discounting curves"

$$z = ke^{-\delta t},$$

where k is a positive parameter. Choose the highest curve from this family that meets the price contour $p = p(t)$, and let T_∞ denote the point of contact. By extracting (and selling) all the reserves R at time $t = T_\infty$, the mine owner clearly maximizes his present value.

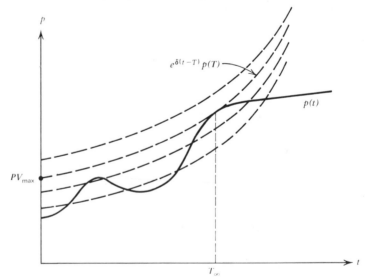

Figure 5.9   Maximization of Eq. (5.9) when $q_{max} = +\infty$.

Notice that with $q_{max} = +\infty$, the problem is reduced simply to maximizing the expression

$$e^{-\delta t}p(t) \cdot R \qquad (5.10)$$

with respect to $t \geq 0$. By differentiation we have

$$\frac{\dot{p}(T_\infty)}{p(T_\infty)} = \delta \qquad (5.11)$$

for the optimal extraction date T_∞. Various versions of this equation are fundamental in exhaustible-resource economics and also arise in many renewable-resource problems (see Chapter 8). For example, Eq. (5.11) is the special case obtained by setting the growth function $F(x) \equiv 0$ in our previous Eq. (3.10) for the singular path in the linear nonautonomous fishery model. The interpretation of Eq. (5.11) is slightly different, however. [In fact we can interpret Eq. (5.11) as defining a *vertical* singular path $t = T_\infty$ that can only be followed by using an impulse control at time T_∞.]

What happens to this solution when $q_{max} < \infty$? Clearly some sort of blocked interval arises (see Section 2.7), with extraction concentrated near the optimal time T_∞ determined from Eq. (5.11). The maximum principle can be used to deduce the exact solution. The Hamiltonian for Eqs. (5.6) and (5.9) is

$$\mathcal{H} = e^{-\delta t}p(t)q(t) - \lambda(t)q(t)$$
$$= [e^{-\delta t}p(t) - \lambda(t)]q(t). \qquad (5.12)$$

By the adjoint equation,

$$\frac{d\lambda}{dt} = -\frac{\partial \mathcal{H}}{\partial x} = 0,$$

so that $\lambda(t) = \lambda =$ constant. Assuming that the reserves will be exhausted at some unspecified time T, we also have the transversality condition in Eq. (4.20)

$$\mathcal{H}_T = [e^{-\delta T}p(T) - \lambda]q(T) = 0.$$

Hence

$$\lambda = e^{-\delta T}p(T). \qquad (5.13)$$

[The control $q(t)$ switches from q_{max} to 0 at $t = T$, but it is always true that the maximized Hamiltonian $\mathcal{H}$ is a continuous function of t (see Pontryagin et al.) Hence Eq. (5.13) follows, even if $q(T)$ is considered to be 0.] Because $q(t)$ must maximize the Hamiltonian given by Eq. (5.12),

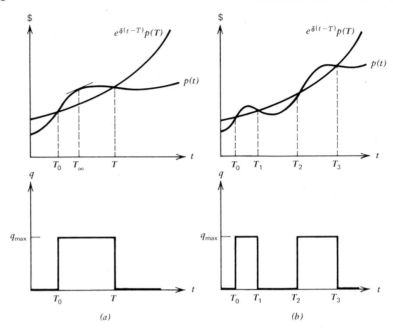

Figure 5.10 Optimal production schedules for an exhaustible resource stock.

we have

$$q(t) = \begin{cases} 0 & \text{when } p(t) < \lambda e^{\delta t} = e^{\delta(t-T)}p(T) \\ q_{max} & \text{when } p(t) > \lambda e^{\delta t} = e^{\delta(t-T)}p(T) \end{cases} \tag{5.14}$$

The solution is indicated in Figure 5.10a. The discount curve $z = e^{\delta(t-T)}p(T)$ intersects the price contour $z = p(t)$ at two points $T_0 < T$. According to Eqs. (5.14), $[T_0, T]$ is precisely the interval during which extraction occurs. The length of this extraction period must satisfy

$$\int_{T_0}^{T} q_{max}\, dt = q_{max}(T - T_0) = R.$$

Clearly as $q_{max} \to +\infty$ the interval reduces to the single point $t = T_\infty$. In the case of a more sinuous price contour (see Figure 5.10b) several periods of production may be required, so that

$$\int_{T_0}^{T_1} q_{max}\, dt + \int_{T_2}^{T_3} q_{max}\, dt + \cdots = q_{max} \Sigma \Delta T_i = R.$$

The Fundamental Price Rule

An apparent implication of the foregoing theory is that generally speaking the real price of exhaustible resources (in a competitive market) can be expected to increase over time at the rate of interest δ. [The theory is easily adapted to cover the case in which $\delta = \delta(t)$ varies over time (cf. Chapter 3, Exercises 1 and 2).] For if $p(t)$ increases at a lower rate mine owners will attempt to dispose of their stocks immediately, whereas if $p(t)$ increases more rapidly than δ owners will tend to hoard stocks. Forces of market equilibrium will then tend to adjust real price levels over time in such a way that Eq. (5.11) becomes an identity:

$$\frac{\dot{p}(t)}{p(t)} \equiv \delta(t). \tag{5.15}$$

This result, which has been called the fundamental theorem of exhaustible-resource economics (see Hotelling, 1931; Gordon, 1967), unfortunately has very little econometric evidence either to support or to refute it. When finite demand elasticities are considered, the rule must be severely modified.

Monopoly: A Nonlinear Model

Next suppose that the exhaustible-resource stock is controlled by a monopolist who faces a demand curve $p = P(q)$ where $q = $ rate of extraction. If extraction costs can be neglected, the monopolist's objective is

$$J\{q\} = \int_0^\infty e^{-\delta t} q P(q)\, dt, \tag{5.16}$$

which is to be maximized subject to the usual conditions:

$$\frac{dx}{dt} = -q, \quad x(0) = R, \quad x(t) \geq 0, \quad q(t) \geq 0.$$

The condition $x(t) \geq 0$ is equivalent to

$$\int_0^\infty q(t)\, dt \leq R. \tag{5.17}$$

We begin by assuming all reserves R will be exhausted at some finite time T to be determined. [The validity of this assumption depends on the demand curve $P(q)$.] Thus $x(T) = 0$ and $q(t) = 0$ for $t > T$, where T is unspecified.

Our Hamiltonian is

$$\mathcal{H} = e^{-\delta t} q P(q) - \lambda q. \tag{5.18}$$

The maximum principle implies that (because $\mathcal{H}$ is nonlinear in q)

$$\frac{\partial \mathcal{H}}{\partial q} = e^{-\delta t} MR(q) - \lambda = 0, \tag{5.19}$$

where $MR = (d/dq)[qP(q)]$ denotes marginal revenue. From the adjoint equation we obtain $\lambda = $ constant, so that

$$\lambda = e^{-\delta T}MR[q(T)]. \tag{5.20}$$

Moreover the transversality condition $\mathcal{H}_T = 0$ yields

$$\lambda = e^{-\delta T}P[q(T)].$$

Consequently we have

$$P[q(T)] = MR[q(T)] = \frac{d}{dq}[qP(Q)]_{q=q(T)} = P[q(T)] + q(T)P'[q(T)],$$

which implies that

$$q(T) = 0, \tag{5.21}$$

except possibly in the case $P'(0) = -\infty$. Since $MR(0) = P(0)$, Eq. (5.20) now becomes

$$\lambda = e^{-\delta t}P(0). \tag{5.22}$$

Thus at $t = T$ the shadow price $\mu(T)$ simply equals the *choke price* $P(0)$, which is the highest price that can be obtained for the mineral. The production rate $q(t)$ must satisfy Eq. (5.19):

$$MR[q(t)] = \lambda e^{\delta t} = e^{\delta(t-T)}P(0). \tag{5.23}$$

Equation (5.23) is the monopolistic analog of the basic rule given by Eq. (5.15). The monopolist adjusts the exploitation rate $q(t)$, so that the marginal revenue grows at the rate of discount. For example, in the case in which $P(q) = a - bq$, Eq. (5.23) becomes

$$q(t) = \frac{a}{2b}(1 - e^{\delta(t-T)})$$

$$p = P[q(t)] = \frac{a}{2}(1 + e^{\delta(t-T)}), \tag{5.24}$$

where the exhaustion date T can be determined from the condition

$$\int_0^T q(t)\,dt = \frac{a}{2b}\int_0^T (1 - e^{\delta(t-T)})\,dt = R.$$

This transcendental equation can be solved numerically.

More generally if ϕ_M denotes the inverse function to the decreasing function $MR(q)$, then Eq. (5.23) implies that

$$q(t) = \phi_M(p_0 e^{\delta(t-T)}), \text{ where } p_0 = P(0).$$

Hence T must satisfy

$$\int_0^T \phi_M(p_0 e^{\delta(t-T)})\, dt = \int_0^T \phi_M(p_0 e^{-\delta u})\, du = R.$$

Because $\phi_M(p_0 e^{-\delta u})$ is an increasing function of u, this equation determines a unique value for T (see Figure 5.11).

Figure 5.12 shows the optimal extraction rate $q(t)$ and the corresponding price $p(t)$ for $T = 100$ years for $\delta = 0.05$ and $\delta = 0.20$. For lower values of T, the appropriate curves have the same graph but start at $t = 100 - T$. Observe that when the exhaustion date T is remote, the monopolist operates near the point $q = a/2b$, $p = a/2$ at which current profits pq are maximized. Only as T becomes imminent is the exhaustion rate significantly reduced and the price correspondingly increased. In particular

$$\frac{\dot{p}}{p} < \delta, \quad \text{but } \frac{\dot{p}}{p} \to \delta \quad \text{as } t \to T.$$

Shadow Price

We recall from Chapter 4 [Eq. (4.34)] that the adjoint variable $\lambda(t)$ can be interpreted as the shadow price of the resource stock at time t (discounted back to time $t = 0$). More precisely $\lambda(t)$ is the shadow price along the optimal trajectory $x^*(t)$. For the foregoing model of exhaustible-resource exploitation, the shadow price is a constant that is independent of time t. According to the adjoint equation $\dot{\lambda} = -\partial \mathcal{H}/\partial x$, this

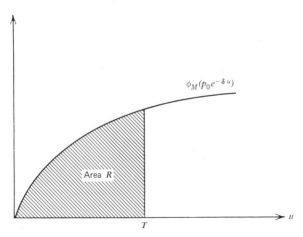

Figure 5.11 Determination of the exhaustion date T.

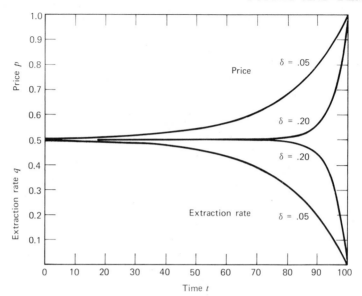

Figure 5.12   Optimal extraction rate $q(t)$ and price $p(t)$ for the monopolist: Eq. (5.24), with $T = 100$ and $a = b = 1.0$.

will be true whenever the Hamiltonian (the total value flow) is independent of the state variable (the capital asset) x.

With regard to the *current* shadow price

$$\mu(t) = e^{\delta t}\lambda(t),$$

we find that for the exhaustible resource model, $\mu(t)$ grows exponentially at the rate δ; that is, $\dot{\mu}/\mu = \delta$. This is the sense in which the basic price formula given by Eq. (5.15) can be considered valid for the monopoly solution and also for the social optimum described as follows.

The Social Optimum

The monopolistic extraction policy is easily compared with the socially optimal policy. Because costs are neglected, the social objective is to maximize total discounted utility:

$$J_{\text{soc}}\{q\} = \int_0^\infty e^{-\delta t} U(q) \, dt, \tag{5.25}$$

where $U'(q) = P(q)$. Minor modifications of the preceding calculation lead to the following analog of Eq. (5.23), for the socially optimal rate of production $q_s(t)$:

$$P[q_s(t)] = e^{\delta(t-T_s)}P(0), \qquad (5.26)$$

where T_s is the socially optimal exhaustion date.

Letting ϕ_s denote the inverse function to $P(q)$, we therefore have

$$q_s(t) = \phi_s(p_0 e^{\delta(t-T_s)}).$$

Also

$$\int_0^{T_s} q_s(t)\, dt = \int_0^{T_s} \phi_s(p_0 e^{-\delta u})\, du = R.$$

Because $P(q) > MR(q)$ for all q, it follows that

$$\phi_s(w) > \phi_M(w) \quad \text{for all } w.$$

Consequently $T_s < T_M$ (see Figure 5.11). Not surprisingly the monopolist *underproduces* (mostly with q near $a/2b$) and therefore practices a *greater* level of conservation than is socially optimal.

However, this result is critically dependent on the hypotheses built into our simple model. For example, if the linear demand curve is replaced by an isoelastic demand curve $P(q) = q^{-\alpha}$, $0 < \alpha < 1$, then the model predicts that the monopolist will precisely follow the socially optimal extraction policy (see Exercise 5). On the other hand if the costs of extraction are considered, the relationship between monopoly and social extraction policies becomes ambiguous.

Cost Effects

In general the costs of extraction $c(x, q, t)$ are expected to depend on the extraction rate $q(t)$, on the level of remaining reserves $x(t)$, and on time t. For mathematical reasons that become apparent later, it is difficult to handle models that involve all three of these independent variables simultaneously. Here we only consider the case in which $c(x, q, t) = c(q, t)$ is independent of the reserve level x (see Exercise 10 for a model in which cost depends on x). The following model applies to the competitive firm facing a nonlinear extraction cost function:

$$J\{q\} = \int_o^\infty e^{-\delta t}[p(t)q(t) - c[q(t), t]]\, dt. \qquad (5.27)$$

This expression is to be maximized subject to the usual conditions:

$$\frac{dx}{dt} = -q(t), \quad x(0) = R, \quad x(t) \geq 0, \quad q(t) \geq 0. \qquad (5.28)$$

Our Hamiltonian is now

$$\mathcal{H} = e^{-\delta t}[pq - c(q, t)] - \lambda(t)q.$$
$$= e^{-\delta t}\{[p - \lambda e^{\delta t}]q - c(q, t)\}. \tag{5.29}$$

The adjoint equation again implies

$$\lambda(t) = \lambda = \text{constant}.$$

According to the maximum principle, the optimal production rate $q = q^*(t)$ must maximize $\mathcal{H}$ over the control set $q \geq 0$ for each time $t \geq 0$. Assuming for the moment that marginal cost $\partial C/\partial q$ is increasing in q, we see that $q^*(t)$ is the unique solution of

$$\frac{\partial C}{\partial q} = p(t) - \lambda e^{\delta t} \tag{5.30}$$

provided this equation has a positive solution. Otherwise $q^* = 0$ (a corner solution); see Figure 5.13.

If the price and cost functions oscillate over time, we obtain an optimal production schedule in which periods of production may alternate with "shutdown" periods, until the resource is ultimately exhausted. This is similar to the linear model described earlier (Figure 5.10b), except that the optimal production rate now varies continuously, rather than switching discontinuously from 0 to q_{max}. [This assumes that $\partial C/\partial q$ is a continuous function of q (see Exercise 3). The preceding analysis assumes that

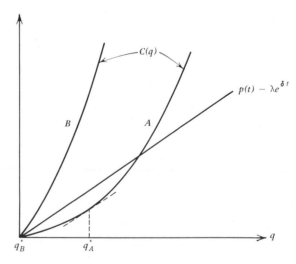

Figure 5.13 The optimal production rate q^* for two cases: A corresponds to Eq. (5.30); B is a corner solution.

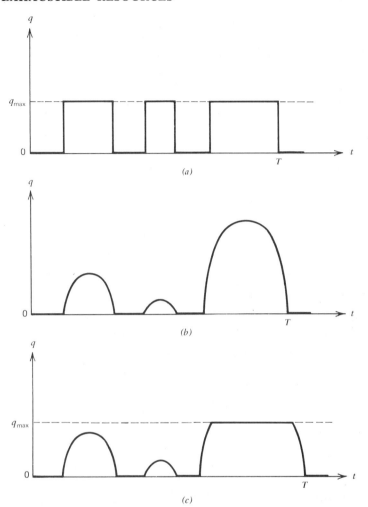

Figure 5.14 Optimal extraction policies: (a) linear model, bang-bang policy; (b) nonlinear model, q_{max} not binding; (c) nonlinear model, q_{max} binding (during third production period).

the upper control constraint q_{max} is either absent or not binding. The case in which q_{max} is binding is not difficult to handle; we simply admit new corner solutions $q = q_{max}$. Figure 5.14 shows the various possibilities.]

To solve Eq. (5.30) for the optimal production rate $q = q^*(t)$, we must first determine the shadow price λ. If the exhaustion date T is finite, it is easy to show that

$$\lambda = e^{-\delta T}[p(T) - \frac{\partial C}{\partial q}(0, T)].\qquad(5.31)$$

This is analogous to Eq. (5.13) and is proved in the same way. Thus if T is known, the entire optimal policy $q^*(t)$ can be deduced. Determination of T, however, is a nontrivial matter. A better approach may be to treat λ as a parameter, use Eq. (5.30) to find $q(t) = q_\lambda(t)$, and then employ an iterative numerical method to determine λ so that

$$\int_0^T q_\lambda(t)\, dt = R.$$

Determination of the adjoint variable (the shadow price) thus becomes progressively more difficult as our optimization problem becomes more complex. Naturally this is to be expected, because the solution of the problem is essentially reduced via the maximum principle to the problem of determining $\lambda(t)$. At least our exhaustible-resource models have the advantage that $\lambda(t)$ is a constant, so that computation is relatively straightforward. The situation is even more complex for nonlinear renewable-resource models.

Nonconvex Cost Functions

Next we consider the case in which the cost function is *nonconvex;* this case is of particular interest in the renewable-resource model described in the next section. For simplicity we consider only the autonomous case, with $p(t) = p$ and $C(q, t) = C(q)$, where $C(q)$ is as shown in Figure 5.15. From the preceding analysis we obtain

$$\lambda = e^{-\delta T}\{p - C'[q(T)]\} = e^{-\delta T}\left\{p - \frac{C[q(T)]}{q(T)}\right\},$$

so that

$$C'[q(T)] = \frac{C[q(T)]}{q(T)}. \tag{5.32}$$

It follows that $q(T) = q^*$, which is the production rate at which marginal cost equals average cost (see Figure 5.15). In general we know that production below this level is never worthwhile because average cost increases for $q < q^*$. In particular if $p < C'(q^*)$ the resource stock will not be exploited at all, because $pq - C(q) < 0$ for all $q > 0$.

The optimal production rate $q(t)$ can be derived from Eq. (5.30), which can now be written in the form

$$p - C'[q(t)] = e^{\delta(t-T)}[p - C'(q^*)]. \tag{5.33}$$

Since $C'(q)$ is increasing for $q > q^*$ Eq. (5.33) implies that (see Figure 5.15)

$$q(t) > q^* \quad \text{for } t < T$$

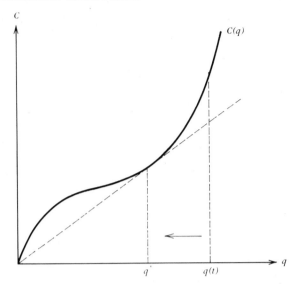

Figure 5.15 Nonconvex cost function: the optimal production rate $q(t)$ decreases toward q^*.

and

$$q(t) \to q^* \quad \text{as } t \to T.$$

Whereas in the convex case we have $q(t) \to 0$ as t approaches the exhaustion date, we now have $q(t) \to q^*$, where q^* is the least economically feasible production rate.

Competitive Exploitation

We can now show that as in Section 5.1, competitive exploitation of a large number of separate mineral deposits leads to a socially optimal extraction policy. However, this result is critically dependent on the assumption that the exploitation involves no externalities. For example, the competitive exploitation of a common pool of petroleum certainly *does* involve significant externalities and generally results in overexploitation and inefficiency.

Consider a mining industry consisting of N firms, the ith firm having a cost function $C_i(q_i)$ and thus an objective functional

$$J_i(q_i) = \int_0^\infty e^{-\delta t} [pq_i - C_i(q_i)] \, dt, \qquad (5.34)$$

where the discount rate δ is assumed given and the price p is to be determined. Equation (5.30) then applies to each firm:

$$\frac{\partial C_i}{\partial q_i} = p - \lambda_i e^{\delta t}, \tag{5.35}$$

where λ_i is the shadow price for firm i. We also have

$$\int_0^\infty q_i(t) \, dt = R_i. \tag{5.36}$$

Finally, market equilibrium requires

$$P = P(Q) = P\left(\sum_1^N q_i\right). \tag{5.37}$$

(Note that the price p in Eq. (5.34) is therefore dependent on q_i, so that Eq. (5.35) is formally incorrect. The assumption of pure competition, however, amounts precisely to neglecting the term $\partial P/\partial q_i$, which is reasonable provided that $q_i \ll Q = \sum_1^N q_i$.)

Equations (5.35)–(5.37) are $2N + 1$ equations for the $2N + 1$ unknowns $q_i(t)$, λ_i, and p. We now show that precisely the same equations arise for the socially optimal exploitation policy. This requires the maximization of

$$J = \int_0^\infty e^{-\delta t}\left[U\left(\sum_1^N q_i\right) - \sum_1^N C_i(q_i) \right] dt,$$

subject to the N state equations

$$\frac{dx_i}{dt} = -q_i(t), \quad x_i(0) = R_i.$$

The Hamiltonian is

$$\mathcal{H} = e^{-\delta t}\left[U\left(\sum q_i\right) - \sum C_i(q_i) \right] - \sum \lambda_i q_i$$

and we immediately obtain $\lambda_i = -\partial\mathcal{H}/\partial x_i = 0$; that is, $\lambda_i = \text{constant}$. Moreover $\partial\mathcal{H}/\partial q_i = 0$ implies

$$\lambda_i = e^{-\delta t}\left[U'\left(\sum q_i\right) - C_i'(q_i) \right]$$

or

$$C_i'(q_i) + \lambda_i e^{\delta t} = U'\left(\sum q_i\right) = P\left(\sum q_i\right).$$

Thus we obtain precisely the same equations we did previously.

The reader should identify the assumption of nonexternalities in the foregoing argument.

Quality

Given several distinct sources of an exhaustible resource, clearly the most profitable sources will be exploited first. The simple model that follows shows why this is true. We consider only two sources x_1 and x_2, with initial reserves $x_i(0) = R_i$. Let $E_i(t)$ denote the effort devoted to extracting ore from source i, and suppose that the resulting rate of production of pure metal equals $\theta_i E_i$, where $\theta_i > 0$ is a coefficient representing the *quality* of the source (assumed constant for each source). To be more explicit, we suppose that

$$\theta_1 < \theta_2, \tag{5.38}$$

or that the second source is more profitable than the first.

We also assume that there is a total effort constraint, so that (with suitably normalized units of effort)

$$E_1 + E_2 \le 1. \tag{5.39}$$

$$E_1 \ge 0, \quad E_2 \ge 0. \tag{5.40}$$

The owner of the two sources faces a given price p, which can be assumed to equal 1, and wishes to maximize

$$J\{E_1, E_2\} = \int_0^\infty e^{-\delta t} [\theta_1 E_1 + \theta_2 E_2] \, dt, \tag{5.41}$$

subject to the above constraints on E_i, and the dynamic equations

$$\frac{dx_i}{dt} = -\theta_i E_i, \quad x_i(0) = R_i, \quad x_i(t) \ge 0 \tag{5.42}$$

for $i = 1, 2$. (The model is easily adapted to account for the case in which there is a constant cost c per unit of effort; this has no effect on the outcome.)

The Hamiltonian of this linear problem is

$$\begin{aligned}
\mathcal{H} &= e^{-\delta t}(\theta_1 E_1 + \theta_2 E_2) - \lambda_1 \theta_1 E_1 - \lambda_2 \theta_2 E_2 \\
&= (e^{-\delta t} - \lambda_1)\theta_1 E_1 + (e^{-\delta t} - \lambda_2)\theta_2 E_2 \\
&= \alpha_1(t) E_1 + \alpha_2(t) E_2.
\end{aligned}$$

The maximum principle requires that (E_1, E_2) must be chosen from the control set U defined by Eqs. (5.39) and (5.40) to maximize this linear expression. It follows that except for special cases (e.g., $\theta_1 = \theta_2$), the optimal control (E_1^*, E_2^*) must always lie at one of the corners of U (see Figure 5.16). Thus $E_1^* = 0$ and $E_2^* = 1$, or vice versa, or $E_1^* = E_2^* = 0$.

Because whenever x_1 and x_2 are both positive a unit of effort devoted to the extraction of x_2 produces greater revenue than one devoted to x_1,

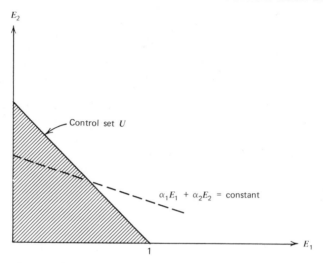

Figure 5.16   The control set given by Eqs. (5.35) and (5.36).

it is intuitively clear that the optimal solution (E_1^*, E_2^*) must satisfy

$$E_1^* = 0, \quad E_2^* = 1 \quad \text{for} \quad\quad 0 \le t \le T_1$$

$$E_1^* = 1, \quad E_2^* = 0 \quad \text{for} \quad\quad T_1 < t \le T_1 + T_2$$

$$E_1^* = 0, \quad E_2^* = 0 \quad \text{for all} \quad t > T_1 + T_2$$

where T_i denotes the time required to exhaust the ith source, using $E_i = 1$. To derive this result via the maximum principle, it is necessary to invoke the *jump condition* (Pontryagin et al., 1962, Chapter 7) pertaining to the state-variable constraints $x_i \ge 0$. We do not pause to work out the details here. A direct proof by reductio ad absurdum is also easily fabricated.

Depletion Allowances

Inasmuch as this book is concerned primarily with renewable resources, our coverage of the exhaustible-resource case is quite superficial. [The reader is referred to Gaffney (1967), Herfindahl and Kneese (1974), and the *Review of Economic Studies* (April 1975) for further details concerning exhaustible-resource economics.] Nevertheless the basic theory given in the preceding section can be used to address various important questions of government policy, such as taxes, depletion allowances, and the like. Let us consider the question of depletion allowances.

According to federal law, a depletion allowance is a tax reduction that is proportional to the annual amount of reserves extracted. It thus plays the role of a negative *severance tax*, a term that refers to a tax proportional to the amount of reserves extracted. Consider the autonomous, competitive model, with price $= p$ and cost $= C(q)$. The modified objective functional is

$$J_\tau\{q\} = \int_0^\infty e^{-\delta t}[(p-\tau)q - C(q)]\, dt, \qquad (5.43)$$

where $\tau =$ constant represents the depletion allowance rate when $\tau < 0$ or the severance tax rate when $\tau > 0$.

From Eq. (5.32) it follows that $q(T) = q^*$, regardless of the value of τ [unless $p - \tau < C'(q^*)$, in which case the optimal extraction policy is $q(t) \equiv 0$]. Equation (5.33) now becomes

$$\frac{(p-\tau) - C'[q(t)]}{(p-\tau) - C'(q^*)} = e^{\delta(t-T)}. \qquad (5.44)$$

To study the effect of the parameter τ, we consider the example $C(q) = cq + \gamma q^2$. Then $q^* = 0$ and Eq. (5.44) can be solved explicitly for $q(t)$:

$$q(t) = \frac{p - \tau - c}{2\gamma}(1 - e^{\delta(t-T)}). \qquad (5.45)$$

First let $\tau = 0$, and consider the family of curves given by Eq. (5.45) for varying values of T (see Figure 5.17a). Precisely one of these curves satisfies the requirement

$$\int_0^T q(t)\, dt = R, \qquad (5.46)$$

which determines the exhaustion date T. Now suppose $\tau > 0$. Each curve of the family given by Eq. (5.45) moves downward (see Figure 5.17b). Obviously the T value determined by Eq. (5.45) must increase. The optimal extraction schedules $q^*(t)$ for these two cases are shown in Figure 5.17c. We see that a severance tax $\tau > 0$ therefore has the effect of spreading out the extraction of the resource and lengthening the period of extraction.

Conversely, the effect of a depletion allowance $\tau < 0$ is to accelerate the extraction of a resource and to shorten the period of extraction. [Ideally a depletion allowance should be proportional to the rate of decrease in the value of the mine, rather than proportional to the extraction rate. Such a depletion allowance would have no effect on the optimal extraction schedule (Hotelling, 1931).]

Since $C'(q)$ is increasing for $q > q^*$, it is easy to see that these results are valid for arbitrary cost functions $C(q)$.

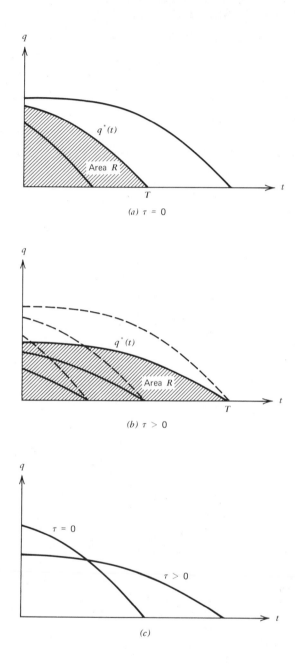

Figure 5.17 The effect of a severance tax $\tau > 0$ is to "spread out" the extraction and postpone the date of exhaustion. A depletion allowance has the opposite effect.

5.3 SUPPLY AND DEMAND IN FISHERIES

In this section we study questions of supply and demand in renewable-resource exploitation, continuing to work within the context of fisheries. We begin with the open-access fishery.

The Equilibrium Supply Curve of the Open-Access Fishery

Again consider the Schaefer model

$$\frac{dx}{dt} = rx\left(1 - \frac{x}{K}\right) - Ex,$$

with the flow of economic rent being given by

$$R = (px - c)E.$$

The conditions for bionomic equilibrium in the open-access fishery are

$$Y = Ex = rx\left(1 - \frac{x}{K}\right)$$

and

$$px - c = 0.$$

Hence we can express the sustained yield Y in terms of the price p:

$$Y = \frac{rc}{p}\left(1 - \frac{c}{pK}\right). \tag{5.47}$$

The graph of this curve, which is the equilibrium supply curve for the open-access fishery (Schaefer model), appears in Figure 5.18. We note that the output Y is zero for $p < c/K$, then increases to MSY at $p = 2c/K$, and subsequently decreases toward zero again as $p \to +\infty$. The supply curve is therefore *backward bending* for $p \geq 2c/K$ as a consequence of the biological overfishing that occurs when effort exceeds the level E_{MSY} that maximizes the sustained yield. Thus Figure 5.18 is simply another way of illustrating the fact that other things remaining equal, biological overfishing takes place when price levels are sufficiently high and becomes more severe as the price level increases (cf. Figure 2.2, where price remains constant but cost varies).

For the more general model:

$$\frac{dx}{dt} = F(x) - h, \quad R = [p - c(x)]h,$$

the open-access supply curve is described by the equations

$$h = F(x), \quad p = c(x).$$

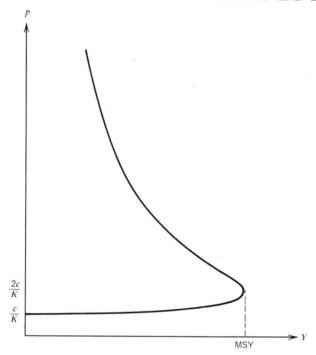

Figure 5.18   Supply curve of the open-access fishery (Schaefer model).

The effects of various combinations of the growth function $F(x)$ and cost functions $c(x)$ are to be described later in this section.

Finite Demand Elasticity

The supply curve in Figure 5.18 specifies the sustained yield or output $Y = h$ corresponding to any given constant price p. Now suppose that the open-access fishery faces a demand curve of finite elasticity, as in Figure 5.19. In this case bionomic equilibrium is established at the point M where the supply and demand curves intersect. This follows from the same argument given in Section 2.1 for the infinitely elastic case: any position on the supply curve below M generates positive rent for the fishermen and therefore causes effort to expand, forcing the system toward M, and so on.

Although the bionomic equilibrium at M fails to generate economic rent, it does result in some positive economic benefits. Because of demand elasticity, consumers' surplus exists (again, see Figure 5.19). But the backward-bending supply curve causes increasing demand to lead to

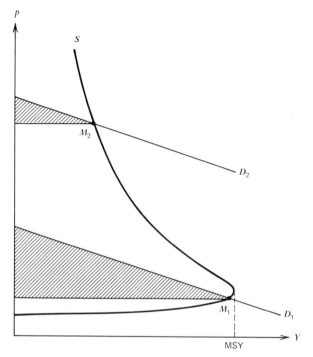

Figure 5.19 Supply and demand in the open-access fishery. The shaded areas represent consumers' surplus.

decreasing amounts of consumers' surplus as biological overfishing becomes more severe. Failure to control the exploitation of the common-property fishery is therefore even more destructive to potential economic benefits in cases where demand is not infinitely elastic.

Bionomic Instability

Still more crimes can be laid at the door of the uncontrolled exploitation of common-property resources. Consider the combination of supply and demand curves illustrated in Figure 5.20. The curve D_2 intersects the equilibrium supply curve at points M_2, M_2', and M_2''. Which of these points is the bionomic equilibrium? It is easy to see that the middle point M_2' is an unstable equilibrium; the open-access fishery will therefore move away from this position and toward either M_2 or M_2''. If the initial position is below M_2', equilibrium will be established at M_2; otherwise equilibrium will be at M_2''. Moreover Figure 5.20 indicates that the entire situation is highly unstable. With demand at D_1 equilibrium is established at M_1, where yield is high and price is low. A relatively minor shift to position D_3,

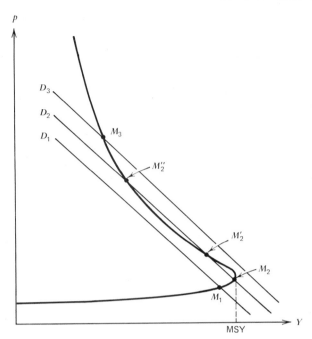

Figure 5.20   Instability of supply and demand in the open-access fishery (Schaefer model).

however, engenders a serious degree of overfishing that greatly reduces sustained yield and increases price.

It is worthwhile to discuss in greater detail the dynamic behavior of an open-access fishery following a critical demand shift of this kind. [A more complete analysis would incorporate an effort-reaction equation such as Eq. (2.4). This question is examined in Chapter 6.] Suppose that the fishery begins in equilibrium at point M_1, but that demand suddenly shifts to D_3. Effort then expands, yield increases temporarily, and price decreases temporarily. As the fish population is reduced, the fishery evolves (perhaps slowly and erratically) towards an eventual new equilibrium at M_3.

The reader may observe a similarity between this model and the description of overfishing "catastrophes" (Section 2.3) arising from depensation effects in the biological growth function $F(x)$. A careful comparison of these two models in fact discloses a remarkable correspondence. In each model an unstable equilibrium separate two stable equilibria, catastrophic "jumps" result from relatively minor shifts in certain parameters, and a "hysteresis" effect delays the return of the equilibrium to its previous position. Yet the two models pertain to quite different causes: depensation is not the same as demand elasticity!

To the mathematician the similarity of these two situations is not surprising: any two-dimensional dynamical system possessing multiple equilibria must behave in essentially this manner. This matter is discussed in greater detail in Chapter 6.

We should also remark here that in a sense the model in Figure 5.20 is far too "precise" to be a realistic description of an actual fishery. In practice random fluctuations in both supply and demand can keep a fishery in a perpetual state of disequilibrium. Our model suggests that such bionomic instability is most likely to occur when demand is reasonably inelastic.

The Discounted Supply Curve

Next we consider the optimally controlled fishery. From our experience with linear models, we conjecture that the theory has many of the same characteristics as the theory of the open-access fishery, particularly in cases where the bionomic growth ratio $\gamma = \delta/r$ is large. This conjecture turns out to be correct, although additional complexities arise because we can no longer equate the social optimum with the sole owner's optimum when demand elasticity is finite.

We begin by letting demand be infinitely elastic (i.e., p is constant). The optimal population level x^* is given by our rule

$$F'(x^*) - \frac{c'(x^*)F(x^*)}{p - c(x^*)} = \delta. \tag{5.48}$$

Solving for the price p we obtain

$$p = c(x^*) - \frac{c'(x^*)F(x^*)}{\delta - F'(x^*)}$$

$$= H_\delta(x^*). \tag{5.49}$$

Equation (5.49) constitutes the definition of the function $H_\delta(x^*)$. The corresponding sustained yield is given by

$$Y = F(x^*). \tag{5.50}$$

Eliminating x^* from Eqs. (5.49) and (5.50), we obtain output Y as a function of price p. This result is the equilibrium supply curve for the optimally controlled fishery, which we refer to as the *discounted supply curve*.

To graph the discounted supply curve for the Schaefer model, we utilize a "four-quadrant" graphical method (see Figure 5.21), which also proves useful for various other cases to be discussed later in this chapter. The S.E. quadrant of this figure contains the logistic growth curve given by Eq. (5.50); the N.W. quadrant contains the graph of Eq. (5.49). In the

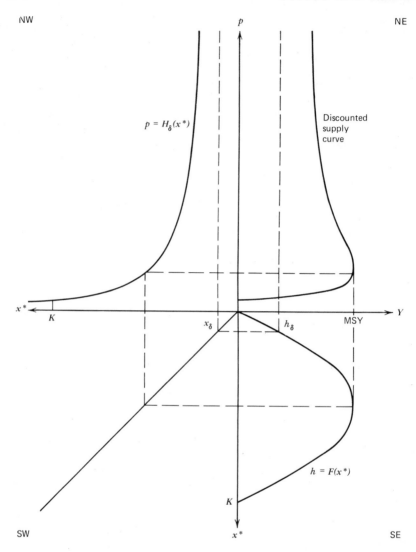

NW p NE

$p = H_\delta(x^*)$

Discounted
supply
curve

x^* K x_δ h_δ MSY Y

$h = F(x^*)$

K

SW x^* SE

Figure 5.21 Discounted supply curve (Schaefer model).

case of the Schaefer model, Eq. (5.49) can be solved explicitly for x^* [cf. Eq. (2.23)]:

$$x^* = \frac{K}{4}\left\{ 1 + \frac{c}{pK} - \frac{\delta}{r} + \sqrt{\left(1 + \frac{c}{pK} - \frac{\delta}{r}\right)^2 + \frac{8c\delta}{pKr}} \right\}.$$

The discounted supply curve, which appears in the N.E. quadrant, is simply the composite of two functions. To graph this curve, we utilize the

45°-transfer line shown in the S.W. quadrant as follows. First we choose a typical value for the variable x^*, using the 45°-line to identify the two x^* axes. Then we complete the rectangle, as shown in Figure 5.21. The S.E. quadrant gives the value of Y, and the N.W. quadrant gives the value of p; thus desired point (Y, p) appears in the N.E. quadrant. The reader should verify this process by checking other x^* values on Figure 5.21. Note that x^* is treated as a parameter that is suppressed in the final graph of the discounted supply curve. However, the four-quadrant diagram immediately identifies the equilibrium population level x^* associated with each equilibrium point (Y, p) on the discounted supply curve.

In general this four-quadrant graphical method is useful in studying problems with three interrelated variables that are subject to two functional relations.

The discounted supply curve just constructed has the following characteristics. For $p < c/K$, there is zero output Y. As p increases from c/K to a value P_{MSY}, output Y increases to MSY. Further increases in p lead to biological overfishing, with Y approaching $Y_\delta = F(x_\delta)$ as $p \to +\infty$, where

$$F'(x_\delta) = \delta.$$

In the case shown $x_\delta > 0$. However, if $\delta > F'(0)$ then $x_\delta = 0$, and the discounted supply curve becomes asymptotic to the vertical p axis.

Thus the discounted supply curve for the optimally controlled fishery (with positive discount rate) is also a backward-bending supply curve similar to the open-access supply curve. Of course the degree of backward bending depends on the rate of discount employed, as indicated in Figure 5.22. For $\delta = 0$ the curve does not bend back at all, but approaches MSY asymptotically. Conversely as $\delta \to +\infty$, the discounted supply curve approaches the open-access supply curve in Figure 5.18.

Optimal Equilibrium (Finite Demand Elasticity)

Now let $P(h)$ denote the demand function for the given fishery resource harvested at rate h. If $U(h) = \int P(h)\, dh$ represents the total social utility of fish consumption (cf. Section 5.1), then the socially optimal fishery management policy requires the maximization of

$$J_{\mathrm{soc}}\{h\} = \int_0^\infty e^{-\delta t}[U(h) - C(x, h)]\, dt, \tag{5.51}$$

where $C(x, h)$ denotes the harvest cost function. As usual the maximization problem is subject to the state equation:

$$\frac{dx}{dt} = F(x) - h(t), \quad x(0) = x_0 \tag{5.52}$$

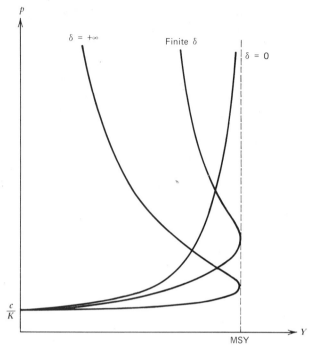

Figure 5.22 Discounted supply curve and its limiting positions.

and to the constraints

$$x(t) \geq 0, \quad h(t) \geq 0. \tag{5.53}$$

As in our earlier work, we continue to assume that $C(x, h)$ is linear in h:

$$C(x, h) = c(x)h. \tag{5.54}$$

The case in which C depends nonlinearly on h is discussed in the following section.

The preceding problem is a generalization of the nonlinear fishery model in Section 4.2. As before we introduce the Hamiltonian expression

$$\mathcal{H} = e^{-\delta t}[U(h) - c(x)h] + \lambda(t)[F(x) - h].$$

The maximum principle implies that

$$\frac{\partial \mathcal{H}}{\partial h} = e^{-\delta t}[U'(h) - c(x)] - \lambda(t) = 0,$$

so that along any optimal trajectory,

$$\lambda(t) = e^{-\delta t}\{U'(h) - c(x)\}. \tag{5.55}$$

(Strictly, this is only true when $h > 0$. This complication is easily handled, however, and we neglect it in the following calculation.) It is left as an exercise for the reader to modify Eqs. (4.24)–(4.26), thereby deriving the equation

$$\frac{dh}{dt} = \frac{1}{U''(h)} \{ [\delta - F'(x)][U'(h) - c(x)] + c'(x)F(x) \} \qquad (5.56)$$

for the optimal harvest rate $h(t)$. Note that this equation reduces to Eq. (4.26) when $c(x) \equiv 0$.

According to the maximum principle, any optimal trajectory $(x(t), h(t))$ must satisfy the nonlinear autonomous system of differential equations [Eqs. (5.52) and (5.56)]. We now wish to determine the equilibria of this system and, if possible, to identify the optimal approach trajectory for the infinite time-horizon case.

Setting $\dot{x} = \dot{h} = 0$ and solving the Eqs. (5.52) and (5.56), we obtain (because $U' = P$)

$$F'(x) - \frac{c'(x)F(x)}{P(h) - c(x)} = \delta, \qquad (5.57)$$

where

$$h = F(x). \qquad (5.58)$$

Once again we recognize our basic optimality rule [Eq. (5.48)], except that the fixed price p is now replaced by the demand price $p = P(h)$ at the optimal equilibrium harvest rate $h = F(x)$. The economic interpretation of Eq. (5.57) is the same as before.

Equations (5.57) and (5.58) can be written

$$\left. \begin{array}{l} P(h) = H_\delta(x) \\ h = F(x) \end{array} \right\}, \qquad (5.59)$$

where $H_\delta(x)$ is defined as in Eq. (5.49). Consequently the optimal output h and the optimal price p are reached at the point where the demand curve and the discounted supply curve intersect (see Figure 5.23). A point (p, h) on the discounted supply curve must satisfy $p = H_\delta(x)$ and $h = F(x)$, whereas a point on the demand curve must satisfy $p = P(h)$. Thus Eqs. (5.59) are satisfied if and only if the point (p, h) lies on both curves.

The analysis of supply and demand in the controlled fishery is similar to supply and demand factors in the open-access case. A shift in the demand curve affects the optimal equilibrium in the same way. What is somewhat surprising is that the optimally controlled fishery may also be subject to multiple equilibria and therefore to the associated effects of instability. In this regard, our model may be considered inadequate inasmuch as costs

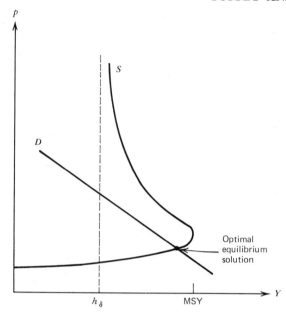

Figure 5.23 Supply and demand equilibrium for the optimally controlled fishery.

that may be associated with an unstable production policy (e.g., costs of shifting capital in and out of the fishery) are not taken into consideration. It is at least worth noting that the open-access fishery seems to be much more susceptible to dynamic instability than the optimally controlled fishery, because the supply curve bends backwards more sharply in the open-access case. In general the possibility of instability can be expected to depend on the rate of discount. When $\delta = 0$, the supply curve does not bend backwards, so that instability does not occur for any demand curve (see Figure 5.22).

The Monopoly Solution

In the case of the monopolist the objective functional becomes

$$J_{\text{mon}}\{h\} = \int_0^\infty e^{-\delta t}[hP(h) - c(x)h]\,dt. \tag{5.60}$$

It is easy to verify that the resulting equilibrium solution is now characterized by the equations

$$\left.\begin{array}{r}MR(h) = H_\delta(x) \\ h = F(x)\end{array}\right\}, \tag{5.61}$$

where $MR(h) = (d/dh)[hP(h)]$ denotes marginal revenue. Consequently the normal static analysis of supply and demand in Section 5.1 extends directly to the *equilibrium* solutions of our dynamic model: the social optimum occurs when discounted supply equals demand, whereas the monopoly optimum occurs when discounted supply equals marginal revenue (see Figure 5.24).

However the backward-bending nature of the discounted supply curve leads to various novel results. Figure 5.24a corresponds to the normal situation: the monopolist underproduces and overprices relative to the social optimum. In Figure 5.24b, however, the situation is reversed: the monopolist overproduces and underprices relative to the social optimum. The apparent generosity of the monopolist is explained by the dynamics of the exploitation policy. During the initial approach to equilibrium, the monopolist follows the normal practice of underproducing. Thus the monopolist's exploitation of the resource is less severe than the socially optimal exploitation rate. Whether in equilibrium the monopolist produces more or less fish per unit time than the social optimum depends on the position of the demand curve relative to the discounted supply curve.

Various possibilities are also shown in Figure 5.25. In Figure 5.25a the socially optimal harvest rate always exceeds the monopolistic rate; in Figure 5.25b the two curves cross one another. The latter case is similar to the case of an exhaustible resource, as shown in Figure 5.25c.

Dynamic Approach Paths

Next we briefly consider the question of the optimal approach to the equilibrium positions just described. This requires an analysis of the solution

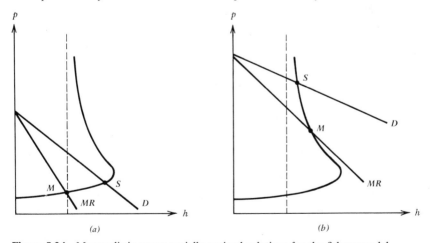

Figure 5.24 Monopolistic versus socially optimal solutions for the fishery model.

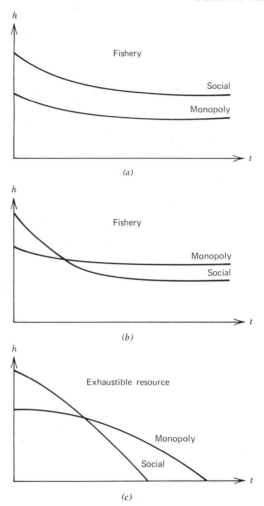

Figure 5.25 Socially optimal versus monopolistic harvest policies: (a) and (b) refer to the fishery model shown in Figure 5.24; (c) refers to an exhaustible resource.

trajectories of the system of differential equations [Eqs. (5.52) and (5.56)]; that is, the *phase-plane* map of this system. The theory of these systems is examined in detail in Chapter 6, and we postpone the full discussion of phase-plane systems until then. Figure 5.26 is a computer-generated graph of the system given by Eqs. (5.52) and (5.56) based on data from the Pacific halibut fishery shown previously in Figure 4.2, but now including the cost data given in Section 2.6. In both cases the optimal equilibrium (x^*, h^*) is a saddle point, and the convergent separatrices

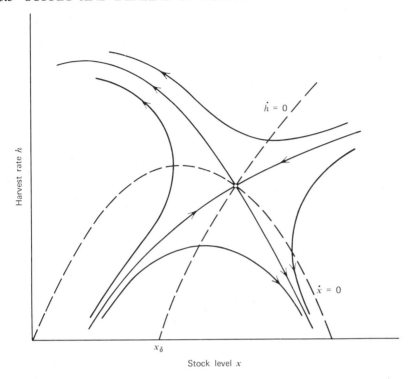

Figure 5.26   Phase-plane diagram of the system given by Eqs. (5.52) and (5.56).

associated with this point constitute the optimal trajectories (depending on whether $x_0 > x^*$ or $x_0 < x^*$) for the infinite time horizon. The effect of including the cost term in this diagram is primarily to reposition the equilibrium and the corresponding trajectories. Note that these separatrix trajectories define an optimal feedback control for our problem: the optimal control h is specified for each stock level x by these curves.

The phase-plane diagram in Figure 5.26 possesses a single equilibrium (x^*, h^*). In other cases the demand curve may intersect the discounted supply curve at several points, each of which is an equilibrium point for the system given by Eqs. (5.52) and (5.56). The phase-plane diagram then becomes considerably more complex; further discussion of this case is postponed to Chapter 6.

Alternative Models

The supply curves in Figures 5.20–5.24 are all based on the Schaefer model. However, Eqs. (5.49) and (5.50) define the discounted supply curve and its limiting case, the open-access supply curve, for models

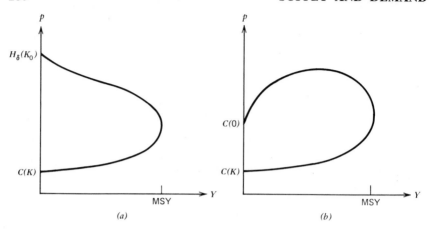

Figure 5.27 Discounted supply curves: (*a*) critical depensation model, with $F'(K_0) < \delta$; (*b*) finite extinction cost, with $F'(0) < \delta$.

employing arbitrary growth functions $F(x)$ and cost functions $c(x)$. Two cases in which the discounted supply curve assumes a somewhat unusual form are illustrated in Figure 5.27. In Figure 5.27*a* the growth function

$$F(x) = r(x - K_0)\left(1 - \frac{x}{K}\right)$$

exhibits critical depensation. When $F'(K_0) < \delta$, the discounted supply curve terminates at a finite price level $p_0 = H_\delta(K_0)$. Figure 5.27*b* shows the logistic growth curve, but employs a cost function for which the cost of extinction $c(0)$ is finite. In this case the discounted supply curve terminates at $p = c(0)$, provided that $F'(0) < \delta$. (This case corresponds to the extinction model discussed in Section 2.8.) It is left as an exercise for the reader to prepare a four-quadrant sketch verifying that these curves are correctly drawn. (Whether the supply curve actually bends around as shown in Figure 5.27*b* depends on the relative values of the parameters.)

Working out the implications of the various demand possibilities is also left to the reader. Obviously extinction of the fishery may be optimal in either of these models; in certain cases the initial stock level will determine whether extinction is optimal. The optimal dynamic approach to extinction must account for transversality conditions, as in the exhaustible-resource case. General results do not appear to be known for this problem.

5.4 NONLINEAR COST EFFECTS: PULSE FISHING

The assumption that the costs of fishing are linearly dependent on the harvest rate h (or alternatively, on the effort rate E) is obviously highly artificial. The additional assumption of a maximum harvest rate h_{max} (or of a maximum effort rate E_{max}) can be considered as introducing a simple (but necessary) nonlinearity into the model. In essence this amounts to supposing that

$$\text{Marginal cost} = \frac{\partial C}{\partial h} = \begin{cases} C(x) & \text{for} \quad 0 \leq h \leq h_{max} \\ +\infty & \text{for} \quad h > h_{max}. \end{cases}$$

In this section we briefly investigate the effects of introducing specifically nonlinear cost functions into our fishery model. Both the case of increasing marginal cost ($\partial C/\partial h > 0$) and the case of nonincreasing marginal cost ($\partial C/\partial h \gtrless 0$) are considered here. The latter case is particularly interesting, because it can lead to a new form of bionomic instability called "pulse fishing." This case of nonincreasing marginal cost is also of considerable interest from the mathematical point of view.

Increasing Marginal Cost

First let us assume that the cost of fishing is given by a function of the form

$$C(x, h) = c(x)\phi(h), \tag{5.62}$$

where

$$\phi(h) > 0, \quad \phi'(h) > 0, \quad \text{and} \quad \phi''(h) > 0. \tag{5.63}$$

Thus marginal costs are an increasing function of the harvest rate h. For example, such a cost function results from a production function of the form given in Eq. (2.8):

$$h = Q(E, x) = ax^{\alpha}E^{\beta},$$

provided that $\beta < 1$. If cost $C(x, h)$ is proportional to effort, the above relation implies that

$$C(x, h) = kx^{-\lambda}h^{1/\beta}, \tag{5.64}$$

where k and λ are positive constants. Provided that $\beta < 1$, this function satisfies Eqs. (5.62) and (5.63).

The objective functional for socially optimal management is now

$$J\{h\} = \int_0^{\infty} e^{-\delta t}[U(h) - C(x, h)] \, dt. \tag{5.65}$$

From previous calculations it can be easily verified that any optimal equilibrium solution (x^*, h^*) must satisfy

$$F'(x^*) - \frac{c'(x^*)\phi(h^*)F(x^*)}{U'(h^*) - c(x^*)\phi'(h^*)} = \delta, \tag{5.66}$$

where

$$h^* = F(x^*). \tag{5.67}$$

[See Exercise 8 for a simple general formula that contains Eq. (5.66) as a special case.]

The corresponding discounted supply curve is obtained by writing $p = U'(h^*)$ and solving Eq. (5.66) for p:

$$p = c(x^*)\phi'(h^*) - \frac{c'(x^*)\phi(h^*)F(x^*)}{\delta - F'(x^*)}. \tag{5.68}$$

One instance of this discounted supply curve is shown in Figure 5.28. In this diagram fishing costs are assumed to be only slightly sensitive to the stock level x; that is, $|c'(x)|$ is assumed to be small. Thus the expression on the right side of Eq. (5.68) is a small perturbation of the function $p = c(x)\phi'(h) = c(x)\phi'[F(x)]$. The latter function is shown by the solid curve in the N.W. quadrant of Figure 5.28; the dashed curve in the N.W. quadrant is the perturbation. (Note that the unperturbed curve together with the marginal cost curve MC, $p = \phi'(h)$ and the growth curve $h = F(x)$ are related by the four-quadrant composite construction.) The discounted supply curve appears as the dashed curve in the N.E. quadrant of the figure.

In this instance the supply curve reverses direction in such a manner that multiple equilibria may arise even for infinitely elastic demand curves. It is never worthwhile to harvest where $MC > p$, so the harvest rate is "choked off" at point h', which is not large enough to cause biological overfishing. But once overfishing has occurred (no matter what the cause), the fishery may become "trapped" at the other stable equilibrium point with h near h_δ.

Although this appears to be a rather special case, it does suggest that bionomic instability may arise from cost factors as well as from demand factors. When both nonlinearities are present, instability is even more likely to occur.

Finally because the open-access fishery can no longer be identified as the limit as $\delta \to \infty$ of the optimally controlled fishery (it is left to the reader to determine why not), it is unclear which case would tend to be more unstable relative to cost nonlinearities.

Existence of Optimal Controls

Here we arrive at a rather subtle mathematical point concerning the *existence* of optimal controls. Until now we have side-stepped this issue and more or less assumed that our control problems possess optimal

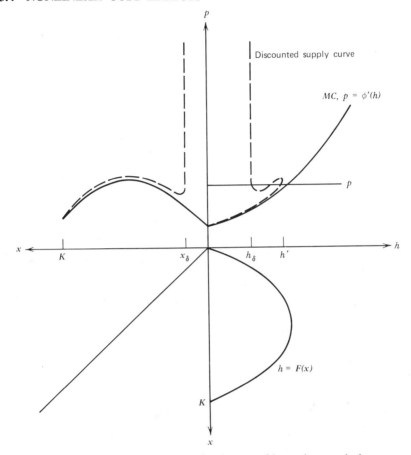

Figure 5.28   Discounted supply curve for the case of increasing marginal cost.

solutions. Except for the discussion of linear theory in Section 2.7, we have made no attempt to justify this assumption. Indeed proof of the existence of optimal controls is frequently the most difficult aspect of control theory. The reader is reminded that the maximum principle is only a *necessary* condition for optimality. Control functions that satisfy all the conditions of the maximum principle are not necessarily optimal, as the following example shows.

An Example

Consider the problem of minimizing (not maximizing!) the integral

$$J\{u\} = \int_0^1 [x^2 + (1-u^2)^2] \, dt, \qquad (5.69)$$

subject to

$$\frac{dx}{dt} = u, \quad x(0) = x(1) = 0. \tag{5.70}$$

The Hamiltonian is

$$\mathcal{H} = x^2 + (1 - u^2)^2 + \lambda u, \tag{5.71}$$

and the necessary conditions can be written

$$\frac{\partial \mathcal{H}}{\partial u} = -4u(1 - u^2) + \lambda = 0 \tag{5.72}$$

and

$$\frac{d\lambda}{dt} = -\frac{\partial \mathcal{H}}{\partial x} = -2x. \tag{5.73}$$

An obvious solution to Eqs. (5.70), (5.72), and (5.73) is $x(t) = \lambda(t) = u(t) \equiv 0$. This yields

$$J\{u\} = 1.$$

Obviously $J\{u\}$ has no maximum (*Proof:* take $u = M$ for $0 \le t < \frac{1}{2}$, $u = -M$ for $\frac{1}{2} \le t \le 1$). Therefore this solution must be a minimum, if anything. But now consider the control

$$u_n(t) = \frac{\sin 2n\pi t}{|\sin 2n\pi t|}, \tag{5.74}$$

or $u_n(t) = +1$ for $0 < t < 1/n$, $u_n(t) = -1$ for $1/n < t < 2/n$, and so on. [The value of $u_n(t)$ at the corner points k/n can be assigned arbitrarily without affecting the solution.] The corresponding state variable $x = x_n(t)$ is the saw-toothed curve in Figure 5.29, and we see that

$$J\{u_n\} = \int_0^1 x_n^2 \, dt = \frac{1}{2n}.$$

Obviously $J\{u\} > 0$ for all u, so we conclude that

$$\inf J\{u\} = 0.$$

We see therefore that whereas the maximum principle yields the solution $u = 0$, this is not an optimal control. In fact there is no optimal control in the usual sense, although a *minimizing sequence* $\{u_n(t)\}$ does exist, with the property that $J\{u_n\} \to 0$. It is also intriguing that the state-variable solution $x = 0$ resulting from the maximum principle is the limit of the minimizing sequence $x_n(t)$.

The properties of this "solution" turn out to be representative of

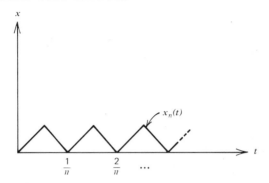

Figure 5.29 A chattering solution.

optimal control (minimizing) problems such that inf $J\{u\}$ is finite, but for which no optimal control exists in the ordinary sense. [Many further examples are given by Young (1969).] The sequence given by Eq. (5.74) is called a minimizing sequence of *chattering controls* for reasons that should be apparent.

A Convexity Condition

More generally, to return to our usual formulation, consider the problem of maximizing the functional

$$J\{u\} = \int_{t_0}^{t_1} g(x, t, u) \, dt, \tag{5.75}$$

subject to the state equation

$$\frac{dx}{dt} = f_0(x, t) + f_1(x, t)u, \tag{5.76}$$

with given initial and terminal values. Observe that Eq. (5.76) is *linear* in the control variable, a condition met by all the control problems treated in this book.

Assume that $J\{u\}$ is bounded for $u \in U$, where the control set U is closed and convex. Then, omitting numerous technical details, [these details are ably discussed by Berkovitz (1974, particularly Corollary 5.1, p. 67)] we can state that *if the integrand $g(x, t, u)$ is concave downward as a function of u, then an optimal control* (in the usual sense) *exists.* Conversely if $g(x, t, u)$ is not concave, then chattering controls may occur.

How do these assertions apply to the objective function given in Eq. (5.65), in which

$$g(x, t, u) = e^{-\delta t}[U(h) - c(x)\phi(h)]?$$

In the case of increasing marginal cost we have

$$\frac{\partial^2 g}{\partial u^2} = e^{-\delta t}[U''(h) - c(x)\phi''(h)] < 0,$$

so that g is concave in u. Thus we are justified in our original assumption that an optimal control exists. But if marginal cost is not increasing, this assumption is no longer justified. We turn now to this case.

Decreasing Marginal Costs

For simplicity we now suppose that marginal cost is decreasing:

$$\phi''(h) < 0. \tag{5.77}$$

In practice this cannot possibly hold for all $h > 0$, so that we assume a maximum harvest rate:

$$0 \leq h(t) \leq h_{\max}. \tag{5.78}$$

Let price be constant $P(h) \equiv p$. Now by formally applying the maximum principle, we again obtain Eqs. (5.66) and (5.67) for optimal equilibrium solutions. Assume that $h_{\max} > h^*$. Then it is easy to see that the equilibrium solution $x = x^*$, $h = h^*$ cannot possibly be optimal!

Consider Figure 5.30. The cost per unit time of harvesting at the fixed rate h^* is $c(x^*)\phi(h^*)$. However, from Figure 5.30 it is clear that a suitable average of harvest rates $h = 0$ and $h = h_{\max}$ provide the same average rate of harvest (h^*), but at a lower cost. Explicitly let

$$h(t) = \begin{cases} h_{\max} & \text{for} \quad 0 \leq t < \alpha\varepsilon \\ 0 & \text{for} \quad \alpha\varepsilon < t < \varepsilon, \end{cases} \tag{5.79}$$

and let $h_\varepsilon(t)$ be continued periodically for all $t \geq 0$. Here ε is a small, positive number (the period of chattering) and α is a constant to be determined.

The change in the population level during a single chatter is approximately

$$\Delta x^* \approx F(x^*)\varepsilon - h_{\max}\alpha\varepsilon.$$

The equilibrium condition $\Delta x^* = 0$ implies that

$$\alpha = \frac{F(x^*)}{h_{\max}} = \frac{h^*}{h_{\max}}.$$

The total cost for one chatter is

$$c(x^*)\phi(h_{\max}) \cdot \alpha\varepsilon + 0 \cdot (1 - \alpha)\varepsilon = \frac{c(x^*)\phi(h_{\max})h^*\varepsilon}{h_{\max}}. \tag{5.80}$$

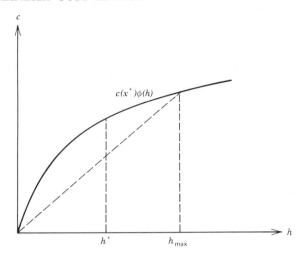

Figure 5.30 Decreasing marginal costs and chattering.

From the concavity of $\phi(h)$ we have

$$\frac{\phi(h_{\text{max}})}{h_{\text{max}}} < \frac{\phi(h^*)}{h^*},$$

so that the total cost per chatter is less than

$$c(x^*)\phi(h^*)\varepsilon.$$

In other words the cost rate per unit time for the chattering control given by Eqs. (5.79) is lower than the cost per unit time for the constant control $h = h^*$, as asserted above. The equilibrium solution derived from the maximum principle is therefore nonoptimal.

But what is really happening here? Recall from Section 5.1 that the profit-maximizing firm never operates at a production rate Q for which marginal cost is falling. But in the present case the biological nature of the resource *forces* the firm to use an average production rate for which marginal cost is indeed falling. This average rate can be achieved at lower cost by means of a chattering production policy. For the model formulated here, the optimal policy requires an infinite rate of chattering; if costs of switching from 0 to h_{max} are included in the model, this is no longer true.

Note that the above argument applies to *any* equilibrium harvest rate h^* between 0 and h_{max}. In fact it is easy to see that the harvest rate h^* determined by Eqs. (5.66) and (5.67) is not appropriate for the chattering case. Instead the optimal stock level x^* is now given by

$$F'(x^*) - \frac{c'(x^*)\bar{\phi}(h_{max})F(x^*)}{p - c(x^*)\bar{\phi}(h_{max})} = \delta,$$

where

$$\bar{\phi}(h) = \phi(h)/h.$$

The corresponding optimal average harvest rate is $h^* = F(x^*)$.

This formula follows naturally from the fact that the maximum principle remains valid for the *average* of a chattering control (see Warga, 1962). The formula also makes good economic sense, because we know that the cost per unit time with chattering is equal to $c(x)\bar{\phi}(h_{max})F(x)$ [see Eq. (5.80)].

The optimal approach to equilibrium is also clear now; it is the bang-bang approach

$$h(t) = \begin{cases} h_{max} & \text{if } x > x^* \\ 0 & \text{if } x < x^*. \end{cases}$$

Pulse Fishing

The modern distant-water fishing fleets of Russia and Japan—and more recently those of other countries such as Korea—are highly capital-intensive, consisting of large factory vessels accompanied by numerous trawlers. These fleets heavily exploit a given area, often removing fishes of several species and then move to a new area. This practice, termed *pulse fishing*, has been highly criticized by fishery biologists, who fear that pulse fishing may severely damage marine ecosystems.

The underlying motivation for pulse fishing remains somewhat mysterious. It has been observed that under certain circumstances pulse fishing can result in a greater average yield than the maximum sustainable yield. [See Clark et al. (1973); Hannesson (1975); and Pope (1973). Pope reports that a computer-simulation prediction that pulse fishing could increase average yield came as a "complete surprise." An analytic discussion of this possibility appears in Section 8.7.] In most cases, however, yield improvement is rather small.

The analysis in this section strongly suggests that pulse fishing is primarily an economic rather than a biological phenomenon. Given the large fixed capacity h_{max} of factory vessels, it is clear that marginal fishing costs are likely to be a decreasing function of h for $h \leq h_{max}$. Hence pulse fishing may be more efficient than sustained-yield fishing. Since the opportunity cost of moving the fleet from one location to another is significant, the optimal policy for any given location is a rather "sluggish" chattering. (See also Sections 7.7 and 8.7.)

5.5 SUMMARY AND CRITIQUE

In this chapter we have incorporated supply and demand questions into our study of renewable-resource economics. Our discussion was broadened to include the exhaustible-resource case for the sake of general interest as well as to illustrate the techniques of optimal control theory in a fairly simple setting.

Although nonlinear demand and cost functions seriously complicate the theory, the general results obtained in this chapter are quite intelligible. For the fishery model, we have seen that both open-access and discounted equilibrium supply curves can easily be constructed. These curves act just like the standard supply curves used in static analysis, except that they are always backward bending unless the discount rate is zero.

Perhaps not quite as expected are our findings concerning two quite different kinds of bionomic instability. The first type, which might be called "demand instability," arises from multiple intersections of the supply and demand curves. The second type of bionomic instability, which we have associated with pulse fishing, arises from a conflict between biological growth and economic efficiency requirements.

Finally we have touched briefly on the question of optimal approach paths. The linear model appears as a special dividing case in the sense that when linearity is replaced by concavity [of the objective $g(x, t, h)$], we obtain continuous asymptotic approach paths to equilibrium solutions, whereas the convex case produces bang-bang approaches to ultimate chattering controls.

Many of our early criticisms in Sections 1.3 and 2.9 remain valid for the present analysis. Our biological models are still greatly oversimplified, and the uncertainty and inadequacy of data is not considered. In the latter respect it should be noted that bringing demand curves into consideration obviously further complicates the econometric problem. Demand elasticity estimates have been attempted for a few fisheries (cf. Anderson, 1973, quoting Bell et al., 1970).

A serious analytical deficiency in Chapter 5 lies in our conclusion that bionomic instability may result from optimal fishery exploitation policies. There are several obvious reasons why instability itself is often undesirable on social, biological, and economic grounds. Our models completely ignore this question, although they do indicate that in many cases bionomic instability might have to be explicitly controlled if "free-market" conditions failed to do so.

EXERCISES

1. Show that the demand curve $p = aQ^\alpha$ ($\alpha < 0$) has constant elasticity $e = 1/\alpha$, and prove the converse. Also show that these isoelastic

demand curves have the property that

$$\frac{\text{marginal revenue}}{\text{price}} = \text{constant}.$$

2. Show that at the output level Q that maximizes total revenue, we must have $e = 1$. Why are isoelastic demand curves with $\alpha \neq 1$ unrealistic?

3. Assume a piecewise-linear cost curve, as indicated in the figure at right. Show that in the static case with $p = $ constant, production always occurs at one of the vertices Q_i. Then prove the same result via the maximum principle for the mining model given by Eq. (5.27).

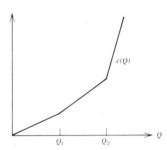

4. Discuss the effect of the discount rate on the nonlinear mining model given by Eq. (5.16) for the case $P(q) = a - bq$. In particular, (a) show that there is no optimal control when $\delta = 0$, and (b) investigate the case $\delta \to +\infty$.

5. Consider an isoelastic, nonautonomous demand function

$$P(q, t) = f(t)q^{\alpha}.$$

Show that the monopoly solution to the exhaustible resource problem (zero extraction costs) is also socially optimal. (This problem is almost trivial: compare the objective functionals.)

6. For each (sustained) level of output Y, show that the open-access supply curve represents the *average* cost of output C/Y. (The curve has two branches, because each level of output $<$ MSY can be achieved at two different cost levels.) Similarly show that the discounted supply curve for $\delta = 0$ represents the *marginal cost dC/dY.*

7. Construct a four-quadrant diagram to verify that Figure 5.27 is correct. (This figure is incorrectly drawn in Herfindahl and Kneese, 1974, p. 158.)

8. Consider a general objective functional of the form

$$J\{h\} = \int_0^\infty e^{-\delta t}\pi(x, h)\,dt,$$

as usual subject to the biological growth equation $\dot{x} = F(x) - h$. Show that optimal equilibrium solutions satisfy

$$h = F(x) \quad \text{and} \quad F'(x) + \frac{\partial \pi / \partial x}{\partial \pi / \partial h} = \delta.$$

This equation contains Eqs. (2.16), (5.17), (5.61), and (5.66) as special cases. (Of course the nonautonomous case $\pi = \pi(x, t, h)$ is not covered, for the simple reason that optimal equilibrium solutions do not exist.)

9. If μ denotes the shadow price $\mu = \lambda e^{\delta t}$ for the problem in Exercise 8, show that along an optimal path we have

$$F'(x) + \frac{\partial \pi / \partial x}{\partial \pi / \partial h} = \delta - \frac{\dot{\mu}}{\mu}.$$

The optimal approach path can be considered as a modification of the optimal equilibrium solution, resulting from time changes in the shadow price of the resource. [Unfortunately this interpretation does not help to solve the problem, because the above is a differential equation for the unknown function $\mu(t)$ that involves two further unknown functions $x(t)$ and $h(t)$. As we mention in this chapter, numerical integration is required to solve the dynamic problem.]

10. Develop and discuss an exhaustible-resource model in which extraction costs depend on cumulative extraction (or equivalently on the remaining reserves).

BIBLIOGRAPHICAL NOTES

The introduction to supply and demand theory given in Section 5.1 is extremely superficial. The reader may wish to refer to Samuelson (1973), Herfindahl and Kneese (1974), or any of several other sources for further details.

One of the earliest works on exhaustible resource economics is the paper by Gray (1914). A brilliant analysis based on variational techniques appears in the paper by Hotelling (1931), which is still a basic reference in the field. More recently the volume edited by Gaffney (1967) is worthwhile. Some of Hotelling's results were reinterpreted by R. L. Gordon (1967), but see the important correction due to Goldsmith (1974). Burt and Cummings (1970) utilize the maximum principle, but rather inconclusively; see also Smith (1968). An excellent summary of the subject in nonmathematical terms is given by Solow (1974). In view of the rapid current expansion in the literature on exhaustible-resource economics, we do not attempt to survey the last two or three years here.

Nonlinear dynamic models of renewable-resource exploitation have been published by Quirk and Smith (1970) and Plourde (1971). A simplified treatment appears in Clark and Munro (1975).

The backward-bending open-access supply curve is discussed by Copes (1970, 1972); the corresponding bionomic instability is described by Anderson (1973). Quirk and Smith (1970) and Plourde (1971) also observe the possibility of multiple equilibria in the case of time discounting. Similar difficulties have already been observed in models of economic growth (Leviatan and Samuelson, 1969; Brock, 1973).

Scale efficiencies, chattering controls, and pulse fishing are discussed by Lewis and Schmalensee (1976).

6

DYNAMICAL SYSTEMS

An autonomous system of ordinary differential equations is called a "dynamical system." We have already encountered several of these systems in previous chapters, mostly by applying the maximum principle to control problems. Many more examples are still to be introduced.

Any system of ordinary differential equations that are of arbitrary order, autonomous or nonautonomous, can be reduced by simple substitution to a dynamical system. Thus a two-dimensional, nonautonomous system $\dot{x}_i = F_i(x_1, x_2, t)$ ($i = 1, 2$) can be transformed into a three-dimensional, autonomous system simply by defining a new dependent variable $x_3 = t$. Unfortunately, however, the treatment of autonomous systems in three or more dimensions is much more difficult than in two dimensions.

In this chapter we describe some of the more elementary aspects of dynamical systems, concentrating largely on the two-dimensional case.

6.1 BASIC THEORY

The autonomous system of first-order differential equations

$$\frac{dx_i}{dt} = F_i(x_1, x_2, \ldots, x_n) \quad i = 1, 2, \ldots, n \tag{6.1}$$

or, in vector notation

$$\frac{dx}{dt} = F(x),$$

is called a *dynamical system*. The term *autonomous* refers to the fact that the functions F_i do not depend explicitly on the time variable t.

The term "dynamical system" originates in classical mechanics. Thus

179

the linear motion of a single particle (without forcing) satisfies an equation of the form

$$\frac{d^2x}{dt^2} + a_1(x)\frac{dx}{dt} + a_2(x) = 0.$$

Letting $dx/dt = v$, we can rewrite this equation as a system of first-order equations:

$$\frac{dx}{dt} = v$$

$$\frac{dv}{dt} = -a_1(x)v - a_2(x),$$

thus obtaining a dynamical system in the plane. The x–v plane is often referred to as the *phase plane*.

Returning to the general dynamical system given by Eq. (6.1), we assume without further specification that the functions $F_i(x_1, \ldots, x_n)$ are sufficiently smooth in the sense that partial derivatives of a high enough order exist and are continuous. Unless stated to the contrary, we assume that each $F_i(x_1, \ldots, x_n)$ is defined for all points $(x_1, \ldots, x_n) \in R^n$. In biological or economic applications, however, we often consider only $x_i \geq 0$, at least for some of the variables x_i.

The following fundamental theorem on existence and uniqueness of solutions is proved in standard texts on ordinary differential equations (e.g., Birkhoff and Rota, 1969).

THEOREM.

Eq. (6.1) possess a unique solution $x_i = x_i(t)$, $i = 1, \ldots, n$ [vector notation, $x = x(t)$] that satisfies the given initial conditions

$$x_i(0) = x_i^0, \quad i = 1, \ldots, n \quad [x(0) = x^0]. \tag{6.2}$$

The solution $x(t)$ is either defined for all $t \geq 0$ or

$$\|x(t)\| = \left\{ \sum x_i^2(t) \right\}^{1/2} \to \infty, \text{ as } t \to t^* < +\infty.$$

As an example of the case where $\|x(t)\| \to +\infty$, as $t \to t^*$, consider the single equation

$$\frac{dx}{dt} = x^2, \quad x(0) = 1.$$

The solution is $x(t) = 1/(1-t)$, which approaches $+\infty$ as t approaches 1 from below.

Any solution $x = x(t)$ of Eqs. (6.1) is called a *trajectory* or an integral curve of the system. If $x = x(t)$ is any trajectory and t_0 is any constant, then the curve $x = x(t + t_0)$ is also a trajectory of the same system, because $dx/dt = dx/d(t + t_0)$. This trajectory, which is obtained by a translation of the t-axis, is considered to represent the same *geometrical trajectory* as $x = x(t)$. The curves $x(t)$ and $x(t + t_0)$ have the same graphs as curves in R^n. With this convention, we have the following result.

THEOREM.

Through each given point $x^0 \in R^n$, there passes a unique geometrical trajectory of the dynamical system given in Eqs. (6.1). Consequently no two distinct geometrical trajectories can intersect at any point.

Proof.   Note that Theorem 1 implies the existence of a trajectory through x^0. Let $x^1(t)$ and $x^2(t)$ denote two such trajectories, with

$$x^k(t_k) = x^0, \quad k = 1, 2.$$

Of course if $t_1 = t_2$, then Theorem 1 implies that $x^1(t) \equiv x^2(t)$. Otherwise define new functions

$$\tilde{x}^k(t) = x^k(t + t_k).$$

Then for $k = 1, 2$, we have

$$\frac{d\tilde{x}^k}{dt} = \frac{d\tilde{x}^k}{d(t + t_k)} = F[x^k(t + t_k)] = F[\tilde{x}^k(t)]$$

and

$$\tilde{x}^k(0) = x^0.$$

Hence Theorem 1 implies that $\tilde{x}^1(t) \equiv \tilde{x}^2(t)$; that is,

$$x^2(t) = \tilde{x}^2(t - t_2) = \tilde{x}^1(t - t_2) = x^1(t + t_1 - t_2).$$

By our convention this means that $x^1(t)$ and $x^2(t)$ represent the same geometrical trajectories. QED

EXAMPLE 1.

$$\left. \begin{aligned} \frac{dx}{dt} &= y^2 \\ \frac{dy}{dt} &= x^2 \end{aligned} \right\}. \tag{6.3}$$

We can construct the geometrical trajectories of this system by noting that

$$\frac{dy}{dx} = \frac{x^2}{y^2}$$

or that

$$y^2 \, dy = x^2 \, dx.$$

Hence

$$y = (x^3 + C)^{1/3}, \tag{6.4}$$

where C is an arbitrary constant of integration. The curves given by Eqs. (6.4) are shown in Figure 6.1.

Note that $x(t) \equiv y(t) \equiv 0$ is also a solution of the system given by Eqs. (6.3). This is called an *equilibrium solution*, and the point $(0, 0)$ is called an *equilibrium point* of the dynamical system [Eqs. (6.3)]. For the general case of Eqs. (6.1), any point $x_0 \in R^n$ such that $F(x_0) = 0$ is an equilibrium point, and the corresponding constant solution is an equilibrium solution.

Also note that $dx/dt = 0$ for every point $(x, 0)$ on the x-axis, and $dy/dt = 0$ for every point $(0, y)$ on the y-axis. The x and y axes are called *isoclines* of the

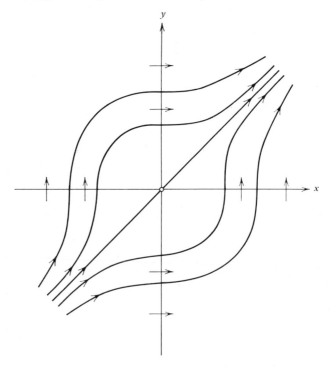

Figure 6.1 Trajectories of the system given by Eqs (6.3).

system in Eqs. (6.3). The trajectories in Figure 6.1 cross the x-isocline vertically and the y-isocline horizontally. The equilibrium point occurs at the intersection of the isoclines.

The reader should realize that although we obtain an explicit geometrical description of the family of all trajectories, we do not obtain the trajectories themselves. For example, it is clear that along any trajectory both $x(t)$ and $y(t) \to +\infty$ as t increases, but it is not clear whether $t \to +\infty$ or $t \to t^* < +\infty$ along a given trajectory. (Because the velocity at a point (x, y) behaves like $x^2 + y^2$, it can be shown that in fact $t \to t^* < \infty$, except for the trajectory $y = x$ in the third quadrant.)

To solve Eqs. (6.3) explicitly, we may try eliminating y as follows:

$$\frac{d^2x}{dt^2} = 2y\frac{dy}{dt} = 2\left(\frac{dx}{dt}\right)^{1/2} \cdot x^2$$

$$\left(\frac{d^2x}{dt^2}\right)^2 = 4x^4\frac{dx}{dt}.$$

This nonlinear, second-order equation is not solvable by any standard textbook method, however. ∎

Even the simplest dynamical system then is often not solvable in an explicit form (although effective numerical techniques do exist). In most cases we cannot even expect to obtain a geometrical description of the trajectories (see Example 1). For this reason, much effort is devoted to the qualitative theory, in which the topological nature of the system of trajectories is studied. By far the most complete results have been obtained for the case $n = 2$, to which we now turn.

6.2 DYNAMICAL SYSTEMS IN THE PLANE: LINEAR THEORY

Consider the system

$$\left.\begin{aligned}\frac{dx}{dt} &= ax + by + p \\[2mm] \frac{dy}{dt} &= cx + dy + q\end{aligned}\right\}, \tag{6.5}$$

where a, b, c, d, p, and q denote constants. Except when the determinant

$$\begin{vmatrix} a & b \\ c & d \end{vmatrix} = ad - bc$$

vanishes, this system has a unique equilibrium point (x_0, y_0), where

$$\left.\begin{aligned} ax_0 + by_0 &= -p \\ cx_0 + dy_0 &= -q \end{aligned}\right\}.$$

By the translation of axes

$$x' = x - x_0$$
$$y' = y - y_0,$$

we can place the equilibrium point at the origin and obtain the following homogeneous linear system (we drop the primes on x' and y' for simplicity):

$$\left.\begin{array}{l} \dfrac{dx}{dt} = ax + by \\[2mm] \dfrac{dy}{dt} = cx + dy \end{array}\right\}. \qquad (6.6)$$

Instead of beginning with the general case here, we consider several examples that have been chosen as prototypes.

EXAMPLE 2.

$$\left.\begin{array}{l} \dfrac{dx}{dt} = ax \\[2mm] \dfrac{dy}{dt} = by \end{array}\right\}. \qquad (6.7)$$

Trivally the solutions are

$$x(t) = x(0)e^{at}, \quad y(t) = y(0)e^{bt}.$$

Hence

$$\left[\frac{x(t)}{x(0)}\right]^{1/a} = \left[\frac{y(t)}{y(0)}\right]^{1/b}. \qquad (6.8)$$

If $a > b > 0$, these trajectories are as shown in Figure 6.2a. The trajectories are defined for all t; they move toward infinity as $t \to +\infty$ and toward 0 as $t \to -\infty$. The equilibrium point $(0, 0)$ is called an *unstable node* in this case.

When a, $b < 0$, the direction along the trajectories is reversed (Figure 6.2b) and the origin is a *stable node*.

In either case, the semiaxes are trajectories called the *arms* of the node.

Next suppose that $b < 0 < a$. The trajectories then are as shown in Figure 6.3. Except for the semiaxes, all trajectories "begin" at infinity ($t = -\infty$) and "end" at infinity ($t = +\infty$). The semiaxes are also trajectories and are called *separatrices*. The case $a < 0 < b$ is similar, except that the direction is reversed along the trajectories. In either case the origin is called a *saddle point*. ■

A saddle point is clearly an unstable equilibrium, but unlike an unstable node, two stable trajectories do converge to the equilibrium point in the case of the saddle point. As we observe in Chapters 4 and 5, this fact is important for optimal control problems.

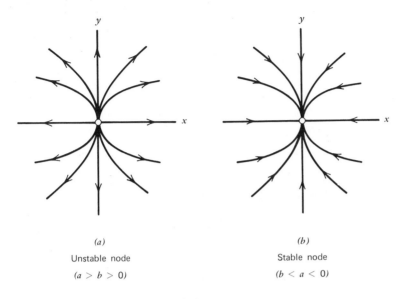

(a)

Unstable node

$(a > b > 0)$

(b)

Stable node

$(b < a < 0)$

Figure 6.2   Trajectories of Example 2.

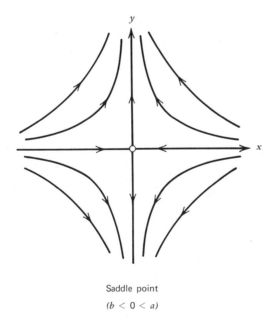

Saddle point

$(b < 0 < a)$

Figure 6.3   Trajectories of Example 2.

185

EXAMPLE 3.

$$\left.\begin{array}{l} \dfrac{dx}{dt}=cy \\[2mm] \dfrac{dy}{dt}=dx \end{array}\right\}. \qquad (6.9)$$

These equations imply

$$\dfrac{d^2x}{dt^2}=cdx.$$

If $cd=\alpha^2>0$, the general solution is

$$x(t)=c_1e^{\alpha t}+c_2e^{-\alpha t}$$

$$y(t)=d\cdot\int x(t)\,dt$$

$$=\dfrac{c_1 d}{\alpha}e^{\alpha t}-\dfrac{c_2 d}{\alpha}e^{-\alpha t}.$$

Hence

$$\dfrac{y(t)}{x(t)}\rightarrow\begin{cases} d/\alpha & \text{as}\quad t\rightarrow+\infty, \\ -d/\alpha & \text{as}\quad t\rightarrow-\infty. \end{cases}$$

The trajectories are as shown in Figure 6.4a; the origin is a saddle point with separatrices $y=\pm(d/\alpha)x$.

The situation is quite different when $cd=-\beta^2<0$, for now the solutions are

$$x(t)=c_1\sin\beta t+c_2\cos\beta t$$

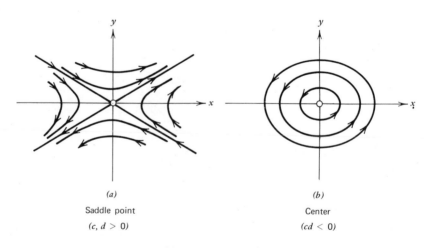

(a)

Saddle point

($c, d > 0$)

(b)

Center

($cd < 0$)

Figure 6.4 Trajectories of Example 3.

and

$$y(t) = -\frac{c_1 d}{\beta} \cos \beta t + \frac{c_2 d}{\beta} \sin \beta t.$$

These are the equations of ellipses centred at the origin (see Figure 6.4b), which is called a *center*. A center is *neutrally stable*. ∎

EXAMPLE 4.

$$\left. \begin{aligned} \frac{dx}{dt} &= -y \\ \frac{dy}{dt} &= x - y \end{aligned} \right\}. \tag{6.10}$$

The solutions are obtained as in Example 3:

$$x = e^{-t/2}\left(c_1 \cos \frac{\sqrt{3}}{2} t + c_2 \sin \frac{\sqrt{3}}{2} t \right)$$

$$y = e^{-t/2}\left(c_1' \cos \frac{\sqrt{3}}{2} t - c_2' \sin \frac{\sqrt{3}}{2} t \right),$$

where c_1' and c_2' are certain constants related to c_1 and c_2, respectively. These curves are spirals converging toward the origin (see Figure 6.5), which is called a *stable focus*. ∎

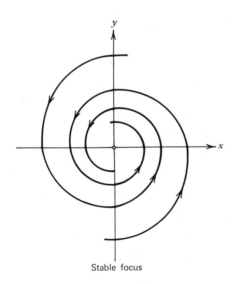

Stable focus

Figure 6.5 Trajectories of Example 4.

With certain exceptional cases to be discussed below, the above examples exhaust the possibilities for the behavior of trajectories of the linear system [Eqs. (6.6)]. To analyze the general case let

$$A = \begin{pmatrix} a & b \\ c & d \end{pmatrix}$$

denote the matrix of coefficients for Eqs. (6.6). Let λ_1 and λ_2 represent the eigenvalues of A; that is, λ_1 and λ_2 are the roots of the equation

$$\det(A - \lambda I) = \begin{vmatrix} a - \lambda & b \\ c & d - \lambda \end{vmatrix} = \lambda^2 - (a + d)\lambda + ad - bc = 0. \quad (6.11)$$

Then the character of the equilibrium point $(0, 0)$ is as described in Table 6.1.

To demonstrate these results we apply the method used in Examples 3 and 4. If $b \neq 0$, we obtain from Eqs. (6.6)

$$\frac{d^2 x}{dt^2} = a \frac{dx}{dt} + b(cx + dy) = (a + d) \frac{dx}{dt} + (bc - ad)x.$$

By direct substitution we then see that $x = e^{\lambda t}$ is a solution if and only if λ satisfies Eqs. (6.10), i.e., if λ is an eigenvalue of A. The general solution is therefore given by

$$x(t) = c_1 e^{\lambda_1 t} + c_2 e^{\lambda_2 t}$$

$$y(t) = \frac{1}{b}\left(\frac{dx}{dt} - ax\right) = c_1' e^{\lambda_1 t} + c_2' e^{\lambda_2 t},$$

where $c_1' = (\lambda_i - a)c_i/b$.

The various cases can now be easily checked, as in Examples 3 and 4.

When $b = 0$, the calculation is somewhat simpler. The eigenvalues are

TABLE 6.1.

Values of λ_1, λ_2	Character of Equilibrium Point
$\lambda_1, \lambda_2 > 0$	Unstable node
$\lambda_1, \lambda_2 < 0$	Stable node
$\left.\begin{array}{l}\lambda_1 < 0 < \lambda_2 \\ \lambda_2 < 0 < \lambda_1\end{array}\right\}$	Saddle point
λ_1, λ_2 complex, Re $\lambda_i > 0$	Unstable focus
λ_1, λ_2 complex, Re $\lambda_i < 0$	Stable focus
λ_1, λ_2 complex, Re $\lambda_i = 0$	Center

$\lambda_1 = a$ and $\lambda_2 = d$, and the general solution (except when $a = d$) is

$$x(t) = c_1 e^{at}$$

$$y(t) = \frac{cc_1}{a} e^{at} + c_2 e^{dt}.$$

It can be easily seen that the origin is either a node or a saddle point (the details of proving this are left to the reader).

Whenever the eigenvalues λ_1 and λ_2 are real and unequal, we know from matrix theory that new coordinates x' and y' can be introduced by means of a linear transformation in such a way that the matrix A becomes diagonalized:

$$A' = \begin{pmatrix} \lambda_1 & 0 \\ 0 & \lambda_2 \end{pmatrix}.$$

Consequently except for the position of the axes, Example 2 is a prototype for all such cases.

If the eigenvalues are equal, however, such a diagonalization may not be possible. Example 5 is a prototype of this situation.

EXAMPLE 5.

$$\left. \begin{aligned} \frac{dx}{dt} &= x \\ \frac{dy}{dt} &= x + y \end{aligned} \right\}.$$

The general solution in this case is

$$x(t) = c_1 e^t$$

$$y(t) = (c_1 t + c_2) e^t.$$

This is different from all previous examples; the trajectories are shown in Figure 6.6. Again the origin is called an unstable node, but it differs from the previous case in that it now has only one pair of arms (the y semiaxes) instead of two. ∎

A final case arises when one of the eigenvalues of A vanishes. In this case A is singular, so that the system in Eqs. (6.6) must be of the form

$$\left. \begin{aligned} \frac{dx}{dt} &= ax + by \\ \frac{dy}{dt} &= q(ax + by) \end{aligned} \right\}.$$

The trajectories are shown in Figure 6.7; all points on the line $ax + by = 0$ are equilibrium points.

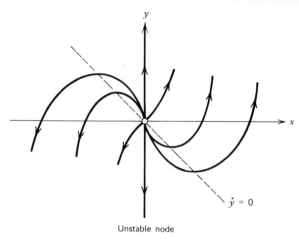

Unstable node

Figure 6.6   Trajectories of Example 5.

Structural Stability

To be a reasonable model of biological or economic phenomena, any dynamical system proposed should be *structurally stable*. By this we mean that the behavior of the system of solution trajectories should not change drastically if the model is subject to a small perturbation. In other words if the predictions of a certain model are critically dependent on the explicit nature of the model, then it is clearly inappropriate to use that

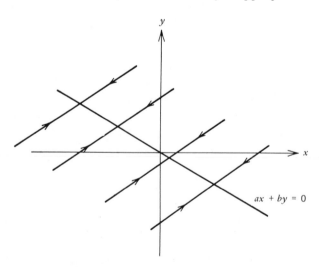

Figure 6.7   The singular case.

model to describe biological or economic phenomena. [A related, but more general, concept is that of the *robustness* of models. If the predictions of a certain model remain qualitatively valid for all other "similar" models, then the given model is considered robust. An important instance to be discussed in Chapter 7 is the fact that (within limits) predictions derived from continuous-time (differential-equation) models often remain valid for the case of discrete-time models.]

Since the eigenvalues of a matrix depend continuously on the matrix elements, it follows that strict eigenvalue inequalities are preserved by small changes in these elements. Again referring to Table 6.1, we conclude that stable and unstable nodes, foci, and saddle points are all structurally stable relative to perturbations of the parameters a, b, c, and d. (Note that structural stability applies to the *system as a whole*, whereas stability refers to the equilibrium point and its relation to the trajectories. The two concepts are quite distinct.) On the other hand centers are structurally unstable: arbitrarily small perturbations may destroy the equality Re $\lambda = 0$, and thus transform a center into either a stable or an unstable focus. The special cases $\lambda_1 = \lambda_2$ or $\lambda_i = 0$ (see Figures 6.6 and 6.7) are also structurally unstable.

6.3 ISOCLINES

Consider a plane, autonomous system that is either linear or nonlinear:

$$\left.\begin{array}{l} \dfrac{dx}{dt} = F(x, y) \\[2mm] \dfrac{dy}{dt} = G(x, y) \end{array}\right\}. \tag{6.12}$$

The curves $F(x, y) = 0$ and $G(x, y) = 0$ are called the *isoclines* for this system. [In some cases (e.g., $F(x, y) = x^2 + y^2 + 1$) isoclines may not exist.] Any trajectory must cross an F isocline vertically and a G isocline horizontally, except for the points where the two isoclines intersect, which are of course the equilibrium points of the system. It is left to the reader to verify that all of the graphs in this chapter satisfy these conditions.

In the case of the nondegenerate linear system [Eqs. (6.6)] the isoclines $ax + by = 0$ and $cx + dy = 0$ divide the plane into four *isosectors* (see Figure 6.8). Here we call an isosector S *terminal* if every trajectory of the system in Eqs. (6.6) that begins in S remains within S for all $t > 0$. Thus S is a terminal isosector if and only if the vectors $V = (ax + by, cx + dy)$ are directed inward at all points of the boundary of S (Figure 6.8a).

The following observations are often helpful in determining the character of equilibrium points.

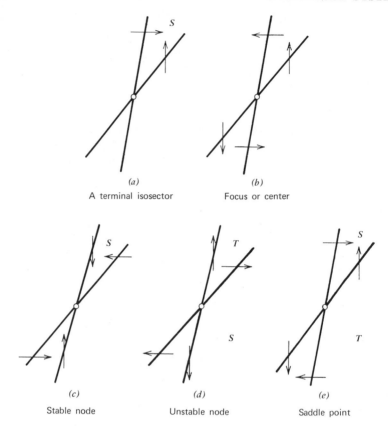

(a)   (b)
A terminal isosector   Focus or center

(c)   (d)   (e)
Stable node   Unstable node   Saddle point

Figure 6.8 Isosectors.

1. There are no terminal isosectors for foci and centers (Figure 6.8*b*).

2. If *S* is a terminal isosector with trajectories approaching the equilibrium point, then the equilibrium point is a stable node (Figure 6.8*c*).

3. If *S* is a terminal isosector with trajectories approaching infinity, then the equilibrium point is either an unstable node or a saddle point. To determine the type of equilibrium point, reverse the direction of the vectors *V* (thus replacing the right sides of Eqs. (6.6) with their negatives). Let *T* be one of the adjacent isosectors to *S*; then *T* becomes a terminal sector for the reversed system. The equilibrium point is (a) an unstable node if the reversed trajectories within *T* converge as $t \to +\infty$ (Figure 6.8*d*), or (b) a saddle point if the reversed trajectories within *T* diverge as $t \to +\infty$ (Figure 6.8*e*).

Consequently if a terminal isosector exists then the character of the equilibrium point can be uniquely determined by simple geometric considerations. If no terminal isosector exists, however, we can only conclude that the equilibrium point is not a saddle point.

We use this method later in studying certain nonlinear problems for which the linearized eigenvalue calculation becomes laborious.

6.4 NONLINEAR PLANE-AUTONOMOUS SYSTEMS

Again consider the system

$$
\left.
\begin{aligned}
\frac{dx}{dt} &= F(x, y) \\
\frac{dy}{dt} &= G(x, y)
\end{aligned}
\right\}.
\tag{6.12}
$$

Let (x_0, y_0) be an equilibrium point, so that $F(x_0, y_0) = G(x_0, y_0) = 0$. Then F and G being assumed smooth we have

$$
\begin{aligned}
\frac{dx}{dt} &= a(x - x_0) + b(y - y_0) + 0[(x - x_0)^2 + (y - y_0)^2] \\
\frac{dy}{dt} &= c(x - x_0) + d(y - y_0) + 0[(x - x_0)^2 + (y - y_0)^2],
\end{aligned}
\tag{6.13}
$$

where $a = \partial F / \partial x (x_0, y_0), \ldots$, and where the terms of $0\{\cdots\}$ are of the order of the square of the distance from (x, y) to (x_0, y_0) as $(x, y) \rightarrow (x_0, y_0)$.

It can be shown that if the linear system obtained by omitting the higher-order terms from Eqs. (6.13) is structurally stable (node, saddle point, or focus), then the nonlinear system Eqs. (6.12) has the same behavior in the neighborhood of (x_0, y_0). [This result is due to Lyapunov; see Birkoff and Rota (1969, p. 135) for the proof.] This fact is useful in discussing the phase-plane diagram of nonlinear systems, because the behavior of trajectories near an equilibrium point can be determined by examining the corresponding linearized system (provided that this system is structurally stable). Thus if λ_1 and λ_2 denote the eigenvalues of the matrix

$$
\begin{pmatrix}
\partial F / \partial x & \partial F / \partial y \\
\partial G / \partial x & \partial G / \partial y
\end{pmatrix}
$$

evaluated at (x_0, y_0), then the character of the equilibrium point (x_0, y_0) is as given in Table 6.1.

EXAMPLE 6 (THE LOTKA-VOLTERRA EQUATIONS FOR A PREDATOR-PREY SYSTEM).

The following dynamical system was proposed by Lotka (1925) and Volterra (1931) as a simple model of a predator–prey interaction:

$$\left.\begin{aligned}\frac{dx}{dt} &= rx - \alpha xy \\[2mm] \frac{dy}{dt} &= -sy + \beta xy\end{aligned}\right\} \tag{6.14}$$

where r, s, α, and β are positive constants. Here $x = x(t)$ represents the prey population, and $y = y(t)$ represents the predator population. Equations (6.14) constitute a model of an ecologically isolated predator–prey system in the sense that the predator population is the only agent that controls the prey and the prey population is the only food source for the predators. Thus if $y = 0$ the prey population grows exponentially, whereas if $x = 0$ the predator population dies at an exponential rate.

There is one nontrivial equilibrium point for the system given by Eqs. (6.14):

$$x_0 = \frac{s}{\beta}, \quad y_0 = \frac{r}{\alpha}.$$

Expanding about this point we obtain

$$\frac{dx}{dt} = -\alpha x_0 (y - y_0) + \cdots$$

$$\frac{dy}{dt} = \beta y_0 (x - x_0) + \ldots.$$

Hence the linearized system has purely imaginary eigenvalues

$$\lambda = \pm i\sqrt{\alpha \beta x_0 y_0} = \pm i\sqrt{rs}.$$

Thus the equilibrium point (x_0, y_0) is a center for the linearized system and consequently either a center or a focus for the original Lotka–Volterra system.

Equations (6.14) can be solved explicitly for the geometrical trajectories:

$$\left(\frac{r}{y} - \alpha\right) dy = \left(-\frac{s}{x} + \beta\right) dx,$$

and hence

$$r \ln y - \alpha y = -s \ln x + \beta x + c \tag{6.15}$$

where c is constant. The family of curves in Eq. (6.15) can be easily graphed by using the four-quadrant method and introducing a dummy variable

$$z = r \ln y - \alpha y = -s \ln x + \beta x + c.$$

The construction (the details of which are left to the reader) shows that the

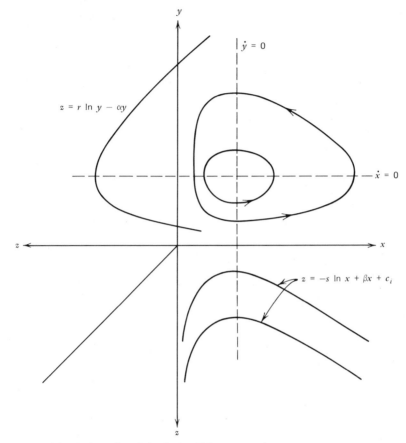

Figure 6.9 Trajectories of the Lotka–Volterra equations.

trajectories are closed orbits, with motion proceeding counterclockwise (see Figure 6.9). The x and y axes are also trajectories. ∎

It is clear that the Lotka–Volterra system is not structurally stable. May (1973) points out that the system is therefore unsuitable as a valid descriptive model of natural predator–prey systems.

Regular population cycles do seem to occur in nature, however, and they require a mechanistic (not a stochastic) explanation. Several types of available models account for structurally stable oscillations. One of the simplest, a discrete-time, single-population model with overcompensation, is described in some detail in Chapter 7. A second model, based on delay-differential equations, does not concern us here, because it seems to be rather artificial and requires an unnecessarily difficult technique (cf. Wangersky and Cunningham, 1957).

Although we have yet to encounter such an example, plane-autonomous systems of differential equations may also possess structurally stable periodic solutions. These solutions, called *limit cycles*, are discussed in the following section.

EXAMPLE 7.   (OPTIMAL FISHERY MANAGEMENT).
Again let us consider the simplified zero-cost nonlinear fishery model in Section 4.2. The maximum principle leads to the dynamical system

$$\frac{dx}{dt} = F(x) - h \tag{6.16}$$

$$\frac{dh}{dt} = \frac{R'(h)}{R''(h)}\{\delta - F'(x)\}, \tag{6.17}$$

where $R(h)$ denotes either total revenue $hP(h)$ or total utility $U(h)$. Assuming first that $R'(h) > 0$ and that $F''(x) < 0$, we can see that the system in Eqs. (6.16) and (6.17) has a unique equilibrium point (x_δ, h_δ) that is determined by

$$F'(x_\delta) = \delta, \quad h_\delta = F(x_\delta).$$

The isoclines are the curves

$$h = F(x), \quad F'(x) = \delta.$$

Examining the isosector diagram (see Figure 6.10) immediately shows that the equilibrium point is a saddle point. The full phase-plane diagram is shown in Figure 4.2.

An interesting variation of this model arises when

$$R'(\hat{h}) = 0$$

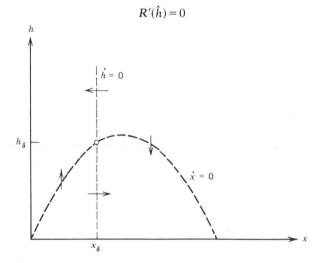

Figure 6.10   Isosectors for the system given by Eqs. (6.16) and (6.17).

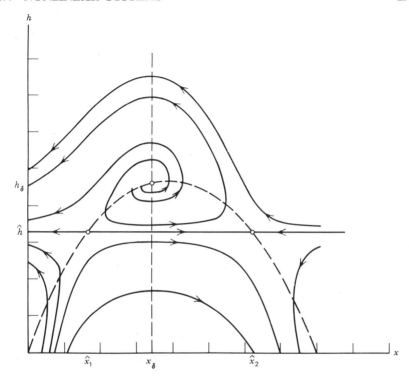

Figure 6.11 Computer-generated phase-plane diagram for the system given by Eqs. (6.16) and (6.17) for the case in which $\hat{h} < h_\delta$.

for some harvest rate $\hat{h} < \max F(x)$. In this case there are additional equilibria at $(\hat{x}_i, \hat{h})$, where $\hat{x}_1, \hat{x}_2$ are the two solutions of $F(\hat{x}_i) = \hat{h}$. An isosector diagram shows that these are saddle-point equilibria and that the point (x_δ, h_δ) now becomes a focus. The full phase-plane diagram appears in Figure 6.11.

Note that the line $h = \hat{h}$ forms separatrices that converge to $(\hat{x}_2, \hat{h})$ but that diverge from $(\hat{x}_1, \hat{h})$. If the initial population level x_0 is greater than $\hat{x}_1$, then these separatrices define the optimal trajectories. This is obvious from the outset, because $h(t) \equiv \hat{h}$ actually maximizes the integrand $e^{-\delta t} R(h)$ for all time t. Thus the optimal policy is independent of the discount rate in this case (but see Exercise 4 at the end of this chapter).

This is by no means an unrealistic case. Many fisheries are (or were) exploited at less than MSY, not primarily because fishing costs rise as the stock level is reduced, but simply because demand chokes off at a level $\hat{h} < \text{MSY}$.

What is the optimal harvest policy when $x_0 < \hat{x}_1$? In this case there exists an optimal separatrix trajectory (not explicitly shown in Figure 6.11) that converges to $(\hat{x}_1, \hat{h})$. Because $F'(x) > \delta$ for $x < x_\delta$, building up the stock level to $\hat{x}_1$ is advisable. Of course the equilibrium at $\hat{x}_1$ is unstable, so that a slight random increase in the stock level results in a shift to the stable equilibrium point at $\hat{x}_2$.

A somewhat different situation arises in this problem when $x_8 < \hat{x}_1$ (again see Exercise 4). ∎

Bifurcation

Let us consider a dynamical system of the form

$$\frac{dx}{dt} = F(x; \varepsilon), \tag{6.18}$$

where ε is a parameter. For simplicity we consider only the case in which x is a two-dimensional vector and ε is a scalar. It is clear that the phase-plane diagram changes as ε varies. What may not be as obvious is the fact that for most values of ε the change in the phase-plane diagram is continuous, whereas for certain isolated ε values this change becomes discontinuous. (By a continuous change from one phase-plane diagram to another, we mean that a continuous transformation of the plane onto itself exists, mapping one diagram into the other. Thus the two diagrams are *topologically equivalent*.) The values of ε at which discontinuous changes in the phase-plane diagram occur are called *bifurcation points*.

Since the study of bifurcation theory is far beyond the scope of this book, we merely illustrate the concept here with the following example.

EXAMPLE 8.

$$\left.\begin{array}{l} \dfrac{dx}{dt} = y \\[2mm] \dfrac{dy}{dt} = x^2 - y - \varepsilon \end{array}\right\}. \tag{6.19}$$

Here ε is a parameter. The isoclines are the x-axis and the parabola $y = x^2 - \varepsilon$. If $\varepsilon < 0$ there are no equilibrium points, whereas if $\varepsilon > 0$ there are two equilibrium points $(\pm\sqrt{\varepsilon}, 0)$. The matrix of the system linearized about $(\pm\sqrt{\varepsilon}, 0)$ is

$$A = \begin{pmatrix} 0 & 1 \\ \pm 2\sqrt{\varepsilon} & -1 \end{pmatrix}.$$

The eigenvalues at $(\sqrt{\varepsilon}, 0)$ are

$$\lambda = -\frac{1 \pm \sqrt{1 + 8\sqrt{\varepsilon}}}{2}$$

so that this is a saddle point. At $(-\sqrt{\varepsilon}, 0)$ the eigenvalues are

$$\lambda = -\frac{1 \pm \sqrt{1 - 8\sqrt{\varepsilon}}}{2},$$

so that this point is a stable node if $\varepsilon < \frac{1}{64}$ and a stable focus if $\varepsilon > \frac{1}{64}$.

The development of the phase-plane diagram as ε passes from negative to positive values is shown in Figure 6.12. All diagrams for $\varepsilon < 0$ are topologically equivalent; indeed because these diagrams contain no equilibrium points, they are equivalent to a family of parallel straight lines (as is obvious from Figure 6.12a). At $\varepsilon = 0$, however, the diagram undergoes a severe change—a bifurcation. This occurs as the two isoclines come into contact and then cross (for $\varepsilon > 0$), producing two equilibrium points. (The case $\varepsilon = 0$ produces a *degenerate* equilibrium point $(0, 0)$ that cannot be classified by means of the linear theory.) Because a stable node is equivalent to a stable focus, we see that all diagrams for $\varepsilon > 0$ are also topologically equivalent. Thus $\varepsilon = 0$ is the only bifurcation point. ∎

Another instance of bifurcation occurs in the nonlinear fishery model in Section 5.3, in cases where the discounted supply curve and the demand curve possess multiple intersections (see Figure 5.20). As the demand curve D shifts upwards in this figure, two bifurcations occur. The first bifurcation transforms the single saddle-point equilibrium at M_1 into three equilibria M_2, M_2', and M_2'', whereas the second bifurcation transforms back to a single equilibrium at M_3. The phase-plane version of the first bifurcation is shown in Figure 6.13; the single saddle-point equilibrium in Figure 6.13a transforms into the three equilibria of Figure 6.13b.

6.5 LIMIT CYCLES

Consider the system

$$\frac{dx}{dt} = y$$

$$\frac{dy}{dt} = -x,$$

(6.20)

which has the circles $x^2 + y^2 = c^2$ as trajectories. Next consider the following perturbation of Eqs. (6.20):

$$\frac{dx}{dt} = y - \varepsilon x(x^2 + y^2 - 1)$$

$$\frac{dy}{dt} = -x - \varepsilon y(x^2 + y^2 - 1),$$

(6.21)

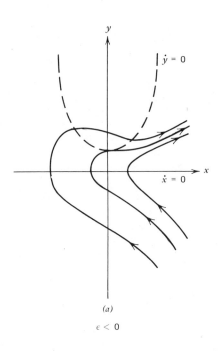

$\dot{y} = 0$

$\dot{x} = 0$

(a)

$\epsilon < 0$

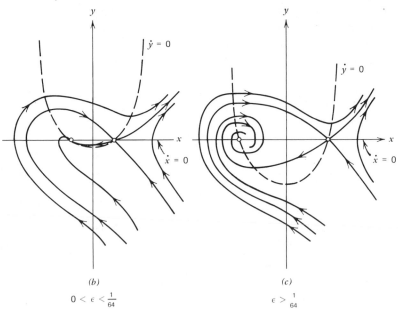

$\dot{y} = 0$

$\dot{x} = 0$

$\dot{y} = 0$

$\dot{x} = 0$

(b)

$0 < \epsilon < \frac{1}{64}$

(c)

$\epsilon > \frac{1}{64}$

Figure 6.12 Trajectories of Example 8.

200

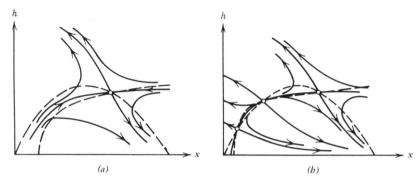

(a)   (b)

Figure 6.13   Bifurcation of supply and demand equilibria in the nonlinear fishery model.

where $\varepsilon > 0$. The circle $x^2 + y^2 = 1$ is a trajectory of this system, but on any other circle $x^2 + y^2 = c^2$ the trajectories of Eqs. (6.21) cross the circle, moving in the direction of $x^2 + y^2 = 1$. Thus every trajectory of Eqs. (6.21) *approaches* the circle $x^2 + y^2 = 1$ (see Figure 6.14). The circle $x^2 + y^2 = 1$ is called a *limit cycle* of the system given by Eqs. (6.21). Let us define this term in general.

A closed trajectory of a dynamical system is called an *orbit* of the system. The motion along orbits is periodic. An orbit is said to be a

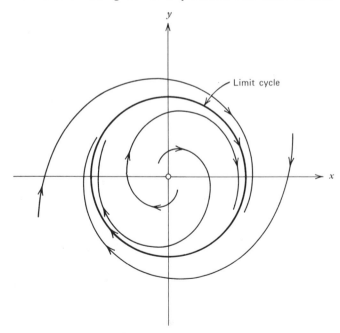

Figure 6.14   Trajectories of the system given by Eq. (6.21), showing limit cycle.

limit cycle if every trajectory that starts at a point close to the orbit converges towards the orbit as $t \to +\infty$. According to the general theory of dynamical systems (cf. Andronov et al., 1973), any orbit that is not one of a family of *concentric* orbits must be either a limit cycle or an *originating cycle* in the sense that all neighboring trajectories diverge from the orbit. An originating cycle is clearly a limit cycle for the time-reversed dynamical system. Note that for $\varepsilon < 0$ the unit circle becomes an originating cycle for the system given by Eqs. (6.21). Also note that $\varepsilon = 0$ is a bifurcation value for this system.

The existence or nonexistence of limit cycles generally cannot be deduced from an inspection of isosectors, as Figure 6.14 shows. Yet it is often desirable to know whether a given dynamical system in the plane possesses a limit cycle, for if this is true the system undergoes periodic oscillations instead of converging to a stationary equilibrium. The following test is often helpful in determining the existence or nonexistence of a limit cycle.

THEOREM (BENDIXON–DU LAC CRITERION).

Consider the dynamical system

$$\frac{dx}{dt} = F(x, y) \tag{6.22}$$

$$\frac{dy}{dt} = G(x, y), \tag{6.23}$$

in which functions $F(x, y)$ and $G(x, y)$ are assumed to be smooth in a given simply-connected region D. Let $B(x, y)$ be a smooth function in D such that the expression

$$\frac{\partial(BF)}{\partial x} + \frac{\partial(BG)}{\partial y} \tag{6.24}$$

does not change sign in D. Then the system given by Eqs. (6.22) and (6.23) has no closed trajectories in D.

Proof. Suppose to the contrary that C is a closed trajectory in D. Then

$$I = \oint_C - BG\,dx + BF\,dy = \iint_{D_0} \left\{ \frac{\partial(BF)}{\partial x} + \frac{\partial(BG)}{\partial y} \right\} dx\,dy \neq 0$$

by hypothesis. Here D_0 denotes the interior of C, and Green's theorem has been employed in the second equality. But on the other hand we have

$$I = \int B\left(-G\frac{dx}{dt} + F\frac{dy}{dt}\right) dt = \int B(-GF + FG)\, dt = 0.$$

This contradiction establishes the validity of the theorem.

EXAMPLE 9 (BIONOMIC EQUILIBRIUM).

The Bendixon–du Lac test can be used to investigate the nature of the bionimic equilibrium of the fishery model in Section 2.3. The dynamical system for the case of the Schaefer model is

$$\frac{dx}{dt} = rx\left(1 - \frac{x}{K}\right) - Ex = F(x, E) \tag{6.25}$$

$$\frac{dE}{dt} = kE(px - c) = G(x, E). \tag{6.26}$$

The region D is the first quadrant $x, E > 0$. Bionomic equilibrium $(\dot{x} = \dot{E} = 0)$ occurs at the point (x_∞, E_∞), where

$$x_\infty = \frac{c}{p}, \quad E_\infty = r\left(1 - \frac{x_\infty}{K}\right).$$

The coefficient matrix of the linearized system is

$$\begin{pmatrix} \dfrac{\partial F}{\partial x} & \dfrac{\partial F}{\partial E} \\[2mm] \dfrac{\partial G}{\partial x} & \dfrac{\partial G}{\partial E} \end{pmatrix} = \begin{pmatrix} r\left(1 - \dfrac{2x_\infty}{K}\right) - E_\infty & -x_\infty \\[2mm] k_p E_\infty & 0 \end{pmatrix} = \begin{pmatrix} -\dfrac{rx_\infty}{K} & -x_\infty \\[2mm] kpE_\infty & 0 \end{pmatrix}.$$

The eigenvalues are

$$\lambda_i = -\frac{rx_\infty}{K} \pm \sqrt{\frac{r^2 x_\infty^2}{K^2} - 4kpx_\infty E_\infty}. \tag{6.27}$$

If the stiffness parameter k in Eq. (6.26) is small, these eigenvalues are both real and negative, and from Table 6.1 we know that the equilibrium point (x_∞, E_∞) is a stable node (see Figure 6.15a). On the other hand if k is large, the eigenvalues become complex with Re $\lambda_i < 0$, and (x_∞, E_∞) becomes a stable focus (see Figure 6.15b).

This eigenvalue calculation pertains only to the behavior of the system in the neighborhood of the equilibrium point. Hence the system may have a limit cycle, in which case the term "bionomic equilibrium" is not appropriate. However, we can now apply the Bendixon–du Lac test, introducing the function

$$B(x, E) = x^{-2} E^{1/kp}. \tag{6.28}$$

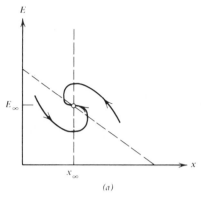
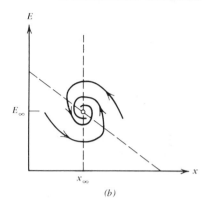

Figure 6.15 Bionomic equilibrium for the open-access fishery (Schaefer model): (a) low stiffness; (b) high stiffness.

A straightforward calculation then shows that Eq. (6.24) is satisfied and that no limit cycles are possible. We do not write the calculation here because a more general case is given in Exercise 5 at the end of this chapter. ■

EXAMPLE 10 (DEPENSATION; BIONOMIC DISEQUILIBRIUM).

Next consider the case in which the growth function $F(x)$ exhibits depensation. (The reader may wish to review Section 2.3, particularly Figure 2.6, at this point). Our dynamical system is

$$\frac{dx}{dt} = F(x) - Ex$$

$$\frac{dE}{dt} = kE(px - c).$$

By the same calculation as above, we see that the eigenvalues at (x_∞, E_∞) are given by

$$\lambda_i = [F'(x_\infty) - E_\infty] \pm \sqrt{[F'(x_\infty) - E_\infty]^2 - 4kpx_\infty E_\infty}. \tag{6.29}$$

Denote by $\bar{x}$ the stock level at which

$$F'(\bar{x}) = \frac{F(\bar{x})}{\bar{x}};$$

due to depensation $\bar{x} > 0$. In the case in which $x_\infty < \bar{x}$ we see that

$$E_\infty = \frac{F(x_\infty)}{x_\infty} < F'(x_\infty).$$

Consequently from Eq. (6.29) we obtain

$$\text{Re } \lambda_i > 0,$$

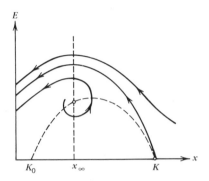

Figure 6.16 Open-access fishery under depensation extinction mode.

which means that the bionomic equilibrium at (x_∞, E_∞) is *unstable*. Observe that this instability occurs precisely in cases in which the total cost curve TC meets the yield-effort (revenue) curve on its lower branch (see Figure 2.6).

Given that (x_∞, E_∞) is an unstable node or focus, what can we say about the trajectories of this system? Consider the trajectory that starts near $(K, 0)$ in Figure 6.16. The isosector analysis immediately implies that this trajectory becomes "trapped," either by hitting the coordinate axis (Figure 6.16) or by cycling around (x_∞, E_∞). In the second case it follows from the instability of the equilibrium at (x_∞, E_∞) that a limit cycle must exist.

This completes the analysis we began in Section 2.3, and shows that whenever the "catastrophe" situation indicated in Figure 2.6 prevails, the model predicts either the extinction of the fishery or a limit-cycle oscillation. ∎

6.6 GAUSE'S MODEL OF INTERSPECIFIC COMPETITION

The following dynamical system was studied both analytically and experimentally by Gause (1935) as a model of competition between two species:

$$\left.\begin{array}{l} \dfrac{dx}{dt} = rx\left(1 - \dfrac{x}{K}\right) - \alpha xy \\[2mm] \dfrac{dy}{dt} = sy\left(1 - \dfrac{y}{L}\right) - \beta xy \end{array}\right\}. \tag{6.30}$$

The symbols r, s, K, L, α, and β denote positive constants. An external resource (food supply) is assumed to exist that supports each population in the absence of the other population, according to a logistic law. However, each population interferes with the other population's use of the resource, as indicated by the negative interaction terms $-\alpha xy$ and $-\beta xy$ in Eqs. (6.30).

Although Gause's equations do not appear to be solvable in closed form, much information can be obtained from a qualitative approach to the system. There are four isoclines:

$$x = 0$$

$$y = 0$$

$$x = \frac{s}{\beta}\left(1 - \frac{y}{L}\right)$$

$$y = \frac{r}{\alpha}\left(1 - \frac{x}{K}\right).$$

These isoclines give rise to three equilibrium points that lie on one or the other of the coordinate axes and in some cases an equilibrium point Q that lies in the first quadrant. (Of course we are only interested in solutions such that $x(t) \geq 0$ and $y(t) \geq 0$.)

Three essentially distinct cases are shown in Figure 6.17. In the first case of *competitive coexistence* (Figure 6.17a), there is a stable node $Q = (x_0, y_0)$, with both x_0 and y_0 positive. In the second case (Figure 6.17b), Q is a saddle point and two stable equilibria (nodes) exist at $(K, 0)$ and at $(0, L)$. The competitive outcome depends on the initial population levels, because one species is ultimately driven to extinction (*competitive exclusion*). In the third case (Figure 6.16c), which is also competitive exclusion, only one stable equilibrium exists, either at $(K, 0)$ or at $(0, L)$. One of the species inevitably wins the competition.

Gause's equations are structurally stable except for certain special cases, one of which is shown in Figure 6.17d. Clearly a small change in the position of the isoclines can transform this diagram, which has a stable node at K, into a diagram of type shown in Figure 6.16b with an added saddle point.

This possibility is significant for cases in which one species in a competitive system is subject to harvesting. Suppose, for example, that a constant effort E is devoted to harvesting species x. Then the first of Eqs. (6.30) becomes

$$\frac{dx}{dt} = rx\left(1 - \frac{x}{K}\right) - \alpha xy - Ex = (r - E)\left(1 - \frac{1}{r - E}\frac{x}{K}\right) - \alpha xy.$$

We can easily imagine a case in which species y is excluded from the natural system given in Eqs. (6.30), but this situation is reversed and species x is excluded once harvesting effort exceeds some critical level (i.e., a bifurcation value). Fishery biologists propose such an explanation

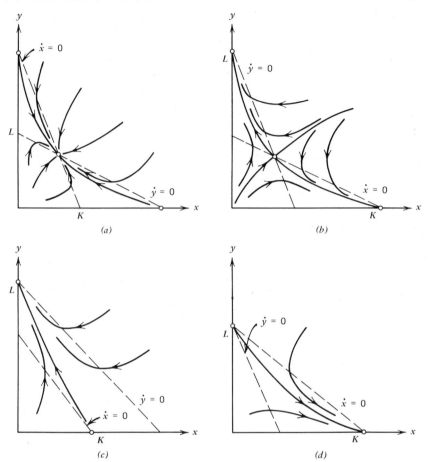

Figure 6.17 Trajectories of Gause's Eqs. (6.30).

for the disappearance of the California sardine (*Sardinops caerula*), which has a competitive population of anchovies (*Engraulis mordax*) (Murphy, 1966).

Economic questions associated with the harvesting of predator–prey and competitive systems are discussed in Chapter 9.

EXERCISES

1. Sketch the phase-plane diagram for the system

$$\left.\begin{array}{l} \dfrac{dx}{dt} = y^2 - 1 \\[2mm] \dfrac{dy}{dt} = x^2 - 1 \end{array}\right\}.$$

2. Compute the eigenvalues corresponding to each of the four equilibrium points in the system in Exercise 1.

3. Consider the linear system

$$\frac{dx}{dt} = -y$$

$$\frac{dy}{dt} = cx - y,$$

where c is a real parameter. Determine the nature of the equilibrium point $(0, 0)$ as a function of c. Show graphically how the phase-plane topology changes with changes in c. What are the bifurcation values of c?

4. Sketch the phase-plane diagram for the system in Example 7, assuming now that $x_\delta < \hat{x}_1$. Also discuss the optimal harvest policy.

5. Show that the system

$$\frac{dx}{dt} = x(a_1 x + b_1 y + c_1)$$

$$\frac{dy}{dt} = y(a_2 x + b_2 y + c_2)$$

has no limit cycles in the first quadrant, provided that

$$\Delta = a_1 b_2 - a_2 b_1 \neq 0$$

and

$$a_1 c_2(b_1 - b_2) + b_2 c_1(a_2 - a_1) \neq 0,$$

by applying the Bendixon–du Lac test with $B(x, y) = x^{\sigma-1} y^{\tau-1}$, where

$$\sigma = \frac{b_2(a_2 - a_1)}{\Delta} \quad \text{and} \quad \tau = \frac{a_1(b_1 - b_2)}{\Delta}.$$

6. Describe the first-quadrant trajectories of the system

$$\frac{dx}{dt} = rx\left(1 - \frac{x}{K}\right) - \alpha xy$$

$$\frac{dy}{dt} = sy\left(1 - \frac{y}{L}\right) + \beta xy,$$

where α, $\beta > 0$. (This model, used by Larkin (1966) to describe a predator–prey system, is discussed further in Chapter 9.)

7. The following model of optimal pollution control is proposed by Wright (1974). This model takes advantage of the particularly simple

features of control problems with "linear dynamics and quadratic cost functional."

Let x denote the amount of a pollutant in the environment, and let $c(t)$ represent the rate of removal of this pollutant (by some unspecified control mechanism). Assume that

$$\frac{dx}{dt} = \alpha x - c(t), \quad x(0) = x_0,$$

where α is a positive constant. Consider the cost functional

$$J\{c\} = \int_0^\infty e^{-\delta t}(ax^2 + bc^2)\, dt,$$

where a and b are positive constants. The problem is to minimize J.

(a) By replacing x with $e^{\delta t/2}x$ and c with $e^{\delta t/2}c$, reduce this problem to the form

$$\frac{dx}{dt} = \alpha x - c \tag{1}$$

$$J\{c\} = \int_0^\infty (Ax^2 + Bc^2)\, dt. \tag{2}$$

(b) From the maximum principle show that

$$\frac{dc}{dt} = \frac{A}{B} x - \alpha c \tag{3}$$

is a necessary condition.

(c) Solve the system given by Eqs. (1) and (3) explicitly, and discuss the optimal solution of the original problem.

BIBLIOGRAPHICAL NOTES

Two recent books that extend far beyond the treatment of dynamical systems in this chapter, are Andronov et al. (1973) and Hirsch and Smale (1974). The former gives a large number of particular applications of the Bendixon–du Lac test.

Structural stability and the related concepts of bifurcation and "catastrophes" are currently under active investigation; see Thom (1972) and Zeeman (1974).

The behavior of dynamical systems proposed as models of ecosystems is analyzed in detail by May (1973). We discuss this subject further in Chapter 9.

7
DISCRETE-TIME
METERED MODELS

The analysis of renewable-resource economics in the preceding chapters is based entirely on the use of continuous-time mathematical models (differential equations). These models involve the particularly restrictive assumption that the response of the population to external forces such as harvesting is *instantaneous*. Thus the possibility of various delay effects cannot be included in models of this kind; the models are without memory, so to speak

However, most natural biological populations are subject to complex dynamic processes that cannot be encompassed by simple continuous-time models. In many fish populations, for example, recruitment to the fishable stock may only occur several years after the spawning of the existing adult population. Furthermore, the entire life history of fish and other organisms is generally subject to strong seasonal or periodic influences.

In trying to model these and other complications, we face the danger that our mathematical models may become too complex to be analyzed and understood. One way to try to overcome this problem is to use the modern computer to *simulate* complex systems. While not underrating the practical value of computer simulation exercises, it should be realized that from the scientific point of view the results of such exercises at best serve as illustrations of a general theoretical framework.

The aim of this book is to proceed as far as we can theoretically by using the simplest possible methods. In this chapter we introduce a new family of mathematical population models, which we refer to as *metered models*. These models consist of two components, the first of which is a

210

first-order *difference equation* or *recurrence scheme*

$$x_{k+1} = F(x_k),$$

relating the population level x at time $t = t_{k+1}$ to the population level at a certain previous time $t = t_k$. (For simplicity we usually suppose that $t_k = k$, so that $t_{k+1} - t_k$ represents one "cycle" or "year" in the population's history.) Discrete-time population models of this kind are frequently employed (for example, in the Pacific salmon fishery), and these models are the subject of extensive study in the biological literature. In spite of their apparent simplicity, these models can exhibit amazingly complex dynamic behavior.

The second component of our metered model is the possibility of constructing the growth function $F(x)$ on the basis of assumed biological processes of birth and mortality that occur during each intervening time interval $t_k \le t \le t_{k+1}$. Thus the short-term dynamics of the model are incorporated in the function $F(x)$, whereas the long-term dynamics are described by the recurrence formula. This metered-model approach is extremely flexible, and can be used to model the effects of such diverse biological processes as predation, cannibalism, overrunning food supplies, and winter die-off.

The metered-model approach also permits an analysis and optimization of harvesting policy. Indeed this method allows us to make a more detailed analysis of the harvesting process than we can readily achieve by using continuous-time models, and thus leads to the explicit formulation of the production function of a fishery.

Finally we remark that the metered models in this chapter do not incorporate an age structure (or "cohort" structure, in the terminology of fisheries). This interesting but difficult problem is examined in Chapter 8.

7.1   A GENERAL, METERED STOCK-RECRUITMENT MODEL

The models in this chapter are based on the simple flowchart model shown in Figure 7.1. In this model the parent stock of the kth generation (P_k) gives rise to a determined number of young (Y_k), which ultimately provide a certain number of "recruits" (R_k). Some of the recruits may be harvested (H_k); the remainder form the parent stock (P_{k+1}) of the next generation. Following the terminology of fishery biologists, we sometimes refer to P_{k+1} as the *escapement*. Note that the model assumes that none of the parent stock P_k survives to be added to the harvestable stock of recruits; this restrictive assumption is relaxed later. The model also assumes that harvesting occurs just prior to the reproduction process.

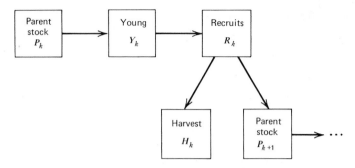

Figure 7.1 A general, metered stock-recruitment model.

Our model implies that R_k is a well-determined function of P_k:

$$R_k = F(P_k). \qquad (7.1)$$

The function $F(P)$ is called the *stock-recruitment relation*. The dynamics of the model under harvesting conditions are governed by the equation

$$P_{k+1} = R_k - H_k = F(P_k) - H_k, \qquad (7.2)$$

or equivalently

$$R_{k+1} = F(P_{k+1}) = F(R_k - H_k). \qquad (7.3)$$

The problem of optimal harvest policies is discussed later in this chapter. First we consider the dynamics of an unexploited population, in which case we have simply

$$P_{k+1} = F(P_k). \qquad (7.4)$$

Given an initial population P_1, Eq. (7.4) determines the future population levels $P_2, P_3, \ldots$, recursively. The equation is a discrete-time analog of the differential equation $dx/dt = F(x)$, because it can be written in the form

$$\frac{\Delta P_k}{\Delta k} = P_{k+1} - P_k = F(P_k) - P_k,$$

so that $F(P) - P$ corresponds to the net growth rate $F(x)$ of the continuous-time model.

The term "metered model" refers to the fact that the value P_k constitutes only a single data point for each time period k.

The dynamic behavior of Eq. (7.4) can be considerably more complex than the behavior of the continuous-time analog

$$\frac{dx}{dt} = F(x).$$

For the preceding equation, equilibrium points $x = K$ are characterized by $F(K) = 0$; such an equilibrium is stable if $F'(K) < 0$ and unstable if $F'(K) > 0$ (see Figure 1.5). Furthermore any solution $x = x(t)$ of this equation converges monotonically to a stable equilibrium point; oscillations cannot occur.

Various possibilities for the discrete-time model are shown in Figure 7.2. Equilibrium points $P = K$ are now characterized by the equation

$$F(K) = K.$$

The 45° line shown in Figure 7.2 contains any of these equilibrium points and also serves as a *transfer line* to move $P_{k+1} = F(P_k)$ back to the horizontal population axis. Thus the sequence $P_1, P_2, P_3, \ldots$, of successive population levels can be determined graphically, as indicated in Figure 7.2a.

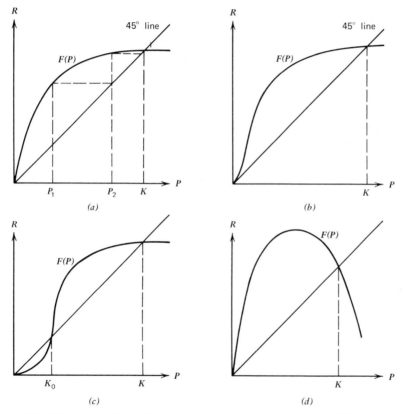

Figure 7.2 Types of stock-recruitment curves: (a) normal compensation; (b) depensation; (c) critical depensation; (d) overcompensation.

In the case in which $F(P)$ is an increasing function, as it is in Figure 7.2a, this sequence converges monotonically toward the equilibrium K. This also holds true for the case of noncritical depensation (see Figure 7.2b). Critical depensation (Figure 7.2c) gives rise to a minimum viable population level K_0, which is an unstable equilibrium for Eq. (7.1). These three cases in which $F(P)$ is always an increasing function of P (although $F''(P)$ may change sign) are analogous to the three earlier cases shown in Figure 1.5.

However, the *overcompensation curve* shown in Figure 7.2d introduces additional complications, because the population sequence $\{P_k\}$ now undergoes oscillations. The oscillations may be simple, with P_k converging to the equilibrium point at K (Figure 7.3a), they may be extremely

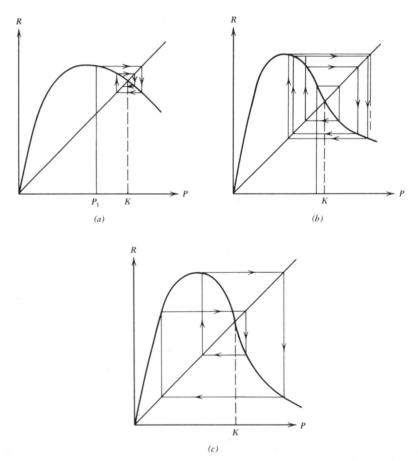

Figure 7.3 Oscillations: (a) stable convergence to K; (b) limit cycle with period 2; (c) limit cycle with period 4.

complex, with limit-cycle periodic oscillations (Figure 7.3b) or they may exhibit an even more irregular behavior pattern (Figure 7.3c).

First let us consider the case of stability. The equilibrium point K for Eq. (7.4) is termed *stable* if $P_k \to K$ (as $k \to \infty$) whenever the initial population P_1 is sufficiently close to K.

The following simple theorem almost gives a necessary and sufficient condition for stability.

THEOREM

Assume that $F(P)$ is continuously differentiable. Let K be an equilibrium point such that $F(K) = K$. Then K is stable, provided that

$$-1 < F'(K) < +1. \qquad (7.5)$$

Conversely K is unstable if $F'(K) > +1$ or if $F'(K) < -1$. The cases of equality are ambiguous.

The rigorous proof of these results is a simple exercise in elementary mathematical analysis. Let us outline the proof that Eq. (7.5) implies stability. Because $F'(P)$ is assumed to be continuous we have $|F'(P)| < 1$ for P sufficiently close to K. Hence by the mean-value theorem, $|F(P) - K| = |F'(P^*)| \cdot |P - K| < |P - K|$. This shows that $F(P)$ is closer to K than P is. Similarly $F[F(P)]$ is closer to K than $F(P)$ is, and so on. Consequently the sequence P, $F(P)$, $F[F(P)], \ldots$, must converge to K; that is, K is stable. The converse cases are similar.

Now what can we say about the behavior of the population sequence in the case in which $F'(K) < -1$? The curves shown in Figure 7.3 have the form

$$F(P) = Pe^{r(1-P)}; \qquad (7.6)$$

the equilibrium position K is now normalized at $K = 1$. This family of curves, called *Ricker curves* (Ricker, 1954), is used in the management of the Pacific salmon (*Oncorhynchus* species) populations. A biologically based derivation of the function given by Eq. (7.6) appears in Section 7.4. The dynamic behavior of the stock-recruitment relation [Eq. (7.1)] for the family of Ricker curves is studied by May (1974) and Oster (1975), who show that:

1. For $0 < r \leq 2$, the point $K = 1$ is a stable equilibrium. This is clear from the above theorem, because

$$F'(1) = 1 - r.$$

2. There exists an increasing sequence $2 = r_1 < r_2 < \cdots$, such that when $r_n < r < r_{n+1}$ the population sequence $\{P_k\}$ undergoes limit-cycle oscillations of period 2^n. (Figure 7.3b shows the case $n = 1$; Figure 7.3c shows the case $n = 2$.)

3. The sequence $\{r_n\}$ approaches a critical value $r^* = 2.6924$ .... For any $r > r^*$, "there exist cycles of every period 2, 3, 4, ..., along with an uncountable number of initial population levels for which the system does not eventually settle into any finite cycle" (May, 1974, p. 645).

The latter situation has been called dynamic "chaos," since the successive population levels $\{P_k\}$ produced under these conditions are to all intents undistinguishable from purely random fluctuations. May (1974) seems to attribute this increasingly chaotic behavior to a large intrinsic growth rate r, but in reality it is not the growth rate itself but the high degree of overcompensation as measured by $F'(K)$ that is significant here. In Section 7.5 we see that in some cases Eq. (7.1) is capable of even more remarkable behavior.

Yield-Effort Curves

As we know from Chapter 1, a given stock-recruitment curve can be used to derive a corresponding yield-effort relation. Recall that this is a static concept, with a constant harvest H (equal to yield Y) being taken from some fixed population of recruits R:

$$H = R - P = F(P) - P. \tag{7.7}$$

We model the harvesting stage by means of a continuous submodel:

$$h(t) = E(t)X(t)$$

$$\frac{dX}{dt} = -h(t), \quad 0 \le t \le T,$$

where $X(t)$ denotes the population size, $h(t)$ is the harvest rate, and $E(t)$ represents the rate of effort applied during the "harvest season" $0 \le t \le T$. (A more detailed analysis is given in Section 7.6.) Because harvesting reduces the recruit population R to the escapement level P we have

$$X(0) = R = F(P) \quad \text{and} \quad X(T) = P.$$

Consequently the total effort E devoted to the harvest is given by

$$E = \int_0^T E(t)\, dt = \int_0^T \frac{h(t)\, dt}{X(t)}$$

$$= \int_P^{F(P)} \frac{dX}{X}$$

or

$$E = \ln\left[F(P)/P\right]. \tag{7.8}$$

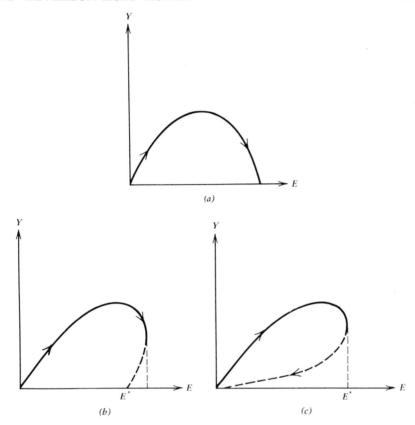

Figure 7.4 Yield-effort curves: (a) normal compensation; (b) noncritical depensation; (c) critical depensation.

Equations (7.7) and (7.8) are parametric equations for the yield-effort curve, having the escapement population P as a parameter. These curves possess the same general characteristics as the yield-effort curves described in Chapter 1 (see Figures 1.6, 1.7, and 1.8). Three examples are illustrated in Figure 7.4; the arrow on each curve indicates the direction of decreasing population parameter P. The implications in terms of stability and other factors are the same as those given in Chapter 1 and are not repeated here.

7.2 THE BEVERTON–HOLT STOCK-RECRUITMENT MODEL

In Sections 7.2–7.4 we describe some standard models of stock-recruitment relationships based on hypotheses concerning the life-history of various fish species. These hypotheses are designed to reflect certain biological mechanisms considered to be responsible for compensation, overcompensation, or depensation. We begin with a simple (but spectacular) model of compensation due to R. J. H. Beverton and S. J. Holt (1957).

Certain fish populations such as the North Atlantic plaice and haddock studied by Beverton and Holt, as well as other demersal (i.e., bottom-feeding) species, have the characteristic that recruitment appears statistically to be virtually unaffected by fishing, at least within "normal" ranges of fishing effort. (In these fisheries overfishing, when it occurs, is not associated primarily with an excessive reduction of the population level, but rather with the capture of immature fish that have failed to grow to their optimal size. This interesting point is examined in greater detail in Chapter 8.) These fish species often possess extremely high fertility rates, with a single female laying as many as 10^6 eggs, only a very small proportion of which survive to become adult fish. Recruitment is usually highly variable, with the occasional appearance of exceptionally large year-classes that are believed to result from unusually favorable environmental conditions.

The Beverton–Holt stock-recruitment model attempts to account for these observations. The model describes a simple mechanism that can lead to extreme population stability (under constant environmental conditions) over a wide range of parent escapement. This model assumes a *density-dependent* relative mortality rate (the equation $N^{-1} dN/dt = -\mu$ is considered to describe a *density-independent* mortality rate; any other relation is referred to as a *density-dependent* relation) of the form

$$\frac{1}{N}\frac{dN}{dt} = -(\mu_1 + \mu_2 N),$$ (7.9)

where $N(t)$ denotes the number of young larval-stage fish alive at time t and where μ_1 and μ_2 may be either positive constants or general, arbitrary nonnegative functions of time t.

Suppose that Eq. (7.9) holds for a fixed interval of time $0 \le t \le T$ (e.g., the larval stage), and write

$$Y = N(0), \quad R = N(T).$$ (7.10)

In the case in which μ_1 and μ_2 are constants, Eq. (7.9) can easily be integrated explicitly over $0 \le t \le T$, and it follows immediately that

$$R = S(Y) = \frac{k_1 Y}{1 + k_2 Y}$$ (7.11)

for certain positive constants k_1, k_2 related to μ_1, μ_2 and T. Beverton and Holt observed more generally that Eq. (7.11) remains valid when the mortality coefficients are arbitrary functions of t.

THEOREM

Let $N(t)$ satisfy Eqs. (7.9) and (7.10), where

$$\mu_i = \mu_i(t) \geq 0.$$

Then there exist constants $k_1 > 0$ and $k_2 \geq 0$ such that Eq. (7.11) holds. If $\mu_2(t) \not\equiv 0$, then $k_2 > 0$.

Proof. We simply note that Eq. (7.9) is a Riccati differential equation, which is known to have a general solution (see Leighton, 1963, p. 237) of the form

$$N(t) = f(t) \frac{u'(t) + kv'(t)}{u(t) + kv(t)},$$

where $f(t)$, $u(t)$, and $v(t)$ are certain functions and k is an arbitrary constant of integration. Because $N(0) = Y$, we obtain by solving for k

$$k = \frac{\alpha + \beta Y}{\gamma + \delta Y}$$

for constants α, β, γ, and δ. Hence

$$R = N(T) = \frac{\alpha' + \beta' Y}{\gamma' + \delta' Y}.$$

Clearly if $N(0) = Y = 0$ then because Eq. (7.9) implies that $N(t)$ is nonincreasing and nonnegative, we have $R = N(T) = 0$. Hence $\alpha' = 0$. This proves that Eq. (7.11) holds.

If $\mu_2 \equiv 0$, then $R = N(T) = N(0)e^{-\mu_1 T} = k_1 Y$, so that $k_2 = 0$. Conversely if $\mu_2(t) > 0$ on any interval $t_1 < t < t_2$, then the following lemma shows that R is a bounded function of Y, so that we must have $k_2 > 0$. Q.E.D.

LEMMA

Suppose that $dN/dt \leq 0$ for $0 \leq t \leq T$ and that

$$\frac{dN}{dt} \leq -\mu N^2 \tag{7.12}$$

for some nontrivial subinterval $t_1 \leq t \leq t_2$. Then $N(T)$ is bounded independently of the initial value $N(0)$.

Proof. Let $N_1 = N(t_1)$, and $N_2 = N(t_2)$. Then Eq. (7.12) implies that

$$\int_{N_1}^{N_2} \frac{dN}{N^2} \leq -\mu \int_{t_1}^{t_2} dt,$$

or

$$\frac{1}{N_1} - \frac{1}{N_2} \le -\mu(t_2 - t_1) = -\sigma < 0.$$

Hence

$$\frac{1}{N_2} \ge \frac{1}{N_1} + \sigma \ge \sigma \tag{7.13}$$

so that $N_2 = N(t_2) \le 1/\sigma$. Since $dN/dt \le 0$ for all t, we conclude that $N(T) \le N_2 \le 1/\sigma$ as required.   Q.E.D.

This lemma is an interesting result in itself, because it asserts that if a density-dependent mortality rate of the form of Eq. (7.9) holds for any arbitrarily brief "critical period" during life-history of the young fish, then there is an absolute upper limit to recruitment that is independent of the size of the spawn. The reader can easily verify that this remains valid if Eq. (7.9) is replaced by

$$\frac{dN}{dt} \le -\mu N^{1+\varepsilon} \quad \text{for some } \varepsilon > 0.$$

Another interesting aspect of the lemma is that the upper-bound $1/\sigma$ [see Eq. (7.13)] equals $1/(\mu \Delta T)$, where $\Delta T = t_2 - t_1$. If $\mu \Delta T$ is subject to stochastic, environmentally induced variations, the same is true for the upper-bound $1/\sigma$. Thus even when escapement is constant, we see that random influences during a brief critical period can seriously affect subsequent recruitment levels. The importance of a critical period of this kind in fish-population dynamics is often noted by biologists (see Beverton and Holt, pp. 44–74).

Returning now to our general stock-recruitment model (Figure 7.1), let us suppose that each female parent lays a fixed number of eggs, of which a fixed proportion hatch to produce young larvae. Then (neglecting the male population) we have

$$Y = \alpha P,$$

where α is a constant, the *fertility* of the species. Hence

$$R = \frac{k_1 Y}{1 + k_2 Y} = \frac{aP}{1 + bP}, \tag{7.14}$$

where a and b are constants.

The function given by Eq. (7.14) is called the *Beverton–Holt stock-recruitment relation*. We note that

$$\left. \frac{dR}{dY} \right|_{Y \to 0} = a \quad \text{and} \quad \lim_{Y \to \infty} R(Y) = \frac{a}{b}.$$

Moreover the Beverton–Holt curve is obviously concave and increasing, being one branch of a hyperbola asymptotic to the line $R = a/b$. In our earlier terminology Eq. (7.14) defines a compensatory stock-recruitment function.

To obtain a positive equilibrium population K we must have

$$a > 1.$$

When $a \approx 1$, the relation in Eq. (7.14) provides a mild degree of population control (see Figure 7.5). But when $a \gg 1$, the control becomes much stronger.

As a numerical example we consider a fish population in which each female lays 10^6 eggs, 10% of which hatch to produce young, so that fertility $\alpha = 10^5$. We presume that there would be a 90% mortality of young if only one parent were to spawn, so that $R = 10^4$ when $P = 1$. Finally we assume a natural equilibrium population $K = 10^9$. Then we have the equations

$$10^4 = \frac{a}{1 + b}$$

$$10^9 = \frac{10^9 a}{1 + 10^9 b}$$

These are easily solved to give

$$a \approx 10^4, \quad b \approx 10^{-5}$$

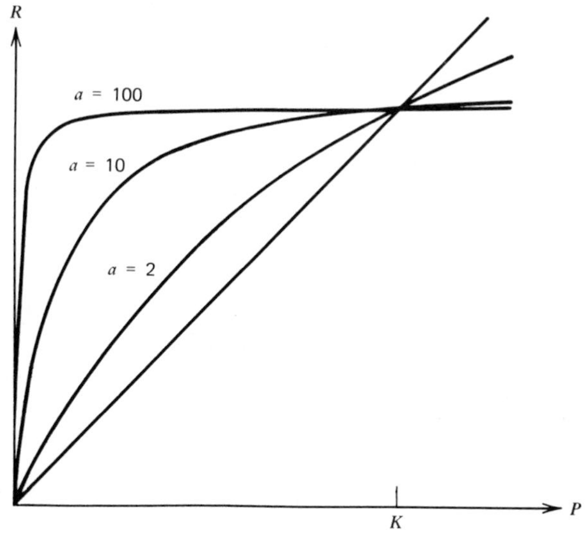

Figure 7.5 Beverton–Holt stock-recruitment curves.

and hence

$$R = \frac{10^4 P}{1 + 10^{-5} P}.$$

Note that a reduction of the parent population K by 99% results in approximately a 1% reduction in recruitment.

The Beverton–Holt model thus provides an elegant and a convincing explanation of the way in which a rather general form of density-dependent pre-recruit mortality can give rise to a situation in which recruitment is virtually independent of the parent stock. The simplistic nature of the model (at least as it is presented here) must be emphasized: our model does not account for temporal or spatial inhomogeneities or ecological complexities such as predator–prey or competitive interactions. Naturally such phenomena can significantly affect population dynamics.

Generalizations

The stock-recruitment function given in Eq. (7.14) is increasing and concave. We now show more generally that (with constant fertility α):

1. The stock-recruitment function $F(P)$ resulting from an arbitrary density-dependent mortality rate

$$\frac{dN}{dt} = -\phi(N, t) \qquad (7.15)$$

is always increasing, assuming $\phi(N, t) \geq 0$ and $\phi(0, t) = 0$.

2. The function $F(P)$ resulting from a time-independent mortality rate

$$\frac{dN}{dt} = -\phi(N) \qquad (7.16)$$

is concave if the positive function $\phi(N)$ is convex.

These results indicate the directions we should follow if we wish to construct models of depensation and overcompensation, as we do in Sections 7.3 and 7.4, respectively.

Assertion (1) is an immediate consequence of the uniqueness theorem for first-order differential equations, [this requires that the function $\phi(N, t)$ satisfy a *Lipschitz-continuity condition* [see Birkhoff and Rota, (1969, p. 20)] which states that Eq. (7.15) possesses at most one solution curve $N = N(t)$ through any given point (N_0, t_0). Consequently two different solution curves $N_1(t)$ and $N_2(t)$ can never intersect. Thus if $Y_1 = N_1(0) > Y_2 = N_2(0)$ (see Figure 7.6), then it must also be true that $R_1 = N_1(T) > R_2 = N_2(T)$ (i.e., that R is an increasing function of Y and therefore of $P = \alpha^{-1} Y$, as claimed).

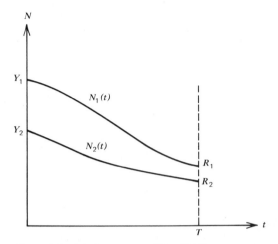

Figure 7.6 Solution curves of Eq. (7.15).

Next assume that Eq. (7.16) holds. Then if $Y = N(0)$ and $R = N(T)$, we have

$$T = \int_0^T dt = -\int_Y^R \frac{dN}{\phi(N)} = \int_R^Y \frac{dN}{\phi(N)}. \tag{7.17}$$

Consider the graph of the function $1/\phi(N)$ shown in Figure 7.7. Given Y, according to Eq. (7.17) the value of R is determined by the condition that

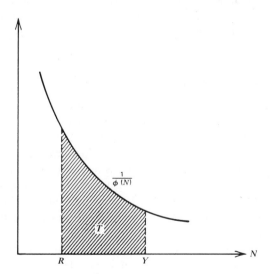

Figure 7.7 The relationship between Y and R, given Eq. (7.18).

the area under the curve from $N = R$ to $N = Y$ is equal to a given constant T.

Because

$$\int_N^Y \frac{dN}{\phi(N)}$$

is a continuous decreasing function of N, we can see that Eq. (7.17) can have at most one solution R. Moreover a necessary and sufficient condition for the existence of a solution R corresponding to every positive initial value Y is that the integral be improper at $N = 0$; that is,

$$\lim_{N \to 0} \int_N^1 \frac{dN}{\phi(N)} = +\infty. \tag{7.18}$$

If this condition does not hold, then positive values of Y exist for which there are no survivors R. This point is examined in greater detail in Section 7.3.

Differentiating Eq. (7.17) with respect to Y we obtain

$$\frac{1}{\phi(Y)} - \frac{1}{\phi(R)} \frac{dR}{dY} = 0, \quad \text{or} \quad \frac{dR}{dY} = \frac{\phi(R)}{\phi(Y)}.$$

Hence

$$\frac{d^2 R}{dY^2} = \frac{[\phi'(R) - \phi'(Y)]\phi(R)}{\phi(Y)^2}. \tag{7.19}$$

If ϕ is convex, then $R < Y$ implies $\phi'(R) < \phi'(Y)$, so that $d^2 R/dY^2 < 0$. This shows that the survivor relation $R = S(Y)$ is concave and therefore that the stock-recruitment relation $R = F(P) = S(\alpha P)$ is also concave. Q.E.D.

7.3 DEPENSATION MODELS

According to assertion (2) in Section 7.2, a convex mortality curve $\phi(N)$ (Figure 7.8a) results in an increasing, concave (compensatory) stock-recruitment function, similar to the Beverton–Holt case. If we wish to model nonconvex depensation curves, therefore, we must consider nonconvex functions $\phi(N)$ (Figure 7.8b). Since a convex function $\phi(N)$ implies that the relative mortality rate $\phi(N)/N$ is an increasing function of N, this condition must be reversed. The question then becomes: what natural phenomena could lead to a relative mortality rate that *decreases* as the population level increases?

An obvious answer to this question would be some form of predation that becomes relatively less effective at higher levels of population. For example, suppose simplistically that predation of young fish occurs at a

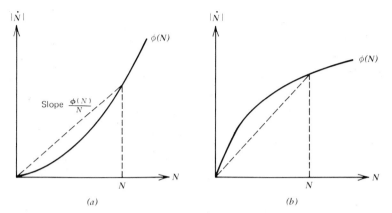

Figure 7.8 Mortality-rate curves: (*a*) convex, implying increasing relative mortality $\phi(N)/N$; (*b*) concave, implying decreasing relative mortality.

constant rate

$$\frac{dN}{dt} = -k, \quad N(0) = Y. \tag{7.20}$$

Then

$$R = N(T) = N(0) - kT = Y - kT. \tag{7.21}$$

If $Y = \alpha P$, the stock recruitment curve $R = \alpha P - kT$ is as shown in Figure 7.9*a*. This curve leads to exponential population growth for $P > K_0 = kT/(\alpha - 1)$.

To incorporate a control mechanism into this model, we consider the

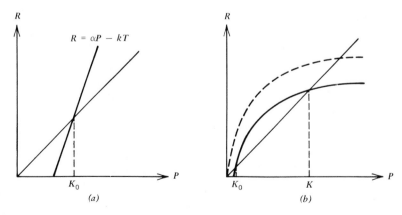

Figure 7.9 Critical depensation curves: (*a*) without control; (*b*) combined with Beverton–Holt type control.

three-stage life history shown below:

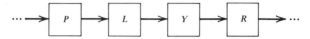

Here L denotes the number of larvae that hatch from eggs laid by the parent stock P. Let the transformation from L to young, adult fish Y be governed by a Beverton–Holt relationship. Assume that the $Y \rightarrow R$ transformation is subject only to a constant predation rate, as above. [An example of this situation occurs in the life history of the salmon. Young salmon fingerlings often negotiate a length of stream, moving down from the spawning areas to a lake where they spend the first year of their lives. The stream may contain a predator population that normally preys on other organisms, but that is capable of devouring a fixed number of salmon fingerlings. A fingerling population that is below this fixed level can be almost completely devoured as it swims downstream. (My thanks to Professor P. A. Larkin, Department of Zoology, University of B.C., for this example.)] Then we have

$$L = \alpha P, \quad Y = \frac{k_1 L}{1 + k_2 L}, \quad R = Y - k_s T.$$

Hence

$$R = \frac{aL}{1 + bL} - c,$$

where a, b, and c are certain positive constants. This curve is shown in Figure 7.9b; it is simply a Beverton–Holt curve that has shifted downward at a fixed distance c.

Note that the mortality-rate function $\phi(N) = k = $ constant of Eq. (7.20) fails to satisfy the condition given by Eq. (7.18). It is easy to verify that in general whenever (7.18) does not hold, critical depensation occurs in the stock-recruitment relation. But as yet we have no model for noncritical depensation. The purpose of the following discussion is to provide such a model.

The Mathematical Theory of Schooling

Certain species of fish, especially *clupeids* (anchovies, sardines, and herrings), are known to form immense, closely packed schools. What is the purpose of these schools? Are the chances of survival for the individual fish increased by swimming with a school? Surely the answer must be affirmative; if it is, this mechanism may lower relative mortality rates for large populations relative to small populations. This consequently would lead to depensation in the stock-recruitment relation.

No advantage of schooling that biologists suggest (food location, navigation, training of young fish, mating) seems to be as important as protection from predation.

If these theories are correct—if the stock-recruitment relationship for certain schooling species is strongly depensatory—then the possibility arises that heavy exploitation can lead to sudden population "crashes" that may not be followed by recoveries even if exploitation is severely reduced. That is precisely what has happened to several of the clupeid fisheries. Usually, however, environmental changes are believed to have contributed primarily to these collapses. (An alternative theory to explain these phenomena based on the Gause model of interspecific competition is described in Chapter 9.)

An interesting analysis due to Brock and Riffenburgh (1963) shows that, perhaps contrary to intuition, the habit of schooling can provide significant protection merely by reducing the rate of detection by predators. This analysis is based on three assumptions:

1. Predators have finite detection ranges.
2. Each predator has a finite "appetite."
3. The number of predators is fixed.

Of these, the first two assumptions clearly are reasonable in general. The third assumption is more restrictive because it rules out standard predator–prey reactions, but it may be realistic in cases where the predators can obtain alternative prey and thus do not depend solely on the schooling species in question.

Let r denote the visual (or other) detection range. In normally clear sea water, r is of the order of 60 meters and is essentially independent of the size of the visual objective. Let N denote the number of prey in a school (assumed spherical), and let c be a constant that defines the average spacing within the school in the sense that a school of N fish has a volume $4\pi c^3 N/3$ and a radius $r_0 = cN^{1/3}$. Then the "visual volume" of the school (i.e., the volume of the region within which a predator can detect the school) equals

$$V = \frac{4\pi}{3}(r_0 + r)^3 = \frac{4\pi r^3}{3}\left(1 + \frac{c}{r}N^{1/3}\right)^3. \tag{7.22}$$

Since predators are assumed to have fixed appetites, we can assume that the rate of predation is proportional to the rate of the detection of schools. The rate of detection is in turn proportional to the visual volume of the school, provided the latter is small in relation to the total volume of sea water over which the predators search. Given these assumptions the

TABLE 7.1. RELATIVE MORTALITY RATES FOR
SCHOOLS OF SIZE N (NORMALIZED TO MAKE $\sigma(10^8) =$
1.0).

N	10^2	10^3	10^4	10^5	10^6	10^7	10^8
$\sigma(N)$	6366	739	100	17	4.4	1.7	1.0

mortality rate of the prey satisfies

$$\frac{dN}{dt} = -kV = -c_1(1 + c_2 N^{1/3})^3 = -\phi(N). \tag{7.23}$$

[Equation (7.23) should be considered valid only for $N \geq a =$ predator appetite. We assume that a is so small that this complication can be neglected, however.]

The relative mortality rate equals

$$\sigma(N) = \frac{\phi(N)}{N} \propto \frac{1}{N}\left(1 + \frac{c}{r}N^{1/3}\right)^3. \tag{7.24}$$

Table 7.1 contains values of $\sigma(N)$ for schools of size 10^2–10^8, under the assumption that $c/r = 0.01$. Because $1/\phi(N)$ represents the relative chances of survival for an individual fish belonging to a school of size N, it is clear that under our assumptions schooling can be an effective survival technique!

Returning to the question of stock-recruitment relationships, we note that because the mortality function $\phi(N)$ in Eq. (7.23) is concave downward, it follows from Eq. (7.19) that the corresponding survival relationship $R = S(Y)$ is convex upward. It can also be shown that $S(Y) \sim cY$ ($c =$ constant) as $Y \to \infty$ (the details are left to the reader). Thus the present model is similar to the case of a fixed predation rate shown in Figure 7.8, except that the stock-recruitment curves in the present model are smooth rather than bent. The depensation may be either critical or noncritical; noncritical depensation occurs if and only if

$$\frac{dR}{dP} = \frac{dR}{dY}\frac{dY}{dL}\frac{dL}{dP} = S'(0)k_1\alpha < 1$$

for $P = 0$.

7.4 OVERCOMPENSATION

We conclude our discussion of stock-recruitment relationships with a brief description of a simple model of overcompensation due to Ricker

(1954, 1958). First recall that overcompensation cannot result from a density-dependent mortality law of the form in Eq. (7.15), because we know that the resulting stock-recruitment curve must increase monotonically.

Ricker assumes that the relative predation rate is proportional to the *initial* larval or egg population:

$$\frac{1}{N}\frac{dN}{dt} = -kL, \quad N(0) = L. \tag{7.25}$$

Ricker's hypothesis is based on the observation that certain species of fish such as salmon habitually cannibalize their eggs and larvae. In this case both the larval population L and the rate of predation are proportional to P, so that Eqs. (7.25) hold.

More generally we may suppose that

$$\frac{1}{N}\frac{dN}{dt} = -[k_1(t) + k_2(t)L], \quad N(0) = L,$$

where $k_1(t)$ and $k_2(t) \geq 0$ are given functions. Consequently

$$Y = N(T) = N(0) \exp\left\{-\int_0^T [k_1(t) + k_2(t)L]\, dt\right\} = aLe^{-bL}.$$

Assuming as usual that $L \propto P$ and that $R \propto Y$, we conclude that

$$R = \alpha Pe^{-\beta P}. \tag{7.26}$$

These are the Ricker stock-recruitment curves we discuss in Section 7.1.

Ricker's model assumes that the larvae are cannibalized or otherwise serve as food. But the larvae themselves require food to grow, and an alternative model of population control (Jones and Hall, 1973) can be based on the utilization of a food supply $F(t)$:

$$\frac{1}{N}\frac{dN}{dt} = -g(F) \tag{7.27}$$

$$\frac{1}{F}\frac{dF}{dt} = -bN. \tag{7.28}$$

Here the relative larval mortality $(1/N)\, dN/dt$ is assumed to depend on the food supply; $g(F)$ is a function that increases sharply as $F \to 0$. Given the initial conditions $N(0) = L$ and $F(0) = F_0$, Eqs. (7.27) and (7.28) can be solved numerically to obtain $Y = N(T)$. An example is given in Figure 7.10. In Figure 7.10a the trajectories $(N(t), F(t))$ are shown for various values of $N(0)$, with F_0 fixed. When $N(0)$ is large, the food supply is rapidly exhausted at the beginning of the larval stage, and subsequently

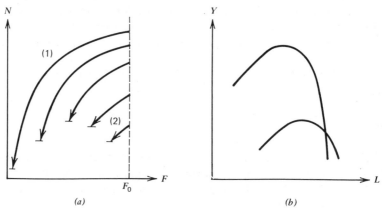

Figure 7.10 Effect of food supply: (a) trajectories (N, F), with F_0 fixed and $0 \le t \le T$; (b) $Y-L$ curves for two different F_0 values (Jones and Hall, 1973).

high rates of larval mortality result (curve 1). Conversely when $N(0)$ is small, there is little reduction in F and consequently little relative loss in N (curve 2). Figure 7.10b shows the relationship between Y and L for two F_0-values. Clearly this model can lead to a stock-recruitment relation that exhibits severe overcompensation.

7.5 A SIMPLE COHORT MODEL

All the discrete-time population models in the preceding sections assume that the parent population P_k dies during spawning and is replaced by the subsequent cohort of recruits $R_k = P_{k+1} = F(P_k)$. We now modify this situation by assuming that the parent stock suffers only limited mortality from one period to the next (see Figure 7.11). The relationship between parent stock P_k and recruits R_k is assumed to be described by the usual stock-recruitment relation

$$R_k = F(P_k). \qquad (7.29)$$

If $\sigma = $ constant represents the survival rate of parents over one period, we then obtain

$$P_{k+1} = \sigma P_k + R_k = \sigma P_k + F(P_k), \qquad (7.30)$$

which may be expressed as

$$P_{k+1} = G(P_k). \qquad (7.31)$$

Thus in this model the total adult population in period $k+1$ is a given function of the population level in the preceding period. [Note that the model assumes that (1) all adults have the same mortality rate $1 - \sigma$, and (2) fecundity is the same for adults of all ages. More complex models in

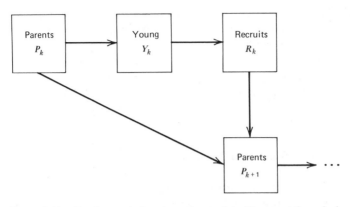

Figure 7.11 Simple population-dynamics model with parental survival.

which both mortality and fecundity are age-dependent have been studied by Leslie (1948); see also Clark (1976).]

Now suppose that the stock-recruitment curve $F(P)$ exhibits strong overcompensation. The corresponding population growth function $G(P) = F(P) + \sigma P$ is then as shown in Figure 7.12. Due to overcompensation, the equilibrium at $P = K$ is highly unstable. Consider, for example, an initial population level P_1 slightly below K, as shown in Figure 7.12. Then

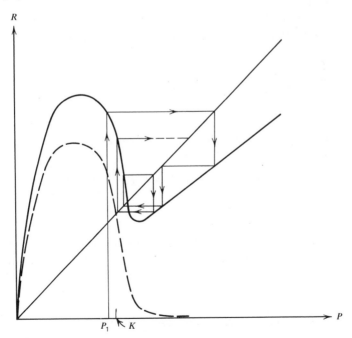

Figure 7.12 Reproduction curve corresponding to Figure 7.11.

$P_2 = G(P_1)$ is much larger than P_1 [a large cohort of recruits $R_1 = F(P_1)$ appears in the adult population at this stage]. In the next period, however, overcompensation reduces recruitment size to an insignificant level $R_2 = F(P_2)$. Thus the population in the next period consists almost entirely of surviving adults σP_2. This process continues for several generations until the population P_k eventually falls below the critical equilibrium level K once more. At this stage another large cohort of recruits is produced, and the process is repeated. Whether the process settles into a regular cyclic pattern or whether the recruitment is "chaotic" cannot be decided without a detailed analysis of an explicit model. Clearly a slight degree of random variability in the system could result in a totally irregular recruitment pattern in any case.

Occasional exceptionally large cohorts typical of certain species such as Lake Erie whitefish (see Table 7.2) dominate some fisheries for many years. Without exception biologists seem to attribute these phenomena to unpredictable environmental fluctuations. While this may indeed be a valid hypothesis, our present model suggests that large-scale, irregular fluctuations in recruitment cannot by themselves be considered conclusive evidence of random effects. Perhaps in general a combination of random effects and deterministic mechanisms is the most believable hypothesis.

An interesting property of growth curves exhibiting overcompensation is the possibility of stabilizing the population and its recruitment by means of a suitable harvest policy. This possibility is somewhat converse to our findings thus far, which indicate that harvesting often leads to increased instability in population dynamics. This question is discussed further in Section 7.7.

A More Complex Model

Further modifications of the life-history model can produce growth curves $P_{k+1} = G(P_k)$ of almost unlimited complexities. In Figure 7.13, for example, we suppose that:

1. The stock-recruitment relationship $F(P_k)$ is of Beverton–Holt type (Eq. 7.14).
2. Recruits and surviving parents form an "adult" stock $A_k = F(P_k) + \sigma P_k$.
3. The adult stock is subject to a survival relation $P_{k+1} = H(A_k)$ that exhibits critical depensation. In this case the overall growth function $G(P_k)$ is given by

$$P_{k+1} = G(P_k) = H[F(P_k) + \sigma P_k].$$

One possible graph of $G(P)$ appears in Figure 7.14. This curve has

TABLE 7.2. AGE COMPOSITION (%) FOR LAKE ERIE WHITEFISH (*COREGONUS CLUPEAFORMIS*) CAUGHT IN 1943–1956 IN YEARLY SAMPLES (LAWLER, 1965). NOTE THE DOMINANCE OF THE 1944 YEAR-CLASS.

Year-Class	Year of Capture													
	1943	1944	1945	1946	1947	1948	1949	1950	1951	1952	1953	1954	1955	1956
1938	38.0	—	2.0	—	0.6	—	—	—	—	—	—	—	—	—
1939	6.0	14.0	2.0	1.0	0.4	—	—	—	—	—	—	—	—	—
1940	9.0	14.0	3.0	1.0	0.4	0.14	0.08	—	—	—	—	—	—	—
1941	3.0	57.0	29.0	2.0	0.4	0.3	0.12	—	—	—	—	—	—	—
1942	—	14.0	63.0	37.0	4.7	0.7	1.7	—	—	—	—	—	—	—
1943	—	—	1.0	38.0	13.0	4.7	3.6	18.0	1.7	0.4	0.9	—	—	—
1944	—	—	—	22.0	75.0	92.3	79.0	14.5	2.0	0.6	0.5	—	—	—
1945	—	—	—	—	3.0	1.6	12.5	45.4	10.5	0.6	0.5	—	—	—
1946	—	—	—	—	1.7	0.14	3.0	22.0	41.0	0.2	0.7	—	—	—
1947	—	—	—	—	—	—	—	—	35.5	1.8	3.4	—	—	—
1948	—	—	—	—	—	—	—	—	9.4	13.5	75.4	—	—	—
1949	—	—	—	—	—	—	—	—	—	82.3	16.1	10.0	—	—
1950	—	—	—	—	—	—	—	—	—	0.4	2.2	36.0	4.2	0.5
1951	—	—	—	—	—	—	—	—	—	—	—	52.0	6.8	4.5
1952	—	—	—	—	—	—	—	—	—	—	—	2.0	42.1	22.5
1953	—	—	—	—	—	—	—	—	—	—	—	—	45.5	58.0
1954	—	—	—	—	—	—	—	—	—	—	—	—	1.2	14.5
Number of Fish in Sample	162	7	140	123	453	1083	1115	55	296	500	403	106	235	200

233

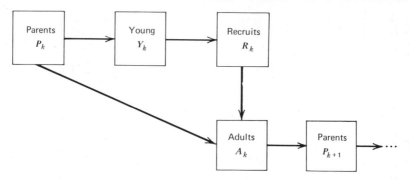

Figure 7.13   Life-history model leading to the growth curve shown in Figure 7.14.

four equilibrium points $P = 0$, K_1, K_0, and K. The equilibria at K_1 and K are stable, whereas the equilibria at 0 and K_0 are unstable. For obvious reasons the point K_0 is called an *escape point*.

We do not pursue this abstract model-construction process further; the purpose of the present discussion is only to illustrate the flexibility of the metered-model approach in population dynamics.

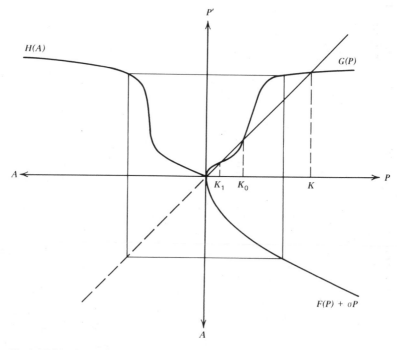

Figure 7.14   Reproduction curve corresponding to Figure 7.13.

7.6 THE PRODUCTION FUNCTION OF A FISHERY

By definition a production function specifies the rate of output of a certain productive process in terms of inputs to the process. The production function of a fishery is usually assumed to be of the form

$$h = h(x, E), \tag{7.32}$$

where the variables E, x, and h have the normal interpretations of fishing effort, fish biomass level, and harvest rate respectively. While our present discussion is limited to the case where E, x, and h are single, real variables, a more detailed study might permit any or all of these variables to be multidimensional. Thus effort E might be broken into its components, capital and labor, and further classified as to form of effort, such as searching for fish, setting and hauling nets, and so on. Similarly the population variable x might contain components for each cohort and for each species subject to capture. Multicohort and multispecies fisheries are discussed in Chapters 8 and 9.

As in any mathematical model, the use of a given production function involves certain underlying assumptions regarding the production process. In this section we briefly discuss some of these assumptions and their implications. We are concerned here with qualitative effects, rather than with detailed, specific-purpose models.

The Schaefer Model

The production function used in the Schaefer model, as well as in other continuous-time models, is

$$h = qEx, \quad q = \text{constant.} \tag{7.33}$$

The basic hypothesis underlying this model is simply that catch per unit effort h/E is proportional to the biomass level, and that this proportionality remains valid for all levels of E and x. This hypothesis in turn is based on several additional assumptions, including:

1. Uniform distribution of the fish population.
2. Nonsaturation of fishing gear.
3. Noncongestion of fishing vessels.

Let us investigate the consequences of relaxing these assumptions.

Gear Saturation

Imagine simplistically that fishing nets of a particular design possess a maximum capacity c_0, so that the biomass of fish captured by one haul of the net is given by

$$h_0 = \max (kx, c_0);$$

that is, the catch h_0 is proportional to the density of the fish stock up to the capacity level c_0. If E denotes the number of nets hauled per unit time we then have [continuing to adopt assumptions (1) and (3) above]

$$h = Eh_0 = kE \max \left(x, \frac{c_0}{k} \right). \tag{7.34}$$

Thus the Schaefer model given by Eqs. (7.33) holds only for $x \leq c_0/k$.

More generally we may assume that gear saturation reduces the catchability of nets in a smooth manner, leading to a production function of the form

$$h = qE\phi(x), \tag{7.35}$$

where $\phi'(x) \geq 0$ and $\phi''(x) \leq 0$. In other words the production function $h(x, E)$ is expected to exhibit decreasing marginal returns to the input factor x as a consequence of ultimate gear saturation.

Congestion

Congestion among fishing vessels is a common occurrence on fishing grounds where the concentration of fish is high. As a model of pure congestion, let us consider a fishing ground A containing a uniform density of fish. We let a denote the area required by a single vessel for the unrestricted operation of its gear. Congestion does not occur if the number of vessels is less than $N_0 = A/a$, but any further increase in the number of vessels simply divides the total harvest into smaller individual catches. (Here we neglect the "packing" problem of fitting n vessels into the fishing area A without overlap, as well as the possibility that congestion may actually lead to a *decrease* in the total harvest rate.) Thus if $E = n =$ number of vessels, we have

$$h(x, E) = qx \max (E, N_0). \tag{7.36}$$

Catch is therefore proportional to effort only for $E < N_0$, the congestion level.

More generally we would expect Eq. (7.36) to be replaced by a relation of the form

$$h(x, E) = qx\psi(E), \tag{7.37}$$

with $\psi'(E) > 0$ and $\psi''(E) < 0$. If both congestion and gear saturation are considered, we obtain

$$h(x, E) = q\phi(x)\psi(E), \tag{7.38}$$

so that the production function satisfies

$$\frac{\partial h}{\partial x}>0, \quad \frac{\partial h}{\partial E}>0, \quad \frac{\partial^2 h}{\partial x^2}<0, \quad \frac{\partial^2 h}{\partial E^2}<0. \tag{7.39}$$

Thus as an explicit result of congestion and gear saturation, the production function of the fishery exhibits decreasing marginal returns to both input factors. These properties are usually assumed to hold for production functions, more or less on an ad hoc basis.

Cost Implications

The equation

$$h = h(x, E) \tag{7.40}$$

can be solved implicitly for the effort E as a function of x and h. If we assume that the cost of fishing is proportional to effort, we then obtain a cost function

$$C = C(x, h).$$

From Eqs. (7.39) it can be easily seen that this cost function satisfies

$$\frac{\partial C}{\partial x}<0, \quad \frac{\partial C}{\partial h}>0, \quad \frac{\partial^2 C}{\partial x^2}>0, \quad \frac{\partial^2 C}{\partial h^2}>0. \tag{7.41}$$

For example, the differentiation of Eq. (7.40) with respect to x (treating E as the dependent variable) gives

$$0 = \frac{\partial h}{\partial x}+\frac{\partial h}{\partial E}\frac{\partial E}{\partial x},$$

so that by Eqs. (7.39)

$$\frac{\partial E}{\partial x}= -\frac{\partial h/\partial x}{\partial h/\partial E}<0.$$

Because $C \propto E$, the first inequality in Eqs. (7.41) is established; the other inequalities follow similarly.

Thus the decreasing marginal returns characteristic of the production function lead to the expected properties for the cost function $C(x, h)$.

Nonuniform Fish Distributions

As far as the fishery is concerned, possible nonuniformities in the distribution of the fish population are of two main types. *Predictable nonuniformities* correspond to normal concentrations of fish, whether they are permanent or vary periodically along regular migration routes. *Unpredictable nonuniformities* correspond to the random movements of the fish population or to the wanderings of individual schools. Let us consider the second type.

As a simple model first suppose that the entire fish population forms into a single school. Once the location of the school is known, the cost of harvesting can be assumed to be simply proportional to the *total* harvest H. Thus

$$\text{Total cost} = c_h H + c_s, \qquad (7.42)$$

where c_h is the unit harvesting cost, H represents the total harvest, and c_s denotes the (expected) searching cost. Note that a continuous-time model is now obviously inappropriate, but that the metered model in Figure 7.1 can be used to good effect (we return to this point later in the discussion). Also a straight cost-accounting model is being used in place of the production-function approach.

Next suppose that the population of recruits R forms into N schools, the number N is possibly dependent on R. The cost of locating these schools obviously depends on the searching technology employed. In an extreme case, such as satellite photography, it may be possible to locate all schools at once at a fixed cost c_s. This is equivalent to the case of a single school, and Eq. (7.42) applies.

The opposite (probably more realistic) case arises when randomly distributed schools are located individually by the fishing vessels. The expected cost of locating one school is then proportional to n^{-1}, where n is the number of schools in existence. Thus the Schaefer unit cost function $c(x) = cx^{-1}$ can be employed, if it is understood that the units of x are considered *standard school biomass*. In this case, however, the *continuous state-variable model* also becomes unrealistic, because x can assume only integer values ($x = 0, 1, 2, \ldots, N$).

The behavior of the open-access fishery under these circumstances is clear: the fish population tends to be exterminated if the value of the last school exceeds the cost of locating it. (Some degree of cooperation may be necessary in sharing the searching cost among the fishing vessels for this to be so.) Of course, a real-world situation is not as simple; in actuality some fish surely escape at the last moment. But if our depensation-schooling model (see Section 7.3) is valid, the few fish that escape may be particularly vulnerable to predation. While schooling behavior may be beneficial to fish when confronted with a natural predator, schooling can be an extremely dangerous habit for fish when confronted with the human predator!

Similar observations apply, often more precisely, to conglomerations (flocks, herds, etc.) of other natural animal populations, such as migratory birds and grazing animals. The extreme vulnerability of migratory bird flocks, for example, resulted in the disappearance of some species, before conservationists established the 1918 International Migratory Bird Treaty between the United States and Canada, and later Mexico.

Predictable Nonuniformities

The case of predictable spatial nonuniformities in the fish population raises many new problems, because the state of the system must now be described in terms of a *distributed parameter* $x = x(p, t)$, where p is a variable representing spatial location and $x = x(p, t)$ now denotes the density of the population at p at time t. A wide variety of models can be conceived, depending on both the recruitment behavior and the movement pattern of the fish. On the economic side, costs may also depend on location p, due to differences in locational distances from port facilities. We do not attempt to discuss these complications here (however see Section 9.4).

One important case exists in which this problem can be reduced to simpler questions. Suppose that the stock-recruitment relationship for the entire population can be adequately described by our simple equation

$$R_{k+1} = F(P_{k+1}),$$

or in the case of harvesting

$$R_{k+1} = F(R_k - H_k).$$

The optimum fishing problem can then be divided into two subproblems: one concerning the most efficient method of achieving any desired total harvest H, and another concerning the choice of an optimal, long-term harvest policy $\{H_k\}$. The latter problem requires the maximization of the expression

$$PV = \sum_{k=1}^{\infty} \frac{\pi(H_k, R_k)}{(1+i)^k}, \qquad (7.43)$$

where $\pi(H_k, R_k) = R(H_k) - C(H_k, R_k)$ represents the net revenue obtained from the harvest H_k and i denotes the annual rate of interest. In formulating this problem (the solution is discussed in Section 7.7), it is presupposed that the cost $C(H_k, R_k)$ represents the *least cost* at which the harvest H_k can be achieved. Thus the first suboptimization problem (cost efficiency) is to be solved prior to the second problem (optimal harvesting).

For example, suppose there are N fishing grounds. Let $H^i = H^i(R^i, E^i)$ denote the production function for the total harvest H^i on the ith ground, as a function of the recruited stock level R^i and effort E^i on the ith ground. The problem is then to determine the least total cost $C = \sum_1^N c_i E^i$ at which a given total harvest $H = \sum_1^N H^i$ can be achieved. This problem can be easily solved by means of Lagrange multipliers. The result is simply

$$\frac{1}{c_i} \frac{\partial H^i}{\partial E^i} = \lambda = \text{constant},$$

where the multiplier λ is chosen so that the constraint $\sum_1^N H^i = H$ is satisfied. However, the solution of the corresponding optimal harvesting problem [Eqs. (7.43)] also requires knowledge of how the total recruitment $R_k = \sum_1^N R_k^i$ is to be distributed among the i fishing grounds following each harvest H_k.

Optimal Fleet Size and Season Length

The preceding analysis shows how to determine the optimal total annual effort E and its distribution among various fishing grounds, given the recruitment levels R^i and the total harvest H. However, the analysis does not suggest how the effort should be distributed in time over the fishing season.

To address this question, we consider any one of the fishing grounds, and let B denote the number of boats to be dispatched to this ground. We also let T denote the corresponding time required for B boats to fulfill the given quota H. We must have

$$T \le T_{\max}, \qquad (7.44)$$

where $T_{\max}$ represents a physical limitation on the length of the fishing season. Neglecting nonuniformities, as well as congestion and saturation effects, we can write

$$\frac{dx}{dt} = -qBx, \quad 0 \le t \le T$$

$$x(0) = R, \quad x(T) = R - H$$

for the fish biomass $x = x(t)$ on the given ground during the fishing season. [The possibility of natural mortality is neglected here; this is an important omission (see Exercise 3).] Therefore,

$$x(T) = Re^{-qBT} = R - H,$$

so that

$$B = \frac{k}{T}, \quad \text{where} \quad k = \frac{1}{q} \log \frac{R}{R-H}. \tag{7.45}$$

Now let $C = C(T)$ denote the cost of operating a single boat for a period of length T. A reasonable model for this function can be simply

$$C(T) = c_0 + c_1 T, \tag{7.46}$$

where c_0 denotes the fixed cost associated with dispatching the vessel and c_1 is the operating cost per unit time. Net revenue for the given fishing ground is then

$$NR = pH - C(T)B$$
$$= pH - \frac{kC(T)}{T}, \tag{7.47}$$

where the constant k depends on the given values of R and H, as in Eq. (7.45). For the cost function [Eq. (7.46)], Eq. (7.47) reduces to

$$NR = (pH - kc_1) - \frac{kc_0}{T}. \tag{7.48}$$

If the fixed cost c_0 is positive, this expression is maximized, subject to the constraint given in Eq. (7.44), when $T = T_{max}$. Thus we reach the conclusion that unless the marginal cost $C'(T)$ increases as the length of the season is extended, *the optimal fleet size B^* is that of the smallest fleet required to capture the quota H in the span of one season.* Note that we must have

$$pH > \frac{kC(T_{max})}{T_{max}} \tag{7.49}$$

for the fishery on the given ground to be viable. Otherwise $NR < 0$ for all $T \leq T_{max}$. On the other hand if $C'(T)$ does increase with T, the rule becomes simply

$$\frac{d}{dT} \frac{C(T)}{T} = 0, \quad \text{or} \quad C'(T^*) = \frac{C(T^*)}{T^*}, \tag{7.50}$$

provided that the solution $T^* < T_{max}$.

The preceding simple model can be used to study the commonly occurring case in which an annual quota H is fixed by a regulating authority, but entry to the fishery remains unrestricted. The net revenue if B boats enter the fishery is given by Eq. (7.47):

$$NR = pH - \frac{kC(T)}{T}, \quad \text{where} \quad T = \frac{k}{B}.$$

If the net revenue is positive, additional boats are attracted to the fishery. Open-access economic equilibrium (not bionomic equilibrium, because the biological equilibrium is established by the regulatory agency) thus becomes established when the fleet size B is such that

$$\frac{kC(T)}{T} = pH. \tag{7.51}$$

The case in which $C(T)$ is given by Eq. (7.46) is shown in Figure 7.15. Note that unless the fishery is not at all viable, the open-access equilibrium involves an excess of boats, which can fulfill the quota in less than the optimal time $T_{\max}$. The greater the price–cost ratio, the more exaggerated this effect becomes.

It has often been observed that both output (yield) and input (effort) must be controlled to accomplish the rational exploitation of a common-property fishery. From our earlier analysis in Section 4.6 we know how a tax on catch can be employed to optimize harvest levels. The present analysis shows that an additional tax or fee must be levied on *capacity* to limit this quantity as well. If we continue to assume that all boats have the same fixed fishing power, we can see immediately that an annual fee per boat equal to

$$\phi = pB^* - C(T^*)$$

ensures that $NR' = NR - B\phi = 0$ for $B = B^*$. This fee leads the open-access (catch-regulated) fishery to adopt an optimal fleet size. If the fishing power is not the same for all boats or if it is subject to change from technological advances, then the optimal fee must obviously reflect these

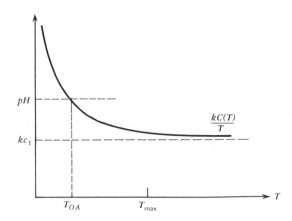

Figure 7.15 Optimal season length ($T_{\max}$) versus open-access season length (T_{OA}).

variations. Furthermore, the optimal fee is sensitive to changes in price and cost levels.

From a practical viewpoint it is clear that severe difficulties are associated with any management program based on taxes, license fees, and the like. Yet ultimately it seems. that some such method must be employed—even if in an approximate, suboptimal form—if marine fisheries are to be managed to the overall benefit of mankind. [The reader is referred to the works of Christy (1973), Crutchfield (1967, 1970) and Scott (1965) in particular for further discussions of these issues.]

7.7 OPTIMAL HARVEST POLICIES

We now turn to the problem of optimal harvest policies for the discrete-time model. The dynamics of the process are described by Eq. (7.3):

$$R_{k+1} = F(P_{k+1}) = F(R_k - H_k). \tag{7.52}$$

The initial recruitment level R_1 is assumed known. Thus if the harvest sequence $\{H_k, k = 1, 2, 3, \ldots\}$ is given, Eq. (7.52) determines the subsequent recruitment levels R_k, $k \geq 2$. The harvests H_k must satisfy the constraints

$$0 \leq H_k \leq R_k = F(P_k). \tag{7.53}$$

Any harvest sequence satisfying these constraints will be called a *feasible* harvest sequence.

We now introduce economic assumptions that are discrete-time analogs of the assumptions made for the linear harvesting model in Chapter 2. (Nonlinear models are discussed in Section 7.8.) First we assume a constant price p for the harvested resource. Then we assume that the instantaneous harvesting cost $C(x, h)$ is linear in h:

$$C(x, h) = C(x)h.$$

The total cost of a harvest H_k taken from a recruit population R_k can therefore be expressed as

$$\begin{aligned} \text{Total cost} = C_T(R_k, H_k) &= C(R_k) + C(R_k - 1) + C(R_k - 2) + \cdots \\ &\quad + C(R_k - H_k + 1) \\ &\approx \int_{R_k - H_k}^{R_k} C(x)\, dx. \end{aligned} \tag{7.54}$$

(As formulated here the model assumes negligible natural mortality during the harvest phase.) (Note that the expression $C(R_k) + \cdots + C(R_k - H_k + 1)$ is simply a Riemann approximating sum for the integral on the right side of this equation.)

The harvest H_k therefore results in a net revenue or economic rent in period k equal to

$$\pi(R_k, H_k) = pH_k - C_T(R_k, H_k) = \int_{R_k - H_k}^{R_k} [p - C(x)]\, dx. \qquad (7.55)$$

The present value of economic rent is then given by

$$PV = \sum_{k=1}^{\infty} \pi(R_k, H_k)\alpha^{k-1} \qquad (7.56)$$

where

$$\alpha = \frac{1}{1+i}, \qquad (7.57)$$

with i denoting the periodic discount rate. We wish to solve the optimal harvest problem; that is, the problem of determining the feasible harvest sequence $\{H_k\}$ that maximizes the present-value expression [Eq. (7.55)], subject to the state equation [Eq. (7.52)].

Not surprisingly the solution to this problem is similar to the continuous-time case in Chapter 2. There exists an optimal parent escapement population P^*, and the optimal harvest policy consists of a most-rapid approach to the equilibrium level P^*. Assuming the validity of this result (the simple proof is given below), we now determine the value of P^*. We suppose that the first harvest H reduces the initial recruitment level R_1 to some level P:

$$H_1 = R_1 - P,$$

and that subsequent harvesting maintains this level:

$$H_2 = H_3 = \cdots = F(P) - P = H.$$

The total present value of this harvest sequence is given by

$$\pi(R_1, H_1) + \sum_{2}^{\infty} \alpha^{k-1} \pi[F(P), H] = \pi(R_1, R_1 - P) + \frac{\alpha}{1-\alpha}\, \pi[F(P), F(P) - P].$$
$$\qquad (7.58)$$

From Eq. (7.55) we have

$$\frac{d}{dP}\, \pi(R_1, R_1 - P) = \frac{d}{dP} \int_{P}^{R_1} [p - C(x)]\, dx = -[p - C(P)],$$

$$\frac{d}{dP}\, \pi[F(P), F(P) - P] = \frac{d}{dP} \int_{P}^{F(P)} [p - C(x)]\, dx$$

$$= F'(P)\{p - C[F(P)]\} - [p - C(P)].$$

Hence the necessary condition for an interior maximum in Eq. (7.58) is that $P = P^*$, where

$$-\{p - C(P^*)\} + \frac{\alpha}{1-\alpha}[F'(P^*)\{p - C[F(P^*)]\} - \{p - C(P^*)\}] = 0.$$

After reordering terms we can rewrite the preceding equation

$$F'(P^*) \cdot \frac{p - C[F(P^*)]}{p - C(P^*)} = \frac{1}{\alpha} = 1 + i. \tag{7.59}$$

This is the discrete-time analog of our optimal escapement formula [Eq. (2.16)] for the continuous-time linear model. The similarity becomes more evident if we rewrite the Eq. (7.59) in the form

$$F'(P^*) - \frac{C[F(P^*)] - C(P^*)}{p - C(P^*)} \cdot F'(P^*) = 1 + i. \tag{7.60}$$

In comparing Eqs. (7.60) and (2.16), it should be noted that the function $F(x)$ appearing in Eq. (2.16) represents *net* productivity, whereas the current function $F(P)$ in Eq. (7.60) represents *gross* productivity (the net productivity is equal to $F(P) - P$).

Although the stock-effect term in Eq. (7.60) (the second term on the left side) is slightly more unwieldy than the corresponding term in Eq. (2.16), it has the same properties; in particular it is positive for all relevant values of P^*. This is a significant conclusion, for it implies that all our earlier analyses of the continuous-time linear model can be reapplied in the discrete-time case, provided the necessary change is made in the marginal stock-effect term. We also expect that the various modifications of Eq. (2.16) discussed earlier possess corresponding analogs for the discrete-time case. This question is examined briefly in Section 7.8.

Rigorous Derivation of the Optimal Harvest Policy

We now wish to prove that the optimal harvest policy is the most-rapid approach policy to the optimal level of escapement P^* given by Eq. (7.59). [The following proof is due to Spence (1973).] We make the following assumptions:

1. $F(x)$ is continuously differentiable, and $F'(x) > 0$ for all x;
2. $C(x)$ is continuous and nonincreasing;
3. Equation (7.59) possesses a unique solution P^*.

We rewrite Eq. (7.59) in the form

$$p - C[F(P^*)] = \frac{(1+i)[p - C(P^*)]}{F'(P^*)}.$$

Letting P_∞ denote the zero-profit level (i.e., $p - C(P_\infty) = 0$) and assuming for now that $P_\infty > 0$, we see that

$$p - C[F(x)] > \frac{(1+i)\{p - C(x)\}}{F'(x)} \tag{7.61}$$

for $x = P_\infty$. Because Eq. (7.59) is assumed to have a unique solution P^*, it follows that Eq. (7.61) must be valid for all x such that $P_\infty < x < P^*$, whereas the reverse inequality must be true for $x > P^*$. These inequalities are basic for our proof.

Because $F'(x) > 0$, the function F has a unique inverse $\psi = F^{-1}$, also defined for $0 \le x \le K$. Let $\{H_k\}$ be an arbitrary feasible harvest sequence and $\{R_k\}$ represent the corresponding sequence of recruitment population levels, as determined by Eq. (7.52). Then

$$R_k - H_k = \psi(R_{k+1}), \tag{7.62}$$

and by Eq. (7.55)

$$\pi(R_k, H_k) = p[R_k - \psi(R_{k+1})] - \{G(R_k) - G[\psi(R_{k+1})]\}, \tag{7.63}$$

where $G(x)$ is an integral of $C(x)$.

By rearranging the present-value expression [Eq. (7.56)] we obtain

$$PV = \sum_{k=2}^{\infty} \frac{V(R_k)}{(1+i)^{k-1}} + p[R_1 - G(R_1)], \tag{7.64}$$

where

$$V(R) = pR - G(R) - (1+i)\{p\psi(R) - G[\psi(R)]\}. \tag{7.65}$$

Now note that by Eq. (7.59)

$$V'(R) = p - C(R) - \frac{(1+i)\{p - C[\psi(R)]\}}{F'[\psi(R)]}$$
$$= 0, \quad \text{if} \quad R = R^* = F(P^*),$$

and moreover that

$$V'(R) \begin{cases} > 0 & \text{for} \quad R < R^* \\ < 0 & \text{for} \quad R > R^*. \end{cases} \tag{7.66}$$

In other words the optimal recruitment level $R = R^*$ maximizes the function $V(R)$.

Now let $\{R_k^*\}$ denote the recruitment population sequence corresponding to the most-rapid approach harvest policy. For $0 \le s \le 1$, define

$$w_k(s) = sR_k^* + (1 - s)R_k \tag{7.67}$$

and

$$I(s) = \sum_{k=2}^{\infty} \frac{V[w_k(s)]}{(1+i)^{k-1}}. \tag{7.68}$$

We now show that

$$I'(s) \ge 0 \quad \text{for all } s, \quad 0 \le s \le 1, \tag{7.69}$$

and moreover that $I'(s) > 0$ unless $R_k = R_k^*$ for every k. This implies that $I(1) > I(0)$; that is,

$$\sum_2^\infty \frac{V(R_k^*)}{(1+i)^{k-1}} > \sum_2^\infty \frac{V(R_k)}{(1+i)^{k-1}}.$$

Thus by Eq. (7.64) the present value produced by the recruitment policy $\{R_k^*\}$ exceeds the present value produced by any other feasible recruitment policy. This proves that the most-rapid approach policy (corresponding to $R_k = R_k^*$) must be uniquely optimal.

The proof of Eq. (7.69) proceeds as follows. First we obtain

$$I'(s) = \sum_2^\infty \frac{V'[w_k(s)] \cdot (R_k^* - R_k)}{(1+i)^{k-1}}.$$

We show that each term in this sum must be ≥ 0. Suppose first that $R_k^* > R_k$. Because $\{R_k^*\}$ is the most-rapid approach path to $R^* = F(P^*)$, it must now be the case that $R_k^* \le R^*$ (see Figure 7.16). Hence Eq. (7.67) implies that

$$R_k \le w_k(s) \le R_k^* \le R^*.$$

Therefore by Eqs. (7.66) $V'[w_k(s)] \ge 0$, so that

$$V'[w_k(s)](R_k^* - R_k) \ge 0.$$

A similar argument shows that $V'[w_k(s)] \le 0$ whenever $R_k^* < R_k$; this completes the proof of Eq. (7.69).

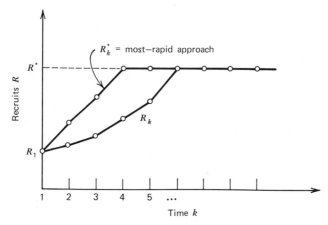

Figure 7.16 If $R_k < R_k^*$, then necessarily $R_k^* \le R^*$.

The reader should observe that the same proof applies in the case in which the constraints in Eq. (7.53) are replaced by

$$H_{\min} \leq H_k \leq H_{\max}, \quad H_k \leq R_k \tag{7.70}$$

provided of course that $0 \leq H_{\min} \leq H_{\max}$. This proof can also be applied to the finite time-horizon problem. In this case the final harvest H_N is such that

$$P_N = P_\infty.$$

Finally if the cost of extinction $C(0)$ is finite then $P_\infty = 0$. If Eq. (7.59) has no solution P^* (but if the fishery is viable; i.e., if $C(K) < p$), then the optimal harvest policy leads to the extinction of the resource stock. It can be shown that this is so whenever

$$p > C(0) \quad \text{and} \quad F'(0) < (1+i)^2$$

[cf. Section 2.8 and Clark (1973b), p. 957]. These results are all analogous to the results in the continuous-time case.

Effects of Overcompensation

The preceding proof depends quite strongly on the assumption that the stock-recruitment curve $F(x)$ is increasing. Certain interesting complications can arise in the opposite case (overcompensation). As an example consider the case shown in Figure 7.17. The equilibrium population K is assumed to be stable [$F'(K) > -1$]. Suppose that the zero-profit level P_∞ is

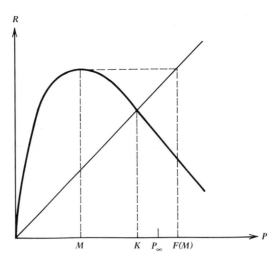

Figure 7.17   A case of overcompensation.

greater than K, but less than $F(M) = \max F(x)$. Given $R_1 = K$, clearly no profit can be made from an initial harvest $H_1 > 0$. But reducing the population to M (for example) in the first period does lead to a recruitment $F(M)$ that can later be harvested at a profit, a potentially desirable policy. The question of optimization is not considered further here, however.

Fixed Costs

The cost expression in Eq. (7.54) implies that the total cost of the harvest H_k approaches zero as $H_k \to 0$. In some cases it may be more realistic to suppose that a positive minimum cost C_0 is incurred whenever it is decided to attempt a harvest, regardless of the size of the harvest actually achieved. This assumption particularly applies to distant-water fishing, where C_0 represents the opportunity cost of moving the fleet to the fishing ground.

Under these circumstances the net revenue expression in Eq. (7.55) must be replaced by

$$\pi(R_k, H_k) = \begin{cases} \displaystyle\int_{R_k - H_k}^{R_k} [p - C(x)] \, dx - C_0, & \text{if} \quad H_k > 0 \\ 0 & \text{if} \quad H_k = 0. \end{cases}$$

What effect will this change have on optimal harvest policies? If a sustained yield $H = F(P) - P$ is obtained each year then the fixed cost is incurred every year, whereas if harvesting occurs only every second year (or less frequently) then the fixed cost is also incurred only every second year (or less). Thus the *average* net annual revenue equals

$$\begin{cases} \displaystyle\int_P^{F(P)} \{p - C(x)\} \, dx - C_0 & \text{if harvested every year} \\ \dfrac{1}{2} \displaystyle\int_P^{F[F(P)]} \{p - C(x)\} - \tfrac{1}{2} C_0 & \text{if harvested every second year.} \end{cases}$$

If C_0 is sufficiently large it is clear that the second expression can exceed the first; indeed the first expression can be negative and the second expression can be positive. In such cases harvesting every other year is superior to a sustained-yield policy.

We conclude therefore that the existence of fixed costs can lead to pulse fishing, in which harvests are obtained only on alternate years or perhaps even less frequently. The reader should observe the similarity between this model and the continuous-time model with decreasing marginal costs $\partial C / \partial h$ in Section 5.4. From the mathematical point of view the mechanism is the same for both models: nonconvexity of the cost function implies that "chattering" controls (i.e., pulse fishing) are less costly on the average than sustained-yield harvests.

Here we do not attempt to work out the optimal pulse-fishing policy, and instead refer the reader to the papers of Jaquette (1974) and Reed (1975). These authors show that the optimal harvest policy is an $s-S$ harvest policy characterized by a rule of the form

$$H_k = \begin{cases} R_k - s & \text{if} \quad R_k \geq S \\ 0 & \text{if} \quad R_k < S. \end{cases}$$

Thus pulse fishing occurs if $F(s) < S$. Note that once s and S have been determined, this is a feedback control law: the current policy depends only on the current state variable R_k. Jaquette (1974) and Reed (1975) prove that this feedback $s-S$ policy is valid for a stochastic stock-recruitment model. (The concept of $s-S$ policies arises in inventory theory, where a large literature exists.)

7.8 THE DISCRETE MAXIMUM PRINCIPLE

In this section we describe the discrete-time analog of the Pontryagin maximum principle, which we then apply to a nonlinear discrete-time fishery model. We then note that both the theory and the results are closely parallel to the continuous-time case.

A discrete-time optimal control problem is equivalent to a problem in nonlinear mathematical programming. Consequently the method of Lagrange multipliers, or more precisely the modern Kuhn–Tucker theory, can be used. In this section we present a formal proof of the discrete maximum principle based on this method. This proof throws additional light on the maximum principle itself, because the adjoint variable λ now appears naturally as a Lagrange multiplier. Indeed the programming approach can be employed to provide a rigorous proof of the discrete-time maximum principle, as well as of the continuous-time Pontryagin principle (although this was not the original method used by Pontryagin). (For further details, see Cannon, Cullum, and Polak, 1970.)

Discrete-Time Optimal Control Problems

For simplicity we again consider only one-dimensional problems here; however, both the notation used and the results stated apply equally to the multidimensional case. Our control problem is formulated in terms of the state equation

$$x_{k+1} - x_k = f_k(x_k, u_k), \quad k = 1, 2, \ldots, N-1, \tag{7.71}$$

where x_k is the state variable and u_k is the control variable. Both the initial and the terminal vaues of the state variable are specified:

$$x_1 = x_1^\circ, \quad x_N = x_N^\circ. \tag{7.72}$$

The control variable is subject to the constraints

$$u_k \in U, \quad k = 1, 2, \ldots N. \tag{7.73}$$

Finally we have an objective functional

$$J\{u\} = \sum_{k=1}^{N} g_k(x_k, u_k). \tag{7.74}$$

[Note that the terminal control u_N does not influence the dynamics via Eq. (7.71), but does affect the objective functional given by Eq. (7.74).] The optimal control problem is to determine the control values u_k, $k = 1$, $2, \ldots, N$, satisfying the constraints in Eqs. (7.73) such that x_N satisfies Eqs. (7.72) and such that the objective functional given by Eq. (7.74) is maximized.

By treating the state Eqs. (7.71) as a family of $N-1$ additional constraints on the objective functional $J\{u\}$, we can consider the preceding control problem as a problem in constrained optimization (i.e., nonlinear programming). In the following discussion we neglect the control constraints in Eqs. (7.73) and consider only the question of interior solutions. In this case the problem can be solved by the classical method of Lagrange multipliers. First we introduce the Lagrangian expression

$$\mathcal{L} = \sum_{k=1}^{N} \{g_k(x_k, u_k) - \lambda_k[x_{k+1} - x_k - f_k(x_k, u_k)]\}, \tag{7.75}$$

where because Eqs. (7.71) do not apply for $k = N$,

$$\lambda_N = 0. \tag{7.76}$$

Necessary conditions for optimality (because x_1 and x_N are given) are then

$$\frac{\partial \mathcal{L}}{\partial x_k} = 0, \quad k = 2, 3, \ldots, N-1,$$

$$\frac{\partial \mathcal{L}}{\partial u_k} = 0, \quad k = 1, 2, \ldots, N,$$

$$\frac{\partial \mathcal{L}}{\partial \lambda_k} = 0, \quad k = 1, 2, \ldots, N-1.$$

[The last condition merely repeats the state Eqs. (7.71).] These equations provide $3N-3$ conditions for the determination of the $3N-3$ unknowns $x_2, \ldots, x_{N-1}; u_1, \ldots, u_N; \lambda_1, \ldots, \lambda_{N-1}$.

Carrying out the first of these differentiations gives

$$\frac{\partial g_k}{\partial x_k} + \lambda_k + \lambda_k \frac{\partial f_k}{\partial x_k} - \lambda_{k-1} = 0,$$

or

$$\lambda_k - \lambda_{k-1} = -\frac{\partial g_k}{\partial x_k} - \lambda_k \frac{\partial f_k}{\partial x_k}, \quad k = 2, \ldots, N-1.$$

The second differentiation yields

$$\frac{\partial g_k}{\partial u_k} + \lambda_k \frac{\partial f_k}{\partial u_k} = 0.$$

If we now introduce the *Hamiltonian* expression

$$\mathcal{H}_k(x_k, u_k, \lambda_k) = g_k(x_k, u_k) + \lambda_k f_k(x_k, u_k), \tag{7.77}$$

these conditions can be written in the form

$$\lambda_k - \lambda_{k-1} = -\frac{\partial \mathcal{H}_k}{\partial x_k}, \quad k = 2, 3, \ldots, N-1; \tag{7.78}$$

$$0 = \frac{\partial \mathcal{H}_k}{\partial u_k}, \quad k = 1, 2, \ldots, N. \tag{7.79}$$

Equation (7.78) is the adjoint equation; Eq. (7.79) is the maximum principle. More generally if the control constraints given in Eqs. (7.73) are considered, the maximum principle is replaced by

$$u_k \text{ maximizes } \mathcal{H}_k(x_k, u, \lambda_k) \quad \text{over} \quad u \in U. \tag{7.80}$$

Naturally the latter case is considerably more difficult to prove.

The similarity between the discrete maximum principle [Eqs. (7.78) and (7.80)] and the Pontryagin maximum principle [Eqs. (4.8) and (4.9)] is evident. The interpretation of Lagrange multipliers as shadow prices is well established in economic analysis (cf. Intriligator, 1971). (In the linear case, the multipliers become the so-called *dual variables*.)

A Nonlinear Fishery Model

Finally we use the discrete maximum principle to derive a formula for optimal equilibrium solutions to the discrete-time fishery model in which the net revenue $\pi(x, h)$ is of a general nature, rather than of the special form given by Eq. (7.55). To do this we must assume that the discrete maximum principle extends to the case of an infinite time horizon, and also that certain essential convexity conditions are satisfied. No attempt is made here to verify these assumptions, however.

In the present notation the state Eq. (7.52) becomes

$$R_{k+1} - R_k = F(R_k - H_k) - R_k,$$

and the objective functional is

$$J = \sum_{k=1}^{\infty} \alpha^{k-1} \pi(R_k, H_k).$$

Thus the Hamiltonian is

$$\mathcal{H}_k = \alpha^{k-1} \pi(R_k, H_k) + \lambda_k [F(R_k - H_k) - R_k].$$

The maximum principle implies that

$$\frac{\partial \mathcal{H}_k}{\partial H_k} = \alpha^{k-1} \frac{\partial \pi}{\partial H_k} - \lambda_k F'(R_k - H_k) = 0,$$

so that in equilibrium we have

$$\lambda_k = \alpha^{k-1} \frac{\partial \pi/\partial H}{F'(R-H)}.$$

Hence

$$\lambda_k - \lambda_{k-1} = (\alpha^{k-1} - \alpha^{k-2}) \frac{\partial \pi/\partial H}{F'(R-H)}.$$

By the adjoint equation

$$\lambda_k - \lambda_{k-1} = -\frac{\partial \mathcal{H}_k}{\partial R_k}$$

$$= -\alpha^{k-1} \frac{\partial \pi}{\partial R} - \lambda_k [F'(R-H) - 1]$$

$$= -\alpha^{k-1} \frac{\partial \pi}{\partial R} - \alpha^{k-1} \frac{\partial \pi}{\partial H} \left\{ 1 - \frac{1}{F'(R-H)} \right\}.$$

Equating these expressions and simplifying we obtain

$$F'(R-H) \frac{\partial \pi/\partial R + \partial \pi/\partial H}{\partial \pi/\partial H} = \frac{1}{\alpha}. \qquad (7.81)$$

If $\pi(R, H)$ is given by Eq. (7.55), we can easily verify that Eq. (7.81) reduces to Eq. (7.59) [note that $R = F(P)$].

EXERCISES

1. Given $P_{k+1} = F(P_k)$ we have

$$P_{k+2} = F[F(P_k)] = F_2(P_k).$$

(a) If $F(P)$ is a Beverton–Holt curve [Eq. (7.14)], show that $F_2(P)$ is also a Beverton–Holt curve (but with different parameters).

(b) Use the four-quadrant method to diagram the growth curves F_2 corresponding to the Ricker function [Eq. (7.6)] for various values of r. Show that F_2 has a single stable equilibrium at $P = 1$ if $r < 2$, but that at $r = 2$ this equilibrium bifurcates into three equilibria, the one at $P = 1$ becoming unstable. If r is located near 2, the two new equilibria for F_2 are stable and correspond to a stable cycle of period two. As r increases further these equilibria also bifurcate, producing a cycle of period four, and so forth. (The process is analyzed in detail by May and Oster, 1975.)

2. Assume that the equation

$$\frac{dx}{dt} = -h(t) = -qE(t)x(t), \quad x(0) = x_0$$

describes the harvesting of a fish population (i.e., natural mortality during the fishing season is negligible). Define the *cumulative catch* $H = .H(t)$ by

$$H(t) = \int_0^t h(t)\, dt.$$

Show that the plot of cumulative catch $H(t)$ versus catch per unit effort $h(t)/E(t)$ is a straight line of slope $-1/q$ and intercept x_0. If catch-effort data are available throughout a single fishing season, both the initial stock level x_0 and the catchability coefficient q can be estimated. The method is due to de Lury (1947).

3. Discuss the effect of a positive density-independent mortality rate M on the optimization of fishing season length. [Modify Eq. (7.48) to agree with the model

$$\frac{dx}{dt} = -(qB + M)x.]$$

4. Most species of shrimps have a life span of about one year. Moreover, recruitment is virtually independent of the previous year's escapement. What is the optimal harvest policy? Through what mechanism, if any, does the open-access shrimp fishery dissipate

economic rent? What form of regulation, if any, is required? (The problem of optimizing the age of capture is discussed in Chapter 8.)

5. A regulatory authority sets both the annual quota H and the total number of vessels licensed for fishing B. Thereafter the price of fish rises dramatically, substantially increasing the fishermen's rents. What happens next? (Assume that both regulations are obeyed.)

BIBLIOGRAPHICAL NOTES

General references on discrete-time population models include Maynard Smith (1968, 1974), May (1973), and Nicholson and Bailey (1935, insect populations). Oscillations resulting from overcompensation are discussed by Hoppensteadt and Hyman (1975), May (1974), Oster (1974), and May and Oster (1975). See Samuelson (1965) for economic applications of discrete-time models.

Stock-recruitment curves and fish population dynamics are discussed by Beverton and Holt (1957), Cushing (1968), Nikolskii (1969), Larkin, Raleigh, and Wilimovsky (1964), Ricker (1954, 1958), and Watt (1968). The International Council on the Exploration of the Sea sponsored a Symposium on Fish Stocks and Recruitment in 1973 (see Parrish 1973).

On the topic of fish schooling, see Breder (1967), Brock and Riffenburgh (1963), Hamilton (1971), and Shaw (1970). The possibility that schooling might lead to depensation is discussed in Clark (1974a).

Only a small amount of the literature is devoted to the form of the production function of the fishery, but see Bradley (1970) and Rothschild (1972). Several economists (Smith, 1969; Turvey, 1964) attempt to model the effect of mesh size on production functions without recognizing the necessity of adopting a more complex cohort model to clarify this concept. We discuss this problem in Chapter 8.

The optimal harvesting problem for discrete-time models is analyzed by Clark (1971, 1972, 1973b), Mann (1970), Plourde (1974), and Spence (1973). The extension to stochastic discrete-time models is undertaken by Jaquette (1972, 1974) and Reed (1974, 1975). A discrete-time model with recruitment delay is considered by Clark (1976).

The discrete maximum principle and its relations to mathematical programming and to continuous optimal control theory are described in detail by Cannon, Cullum, and Polak (1970). See also Fan and Wang (1974) and Intriligator (1971). The closely related techniques of dynamic programming are described by Bellman (1957), and Danø (1975).

8

GROWTH AND AGING

The models of biological population dynamics in the preceding chapters are vastly simplified. With minor exceptions in Chapter 7, these models are all *lumped-parameter models*, with a single real variable representing the size of the population at time t. For practical purposes it is usually convenient to suppose that this variable represents the stock biomass. Thus harvesting affects this biomass level, with an optimal harvest policy normally giving rise to some optimal equilibrium stock level.

Unfortunately the dynamics of many important biological resources cannot be realistically described by means of simple lumped-parameter models. Most fish populations, for example, consist of fish of several different ages; both commercial value and reproductive potential generally depend on the age of the individual fish. These phenomena are often highly significant in determining optimal harvest policies. But including age structure in the analysis introduces significant new mathematical difficulties. Indeed the problem of the optimal harvesting of age-distributed populations remains unsolved.

Given that fish (and all kinds of animals) grow larger and more valuable with age, it is clear that optimal harvesting is concerned with both the age of the fish when they are caught and the number of fish caught per unit time. To gain insight into the problem, it is useful to begin with the special case of a single population of equally aged individuals. The problem then is to determine the optimal age of harvest of the population. This is quite similar to the well-known problem of determining an optimal rotation period in forestry management.

In the first section we begin with the forest rotation problem in its simplest form; this problem has an extensive (and rather controversial) literature. We also study the problem of optimal forest thinning, a subject that has produced very little literature. Here we find another application of optimal control theory.

256

Next we return to fishery management to study in some detail a fishery model due to Beverton and Holt (1957). This model has been applied to a wide variety of commercial species, including the North Sea plaice, the Atlantic haddock, the Atlantic cod, the Yellowtail flounder, and the Peruvian anchovy. The Beverton–Holt model is a mixed discrete-time/continuous-time model that incorporates age structure without resorting to partial differential equations.

8.1 FORESTRY MANAGEMENT: THE FAUSTMANN MODEL

The forestry manager faces a number of interesting biological and economic problems. The basic resource, the forest, possesses various unique biological characteristics. But from the economic point of view, a standing forest is just one particular form of growing capital. Of course, forests serve man in many other important ways, but in the following analysis of forestry economics, we concentrate solely on the factor of timber production.

The commercial value V of a single tree is determined by the volume and the quality of timber the tree can produce. Clearly $V = V(t)$ depends on the age t of the tree. Typically $V(t)$ is graphed as shown in Figure 8.1. Trees of age $t < t^*$ have no commercial value; the value for trees of age

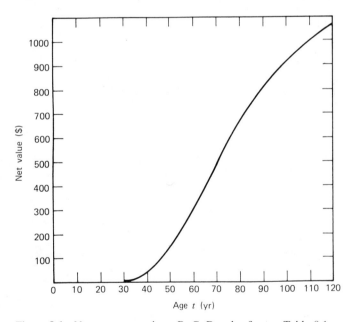

Figure 8.1 Net stumpage values: B. C. Douglas fir; see Table 8.1.

$t > t^*$ increases with age as the volume of usable timber increases. Eventually the tree approaches biological maturity, its growth ceases, and its value reaches a plateau. Ultimately decay sets in and the value of the tree begins to decline to zero. The time scale for this process depends on the species of tree and the habitat in which it grows; many important commercial trees continue to grow and increase in value for several hundred years.

Assuming that the curve $V(t)$ is known, what is the optimal age at which to fell a tree (or more realistically to fell a stand of trees of the same age)? If c denotes the cost of felling, so that $V(t) - c$ represents the net value of the stand, maximizing $PV = e^{-\delta t}[V(t) - c]$ (our basic rule) yields

$$\frac{V'(T)}{V(T) - c} = \delta. \tag{8.1}$$

However, Eq. (8.1) does not encompass an important aspect of the forest rotation problem. Once the trees are removed from a given area, the land is available for new forest growth. Clearly the longer the felling of the existing forest is delayed, the longer it takes to acquire revenues from future harvests. The opportunity cost of utilizing the forest site for the existing stand of trees must be considered.

Let us examine a sequence of times $T_1 < T_2 < \cdots$, with the property that at each time T_k the existing forest is felled and a new forest is planted. Let $t = 0$ represent the time of the first planting. Suppose that soil productivity and all prices and costs remain constant for all future time. If c now denotes the sum of logging and replanting costs, we can express the total present value of all future harvests as

$$PV = e^{-\delta T_1}[V(T_1) - c] + e^{-\delta T_2}[V(T_2 - T_1) - c]$$
$$+ e^{-\delta T_3}[V(T_3 - T_2) - c] + \cdots. \tag{8.2}$$

Then the problem is to choose the times $T_1, T_2, \ldots$, so as to maximize this expression. Note that we are assuming an infinite time-horizon.

The problem of maximizing Eq. (8.2) by the appropriate choice of infinitely many values $T_1, T_2, \ldots$, may seem formidable. But we can conclude immediately from our assumptions that *all rotation periods are of equal length;* that is

$$T_k = kT, \quad k = 1, 2, \ldots. \tag{8.3}$$

To explain this conclusion, consider the problem facing the forestry manager immediately following the first felling at $t = T_1$. Because we assume that all parameters remain constant for all future time, this second rotation problem is identical to the original problem faced at time $t = 0$.

Identical problems have identical solutions, so that $T_2 - T_1$ must equal T_1 (i.e., $T_2 = 2T_1$). The same argument obviously applies to all future rotations.
Substituting Eqs. (8.3) into Eq. (8.2) we obtain

$$PV = \sum_{k=1}^{\infty} e^{-k\delta T}[V(T)-c] = \frac{V(T)-c}{e^{\delta T}-1}. \qquad (8.4)$$

Maximization of this expression with respect to T requires that

$$\frac{V'(T)}{V(T)-c} = \frac{\delta e^{\delta T}}{e^{\delta T}-1} = \frac{\delta}{1-e^{-\delta T}}. \qquad (8.5)$$

Equation (8.5) for the optimal rotation period T is called the *Faustmann formula;* it was derived in 1849 by M. Faustmann, a German forester. Comparison with Eq. (8.1) shows that the rotation aspect introduces an additional factor $1-e^{-\delta T}$ in the denominator of the right side of the equation. Because $1-e^{-\delta T} < 1$, we can see that taking the rotation into account leads to a decrease in the age of cutting, as expected.
Equation (8.5) can be rewritten in the form

$$V'(T) = \delta[V(T)-c] + \delta \frac{V(T)-c}{e^{\delta T}-1}. \qquad (8.6)$$

The first two terms in this equation have the same meaning as in Eq. (8.1); $V'(T)$ is the increase in the net value of the standing forest over a unit time interval $T \le t \le T+1$; $\delta[V(T)-c]$ is the interest that can be earned if the net revenue from cutting $V(T)-c$ is invested at an interest rate δ.
The third term reflects the rotation aspect of the problem. A decision to harvest trees at age T is, as we know, a decision that affects both the existing stand of trees and all future stands on a particular site. The expression

$$\frac{V(T)-c}{e^{\delta T}-1}$$

is the present value of this stream of future revenues; in the forestry literature this expression is called the *site value.* The condition of Eq. (8.6) is that the forest be cut at age T, when the marginal increment to the value of the trees equals the sum of the opportunity costs of investment tied up in the standing trees and in the site.
If the net stumpage value $V(T)-c$ is known, the optimal rotation period as a function of age T for a given forest stand can be easily obtained by employing the graph in Figure 8.2. The family of curves in

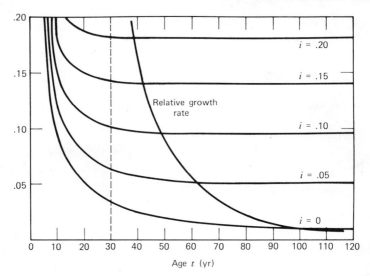

Figure 8.2 Graphical determination of the optimal rotation, where i is the annual discount rate; see Table 8.1.

this graph represent the expression

$$\frac{\delta}{1 - e^{-\delta T}},$$

which occurs on the right side of Eq. (8.5) for various values of the annual discount rate $i = e^{1+\delta}$ ranging from zero to 20% per annum.

Superimposed on this family of curves is the curve with the ordinate

$$\frac{V'(T)}{V(T) - c}$$

corresponding to data published for a typical stand of Douglas fir trees in British Columbia (Pearse, 1967). This data is reproduced in Table 8.1.

The optimal rotation period T as a function of interest rate is given in Table 8.2. This table indicates the corresponding average annual yield.

As in the fishery models in previous chapters, the case of zero discounting here corresponds to the maximization of sustained economic rent (in this instance in terms of average annual yield). To see this, we note that as $\delta \to 0$, by l'Hôpital's rule we have

$$\lim_{\delta \to 0} \frac{\delta}{1 - e^{-\delta T}} = \frac{1}{T}.$$

TABLE 8.1. NET STUMPAGE VALUES FOR A "TYPICAL" STAND OF BRITISH COLUMBIA DOUGLAS FIR TREES (PEARSE; 1967).

Age T (yr)	Net Stumpage Value $V(T) - c$ ($)	Relative Growth Rate $\dfrac{V'(T)}{[V(T) - c]}$ (%)	Average Yield $\dfrac{[V(T) - c]}{T}$ ($/yr)
30	0	+∞	0
40	43	16.6%	1.08
50	143	9.1%	2.86
60	303	5.8%	5.05
70	497	3.5%	7.10
80	650	2.4%	8.12
90	805	1.6%	8.94
100	913	1.1%	9.13
110	1000	0.8%	9.09
120	1075	—	8.93

Thus Eq. (8.5) becomes

$$V'(T) = \frac{V(T) - c}{T} \quad \text{when} \quad \delta = 0. \tag{8.7}$$

This equation means that cutting at age T maximizes the average annual economic yield $[V(T) - c]/T$.

We can see that as the discount rate increases, both T and the average

TABLE 8.2. OPTIMAL ROTATION AND AVERAGE ANNUAL YIELD FOR BRITISH COLUMBIA DOUGLAS FIR TREES (SEE TABLE 8.1).

Annual Discount Rate i	0	0.03	0.05	0.07	0.10	0.15	0.20
Optimal Rotation Age T (yr)	100	70	63	56	49	43	40
Average Annual Yield ($/yr)	9.10	7.10	5.60	4.20	2.80	1.70	1.20

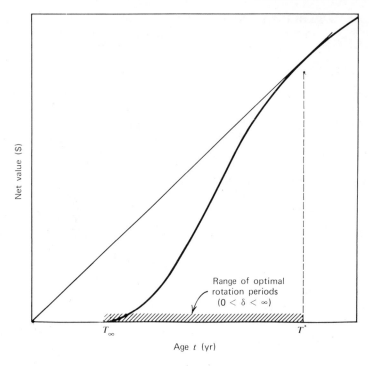

Figure 8.3   Optimal rotation periods.

yield $\{V(T) - c\}/T$ decrease. As $\delta \to +\infty$, we have $T \to T_\infty$, where

$$V(T_\infty) - c = 0.$$

If we imagine that the case $\delta = \infty$ represents the unmanaged, common-property forest, we can see that such a situation results in the dissipation of economic rent; trees are always cut as soon as their market value equals the opportunity cost of logging.

These results are summarized in Figure 8.3, which graphs the data in Table 8.1.

Response to Changes in Demand

The Faustmann model can be used to predict the response of the forest industry to a change in the demand for forest products or in the relative cost of logging operations. Suppose that the value of timber increases from its former level $V(T)$ to a higher level $pV(T)$, where $p > 1$, and assume that this increase is expected to be permanent. The Faustmann formula can then be rewritten

$$\frac{pV'(T)}{pV(T)-c} = \frac{V'(T)}{V(T)-c/p} = \frac{\delta}{1-e^{-\delta T}}. \tag{8.8}$$

Thus T depends only on the cost–price ratio, and an increase in price has the same effect as a decrease in cost.

The left side of Eq. (8.8) is an increasing function of the cost–price ratio c/p. Thus a permanent price increase relative to cost causes the optimal rotation period to decrease (again see Figure 8.2). In a forest that contains stands of various ages, some existing stands become overmature. Therefore the *short-term* effect of a price increase is to increase production as these overmature trees are harvested. The *long-term* effect of the shorter rotation period, however, is the opposite of the short-term effect: the *ultimate* steady-state yield falls if the price rises.

These predictions of the Faustmann model are similar to those of the Schaefer fishery model, although there is one interesting difference. In the Faustmann model, discounting always leads to less than the maximum average annual yield (see Figure 8.3), and this implies that a price increase always leads to a decrease in the ultimate sustained yield from forests of heterogeneous age. In the Schaefer model on the other hand the long-term price response is negative only when the price is high enough to intersect the backward-bending, discounted supply curve. (A model in which the short-term price response is negative and the long-term price is positive is described in Exercise 2.) In the case of the Faustmann model the long-run discounted supply curve is backward-bending over its entire range.

Optimal forest policy is obviously not "myopic" in the sense of the term we introduce in Chapter 3. The determination of an optimal rotation period is strongly dependent on long-term predictions of future prices, costs, and discount rates. Underexploitation of forests can be easily rectified, but overexploitation may not be reversible for many decades. This one-sided irreversibility seems to be characteristic of many resource-management problems.

8.2 A MODEL OF OPTIMAL FOREST THINNING

The Faustmann model is concerned only with the problem of forest rotation. In practice various other forestry practices are often employed in conjunction with clearcutting and replanting. One of the most common practices is *thinning*, a process whereby some trees are removed at various times prior to clearcutting the stand. The main effect of thinning is to improve the growth and the quality of the remaining trees. In some cases ("thinning from below") only the poorer trees are removed, whereas in other cases ("thinning from above") the trees that are removed are of the same overall quality as the average trees in the stand.

In any case thinning in itself is usually profitable, particularly in the case of thinning from above (the case we consider here). We use a simple model of forest growth and thinning that is adapted from a study of Scotch Pine forests in Finland (Kilkki and Vaisanen, 1969). We employ methods of linear optimal control, modifying the dynamic-programming model of Kilkki and Vaisanen to obtain a continuous-time linear model. In this framework the thinning model is formally quite similar in structure to the Schaefer model in Chapter 2, although the thinning model also exhibits several new aspects. (The purpose of introducing this simple model here is not to provide a practical model of optimal forest management, but merely to study the general nature of optimal management policies and to fit the forestry problem into the general capital-theoretic framework adopted in this book.)

The model is based on the forest-growth equation

$$\frac{dV}{dt} = g(t)f(V), \quad V(t_*) = V_*,\tag{8.9}$$

where $V = V(t)$ represents the volume of usable timber in a given forest stand of age $t \geq t_*$. The coefficient $g(t)$ is assumed to be a positive, decreasing function of t. Kilkki and Vaisanen use the specific function

$$g(t) = at^{-b},\tag{8.10}$$

where a and b are positive constants.

The growth function $f(V)$ is positive and concave, with a unique maximum at $V = V_m$. The example used in the preceding reference is

$$f(V) = Ve^{-cV},\tag{8.11}$$

which has a maximum at $V_m = 1/c$ and is concave for $V \leq 2/c$; this growth function is sufficient for our purposes here.

If $h(t) \geq 0$ denotes the rate of removal by thinning, Eqs. (8.9) can be modified to

$$\frac{dV}{dt} = g(t)f(V) - h(t), \quad (t \geq t_*), \quad V(t_*) = V_*.\tag{8.12}$$

The function $h(t)$ is to be treated as the control variable. Rather than adopt an artificial upper-control constraint $h(t) \leq h_{max}$, we allow impulse controls (see Section 2.7). An impulse control that reduces the standing volume to $V = 0$ may be considered a clearcutting operation. Of course the state variable $V(t)$ is constrained to satisfy

$$V(t) \geq 0. \tag{8.13}$$

Let p denote the unit price of usable timber; we assume that p is constant and that it is the same for trees removed by thinning and by clearcutting. (We relax this assumption later in our discussion.) The unit costs of thinning normally vary both with the density of the forest and with the rate of thinning. To retain linearity [note that Eqs. (8.12) are already nonautonomous, so that barring the exclusive use of numerical techniques, we are more or less forced to work with a linear model] we assume that unit costs $C(V)$ do *not* depend on $h(t)$.

We begin with the optimization problem for a single rotation; at this point we neglect the consideration of site value. Our objective functional may then be written

$$PV = \int_{t_*}^{\infty} e^{-\delta t}[p - C(V)]h(t)\, dt. \tag{8.14}$$

We wish to maximize this expression subject to the state equation [Eqs. (8.12)], the control constraint $h(t) \geq 0$, and the constraint given in Eq. (8.13).

By routine calculation the singular solution for this linear, nonautonomous control problem is readily found to be given by

$$f'(V) - \frac{C'(V)f(V)}{p - C(V)} = \frac{\delta}{g(t)}. \tag{8.15}$$

The optimal control consists of a suitable combination of the singular control $h^*(t)$, the bang-bang control $h(t) = 0$, and the impulse controls. The similarity between Eq. (8.15) and the fundamental Eq. (2.16) should be noted here.

First consider the case in which $C'(V) \equiv 0$. Then Eq. (8.15) becomes

$$f'(V) = \frac{\delta}{g(t)}, \tag{8.16}$$

which can be recognized as the usual condition

$$\frac{\partial}{\partial V}[g(t)f(V)] = \delta.$$

By the assumed concavity of f [see Figure 8.4a] this equation has a unique solution $V = V^*(t)$ for each t, $t_* < t < T$, where T is such that

$$g(T) = \frac{\delta}{f'(0)}. \tag{8.17}$$

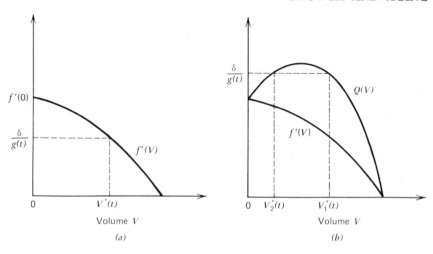

Figure 8.4 Determination of the singular solution $V^*(t)$.

Because $g(t)$ is a decreasing function of t, the singular path $V^*(t)$ also decreases with time, and $V^*(T) = 0$ (see Figure 8.5a). If the initial population V_* lies below the singular curve, then thinning does not begin until some age $t_1 > t_*$. Conversely if $V_* > V^*(t_*)$, an impulse control is applied to thin the stand immediately to the optimal level V^*. Subsequent thinning is adjusted to decrease the volume of standing timber continuously, until the stand is clear of timber at $t = T$. That the optimal solution assumes this unexpectedly continuous form is a consequence of neglecting both the costs of thinning and the subsequent rotations.

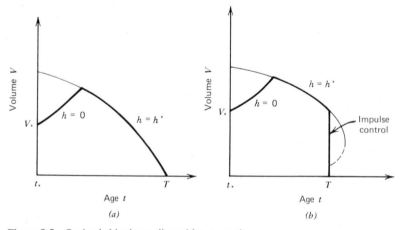

Figure 8.5 Optimal thinning policy without rotation.

The case of zero discounting is somewhat different. Here it is easy to see that no thinning occurs; only a clearcutting impulse control is applied at time T such that $g(T) = 0$ (i.e., when $dV/dt = 0$). [Note that if $F'[V(t_*)] > 0$, then $V(t)$ can never reach a level such that $F'(V) < 0$. Hence it never becomes worthwhile to thin the forest merely to improve the growth rate. This may no longer be the case for thinning from below, however.] Thus in the present model of thinning from above, the occurrence of thinning is a *specific* consequence of time discounting.

Now suppose that $C'(V) \neq 0$. Denote by $Q(V)$ the function that appears on the left side of Eq. (8.15):

$$Q(V) = f'(V) - \frac{C'(V)f(V)}{p - C(V)}.$$

Obviously $Q(V) \geq f'(V)$. If $Q(V)$ is a monotonic function, the problem is similar to the case we just considered; the opposite case is somewhat more interesting, however. Let the graph of $Q(V)$ be as shown in Figure 8.4b. Then for sufficiently large t Eq. (8.15) has two solutions $V_i^*(t)$, $i = 1, 2$. The corresponding singular path is double valued for such t (see Figure 8.5b). It is clear that the lower dashed section of this path is not optimal. The optimal thinning policy must therefore follow the singular path only up to some time $t = T$; at this time an impulse control is then applied to reduce the remaining timber to $V = 0$. (Recall the bang-bang singular nature of all solutions of linear control problems.)

This result is easy to explain. When thinning costs rise as the volume of timber decreases, a stage is reached at which continuous thinning becomes less profitable than a complete clearcutting of the forest. Determining the optimal age T for clearcutting is a blocked-interval problem similar to the problem in Chapter 3. We can well imagine cases in which the singular path is not used and an initial free-growth stage ($h = 0$) is followed directly by clearcutting at $t = T$. Indeed in many forests the practice of thinning is never employed.

Next we turn to the rotation problem with thinning. Let $t = 0$ denote the time of an initial forest clearcutting. Assuming that all parameters remain constant, we know that optimal rotations occur at regular intervals $T, 2T, 3T, \dots$, and that the optimal thinning policy is the same during each rotation. Thus the total present value of thinning and clearcutting all future rotations can be expressed

$$PV = \sum_{k=1}^{\infty} e^{-k\delta T} P_1 = \frac{P_1}{1 - e^{-\delta T}}, \tag{8.18}$$

where P_1 is the present value of the first rotation.

The value P_1 consists of the present value of the thinning revenue given by Eq. (8.14) over $0 \le t \le T$, plus the present value of the clearcutting revenue. Thus if C_T denotes the unit cost of clearcutting at age T we have

$$P_1 = \int_0^T e^{-\delta t}[p - C(V)]h(t)\ dt + e^{-\delta T}(p - C_T)V(T). \qquad (8.19)$$

Substituting in Eq. (8.18) we obtain the desired expression for total present value.

The optimal policy for this problem is significantly affected by the relative magnitudes of the unit costs of thinning $C(V)$ and clearcutting C_T. It seems reasonable to suppose that thinning is always at least as expensive as clearcutting:

$$C_T \le C(V) \quad \text{for all} \quad V > 0. \qquad (8.20)$$

To maximize the objective functional given in Eq. (8.18), we first treat the rotation period T as a parameter and solve the dynamic optimization problem over a fixed interval $0 < t \le T$. We are then left with a straightforward maximization problem with respect to T.

With T fixed our problem is simply to choose the thinning rate $h(t) \ge 0$ so as to maximize the expression P_1 given in Eq. (8.19) subject to the state equation [Eq. (8.12)]. Because the terminal stock level $V(T)$ resulting from the thinning process is unspecified, the transversality condition given in Eqs. (4.45) then applies:

$$\lambda(T) = \frac{\partial}{\partial V}[e^{-\delta T}(p - C_T)V(T)] = e^{-\delta T}(p - C_T). \qquad (8.21)$$

Since the integral in Eq. (8.19) is the same as the integral in the earlier problem [Eq. (8.14)], the singular solution $V^*(t)$ given by Eq. (8.15) is the same for both problems. The value of the adjoint variable $\lambda(t)$ along the singular path is readily seen to be

$$\lambda(t) = e^{-\delta t}\{p - C[V^*(t)]\}. \qquad (8.22)$$

Suppose that thinning and clearcutting costs are identical: $C_T = C[V^*(T)]$. Then Eq. (8.22) implies that $\lambda(T) = e^{-\delta T}(p - C_T)$ if the singular path is followed up to $t = T$. Thus in this case the transversality condition given in Eq. (8.21) is automatically fulfilled by using the singular path for all t together with the usual bang-bang initial approach (see Figure 8.6a). This case is therefore identical to the case of a single rotation considered earlier.

Now suppose that $C_T < C[V^*(T)]$. If the singular path is followed until $t = T$, we have

$$\lambda(T) < e^{-\delta T}(p - C_T)$$

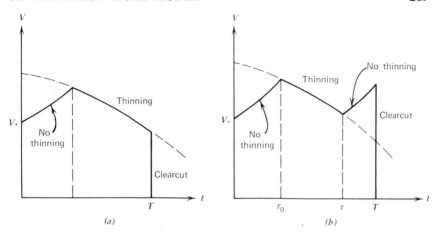

Figure 8.6 Optimal rotation with thinning: (a) thinning costs equal clearcutting costs; (b) thinning costs exceed clearcutting costs.

by Eq. (8.22). To satisfy the transversality condition we must therefore leave the singular path at some time $\tau < T$; that is, thinning must be avoided for $\tau \le t \le T$ (see Figure 8.6b). The reason for this is clear: because timber is more cheaply obtained by clearcutting than by thinning, it is not profitable to thin when clearcutting is imminent.

The simple form of this optimal solution suggests a straightforward technique of determining it numerically. First the singular path $V^*(t)$ and the control $h^*(t)$ are determined via Eq. (8.15). If $V_* < V^*(0)$, the time τ_0 at which thinning begins is determined by solving the growth equation [Eq. (8.9)]. For a given value of τ between τ_0 and T, the value of $V(T)$ is also calculated by solving Eq. (8.9) over $\tau \le t \le T$. The corresponding present value $P_1 = P_1(\tau)$ is then calculated from

$$P_1(\tau) = \int_{\tau_0}^{\tau} e^{-\delta t}\{p - C[V^*(t)]\}h^*(t)\,dt + e^{-\delta T}(p - C_T)V(T).$$

A maximization routine can then be employed to determine the optimal value of τ.

The problem remains of determining the optimal rotation period T so as to maximize the total present value [Eq. (8.18)]. This is also easy to achieve numerically. [See Clark and de Pree (1975).]

8.3 THE BEVERTON–HOLT FISHERIES MODEL

One of the two most commonly used commercial fishery models, the Schaefer model, is examined in detail in Chapters 2, 3, and 5. The second model, due to Beverton and Holt (1957), is the subject of the remainder of this chapter. In the present section we describe the model and outline Beverton and Holt's analysis of it. Basically this analysis amounts to a static yield-effort analysis, as in the standard Schaefer model. Beverton and Holt themselves explicitly recognized the inadequacy of their techniques in determining dynamic factors such as the optimal recovery of an overexploited stock (Beverton and Holt, 1957, Chapter 19).

The dynamic analysis of the Beverton–Holt model in subsequent sections is much more difficult than the dynamic analysis of the lumped–parameter Schaefer model. The results obtained are incomplete, and many interesting questions remain unanswered. Anyone familiar with the complexities of age-structured population models can appreciate the difficulties involved. [See, for example, the models of human population dynamics due to Keyfitz (1968).]

In the Beverton–Holt model the fish population consists of a number of different year-classes or *cohorts*, one resulting from each annual spawning and subsequent recruitment. We begin by discussing a single cohort, letting $N(t)$, $t \geq 0$, denote the *number* of fish of the cohort alive at time t. For simplicity we assume that $t = 0$ corresponds to the time of recruitment of the cohort (i.e., the time at which the cohort first becomes available to the fishing gear). Thus t also represents the age beyond recruitment, which we refer to (somewhat inaccurately) simply as the *age* of the cohort.

The function $N(t)$ is assumed to satisfy the simple differential equation

$$\frac{dN}{dt} = -(M+F)N, \tag{8.23}$$

$$N(0) = R, \tag{8.24}$$

where $M > 0$ is a constant denoting the *natural mortality rate* and $F \geq 0$ represents the *fishing mortality rate*. In this section F is assumed to be constant, and the problem is to determine F in an optimal manner. (In subsequent sections $F = F(t)$ is variable over time and constitutes the control variable for the dynamic optimization problem.) The constant R in Eq. (8.24) denotes the recruitment, which is assumed given.

It should be observed that Eq. (8.23) denotes a tacit assumption that fishing mortality and natural mortality are independent of one another, and also that natural mortality is independent of the stock size N.

Next let $w(t)$ denote the (average) weight of a fish at age t. The specific form of the function $w(t)$ need not concern us, but it is convenient here to suppose that $w(t)$ is bounded and increasing, and that the proportional rate of increase in weight $\dot{w}/w$ decreases with time. The most commonly used example in biology is the *von Bertalanffy weight function*

$$w(t) = a(1 - be^{-ct})^3,$$ (8.25)

where a, b, and c are positive constants.
The total biomass of the cohort is equal to

$$B(t) = N(t)w(t).$$ (8.26)

In the case in which the fishing mortality F equals 0, the "natural biomass" $B_0(t)$ is given by

$$B_0(t) = Re^{-Mt}w(t).$$ (8.27)

By differentiation we then have

$$\frac{dB_0}{dt} = Re^{-Mt}\left\{\frac{dw}{dt} - Mw\right\}.$$

Consequently the natural biomass attains a maximum value at the time (age) $t = t_0$ determined by

$$\frac{\dot{w}(t_0)}{w(t_0)} = M.$$ (8.28)

(In the case in which Eq. (8.28) has no solution $t_0 > 0$, $B_0(t)$ simply decreases for all time $t > 0$.) Because $w(t)$ is bounded, the biomass ultimately approaches zero at an exponential rate. In practice there is usually a maximum age t_{max} for fish, so that $B_0(t) = 0$ for $t > t_{max}$. For analytical simplicity, however, we assume that the exponential decrease in $B_0(t)$ as $t \to \infty$ is a sufficiently close approximation to reality (see Figure 8.7).

Following Beverton and Holt (1957, Chapter 3) we now assume that the fishery can be described in terms of the following parameters:

μ = mesh size;
F = fishing mortality coefficient.

In the static model these parameters are assumed to be constants subject to choice in some optimal fashion.

We suppose that the fishing gear possesses *knife-edge selectivity* in the sense that all fish of age $t \geq t_\mu$ which encounter the nets are captured and all smaller fish escape. Thus in place of Eq. (8.23) we have

$$\frac{dN}{dt} = \begin{cases} -MN & \text{for } 0 < t < t_\mu \\ -(M+F)N & \text{for } t \geq t_\mu. \end{cases}$$ (8.29)

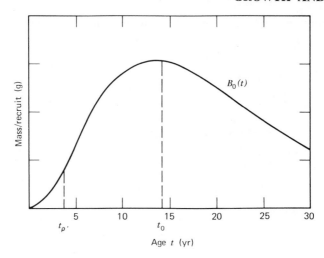

Figure 8.7 The natural biomass curve $B_0(t)$ for plaice: $t_{p'}$ = age of recruitment = 3.72 yrs; t_0 = age of maximum biomass.

Because $N(0) = R$, we deduce that

$$N(t) = \begin{cases} Re^{-Mt} & \text{for} \quad 0 < t < t_\mu \\ Re^{-Mt_\mu}e^{-(M+F)(t-t_\mu)} & \text{for} \quad t \geq t_\mu. \end{cases} \tag{8.30}$$

The total biomass yield obtained from the cohort is equal to

$$Y = Y_\mu(F) = \int_{t_\mu}^{\infty} FN(t)w(t)\, dt$$

$$= RFe^{Ft_\mu}\int_{t_\mu}^{\infty} e^{-(M+F)t}w(t)\, dt. \tag{8.31}$$

Beverton and Holt observe that the expression in Eq. (8.31) is subject to a different interpretation. In addition to representing the total biomass obtained from a single cohort, over its entire life span, this expression also denotes the annual biomass yield in equilibrium obtained from a fish population that consists of cohorts of all possible ages. This is true under the assumption that the fishing mortality coefficient F, the mesh-size parameter μ, and the annual recruitment R remain constant for all time.

The proof of the assertion of Beverton and Holt is immediate. In equilibrium the yield from the ith cohort during one year is equal to the yield of a single cohort during its ith year of life. Summing over i, we see that the total annual yield from all cohorts equals the total lifetime yield from a single cohort.

It must be noted that this result is critically dependent on the equilibrium assumptions. Recruitment R must be the same for all cohorts, and the fishing mortality coefficient F and the mesh size μ must remain constant throughout each cohort's life. Some of the problems involved in a disequilibrium analysis of the Beverton–Holt model are discussed in Section 8.4.

Two examples of sustained yield curves $Y = Y_\mu(F)$ are shown in Figure 8.8; these curves apply to the population of North Sea plaice studied by Beverton and Holt (1957, p. 373). The first curve corresponds to a mesh size such that $t_\mu = 5.0$ yr and exhibits a peak in yield at approximately $F = 0.2$. The second curve corresponds to $t_\mu = 13.5$ yr and increases over the entire range of fishing mortality F.

These curves suggest that such concepts as biological overfishing and maximum sustainable yield are inappropriate with regard to the Beverton–Holt model. Even at infinite levels of effort, the sustainable yield remains positive, because it can be shown that

$$\lim_{F \to \infty} Y_\mu(F) = B_0(t_\mu), \tag{8.32}$$

where $B_0(t)$ is the natural biomass given by Eq. (8.27); see Exercise 6 at the end of this chapter. If we make $t_\mu = t_0$ [see Eq. (8.28)] we can see that the maximum biomass itself can be harvested only by applying an "infinite" fishing mortality. Theoretically the instant the cohort reaches the knife-edge size at age t_0, it is instantly captured in toto.

To obtain a more practical definition of optimal fishing, Beverton and Holt introduce the concept of the *eumetric yield curve*. For each value of F, there exists some mesh size μ that results in the maximum possible sustained yield. The resulting curve $Y = Y_{\text{eum}}(F)$, defined by

$$Y_{\text{eum}}(F) = \max_\mu Y_\mu(F),$$

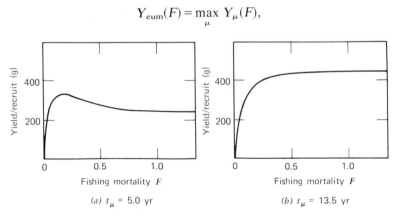

(a) $t_\mu = 5.0$ yr (b) $t_\mu = 13.5$ yr

Figure 8.8 Yield/recruit $\{(1/R)\,Y_\mu(F)\}$ versus fishing mortality F for North Sea plaice.

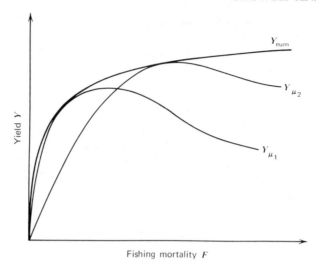

Figure 8.9   Eumetric yield curve as an envelope of the family $\{Y_\mu(F)\}$.

is called the *eumetric yield curve*. This curve is clearly the envelope of the individual yield curves Y_μ (see Figure 8.9).

Now assume, following Beverton and Holt, that the fishing mortality coefficient F is proportional to fishing effort E. Because the rate of landing fish is given by

$$L = FN,$$

this assumption implies that

$$L = qEN,$$

which is the standard catch-per-unit-effort hypothesis we discuss in Chapter 7.

Since the eumetric yield $Y_{\text{eum}}(F)$ represents the greatest sustainable yield that can be obtained at any given level of effort $E = F/q$, it is clear that only points on the eumetric yield curve can be considered "optimal" under the sustained yield assumption. It is equally clear, however, that the eumetric yield curve is probably of little significance for the dynamic optimization problem. Indeed there is no reason to suppose a priori that sustained yield is to be an optimal policy in the dynamic problem. We return to this question later, particularly in Section 8.7.

Continuing for the moment with the static equilibrium analysis, let us try to determine a static optimum. Adopting the usual assumptions that price is constant and that fishing costs are proportional to effort, we obtain a simple cost–revenue diagram (see Figure 8.10). Optimal fishing occurs at E_0, where marginal revenue equals marginal cost. At this point *both* fishing effort and mesh size are optimally determined.

The simplicity of Figure 8.10 leads some authors to the conclusion that the only distiction between the Beverton–Holt model and the Schaefer model is the introduction of a mesh-size parameter μ in the former. Realistically, however, the Beverton–Holt model is vastly more complex than the Schaefer model. If a stock-recruitment relationship is also included, the Beverton–Holt model is almost incomprehensible from the dynamic viewpoint. Some of these difficulties become apparent in Section 8.4.

What outcome should we expect in the common-property case? First we can expect that the fishermen (unless faced by mesh-size regulations) use a small enough mesh size to capture any fish of commercial value. The level of effort can be expected to reach an equilibrium, resulting in the dissipation of economic rent. In this situation overfishing results from two suboptima: effort is excessive, and the mesh size is noneumetric. A regulatory agency may wish to consider the possibility of controlling the mesh size. If a larger mesh size is adopted, a positive rent is eventually generated. If effort remains at E_1 (see Figure 8.11), the equilibrium rent equals AB. However, if effort is unlimited, it expands to a new rent-dissipating level E_2. Although sustained yield is increased by imposing suitable mesh-size regulations, the basic problem of economic inefficiency remains unsolved. Of course social benefits (consumers' surplus) may well

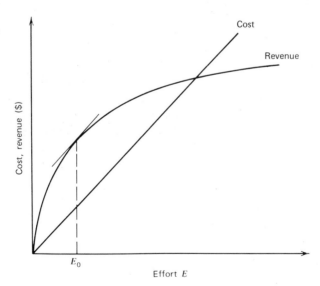

Figure 8.10 Static optimization for the Beverton–Holt model.

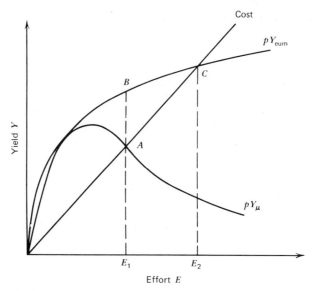

Figure 8.11 Behavior of the common-property Beverton–Holt fishery.

result from the increased supply of fish. (The population biomass is also increased, which is relevant when recruitment is stock-dependent.)

We turn now to the dynamic case.

8.4 DYNAMIC OPTIMIZATION IN THE BEVERTON–HOLT MODEL

We begin with the case of a single cohort, which we assume to be harvested independently of other cohorts. This case is perhaps of limited practical importance, but solving the dynamic optimization problem for a single cohort is helpful in understanding the more interesting case of the dynamic optimization of mixed cohorts.

As before we denote the fishing mortality coefficient by $F = F(t)$. We now permit this coefficient to vary over time subject to the constraints

$$0 \le F(t) \le F_{max}. \tag{8.33}$$

The upper constraint may be assumed finite or infinite, as desired; the latter constraint produces impulse controls.

The number $N(t)$ of fish in the cohort satisfies

$$\frac{dN}{dt} = -[M + F(t)]N, \quad N(0) = R. \tag{8.34}$$

This is the state equation for our dynamic problem; $F(t)$ is the control variable. The objective functional is given by

$$PV = \int_0^\infty e^{-\delta t}[pN(t)w(t) - C]F(t)\, dt, \qquad (8.35)$$

where p is a constant representing the price of fish, $w(t)$ denotes the weight of one fish of age t, and C is a constant cost coefficient. [If the value of the fish depends on age, the expression $pw(t)$ can be replaced by $p(t)w(t)$, with suitable modifications in the ensuing results.] The problem is to determine the optimal solution $F^*(t)$, maximizing the objective functional given by Eq. (8.35). Note that the mesh-size parameter has been suppressed in this formulation, as it is clearly irrelevant for the case of a single cohort. We shall see that $F^*(t) = 0$ for t less than some age t_δ, so that optimal fishing does not capture immature fish, regardless of the gear used.

It is now a routine exercise to determine the singular solution $N^*(t)$ to this linear control problem. This result is

$$N^*(t) = \frac{p^{-1}c\delta}{w(t)[\delta + M - \dot{w}(t)/w(t)]}. \qquad (8.36)$$

If $B^*(t) = N^*(t)w(t)$ denotes the corresponding singular biomass curve we have

$$B^*(t) = \frac{p^{-1}c\delta}{\delta + M - \dot{w}(t)/w(t)}. \qquad (8.37)$$

Before discussing the economic significance of this equation, let us see how singular and bang-bang controls can be combined to obtain the optimal harvest policy.

Figure 8.12 shows the optimal policy in terms of the biomass $B(t)$. The singular path $B^*(t)$ has a vertical asymptote at $t = t_\delta$, where

$$\frac{\dot{w}(t_\delta)}{w(t_\delta)} = M + \delta. \qquad (8.38)$$

Because by assumption $\dot{w}/w$ is a decreasing function of time, Eq. (8.38) has a unique solution t_δ, with $0 \le t_\delta \le t_0$; moreover t_δ decreases with increasing δ. [This should be taken to include the corner solution $t_\delta = 0$, whenever $\dot{w}(0)/w(0) \le M + \delta$.] Recall that the natural biomass curve attains its maximum when $t = t_0$.

Because $\dot{w}/w$ is decreasing, the singular path $B^*(t)$ given by Eq. (8.37) is also decreasing as a function of time. Observe that $\dot{w}(t_0)/w(t_0) = M$, so that we have $B^*(t_0) = p^{-1}C$; that is,

$$pB^*(t_0) = C. \qquad (8.39)$$

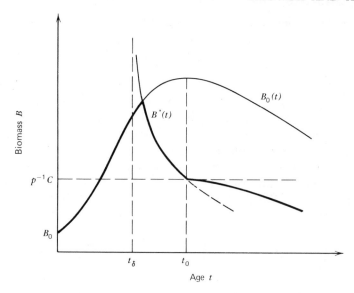

Figure 8.12   Optimal biomass curve for a single cohort.

The optimal biomass level $B(t)$ is denoted by the heavier curve in Figure 8.12. At first harvesting does not occur, and $B(t)$ follows the natural biomass curve $B_0(t)$, $F(t) = 0$. At some time $t_\delta^* > t_\delta$ the natural biomass curve crosses the singular path $B^*(t)$, and a positive fishing mortality $F^*(t)$ (the singular control) is applied to drive $B(t)$ along the singular path $B^*(t)$. When $t = t_0$ fishing is terminated, because for all subsequent t we have in any case $pB(t) < C$, so that no positive revenue can be derived from fishing.

The optimal fishing mortality $F(t)$ can therefore be described by

$$F(t) = \begin{cases} 0 & \text{for} \quad 0 < t < t_\delta^* \\ \text{positive singular control} & \text{for} \quad t_\delta^* \le t \le t_0 \\ 0 & \text{for} \quad \text{all } t > t_0. \end{cases}$$

An expression for the positive singular control can be easily derived, but we do not do so here. [Of course the upper-control constraint $F(t) \le F_{\max}$ could become binding during the singular phase. This would create a blocked-interval problem, as we discuss later in this chapter.]

Although Eq. (8.37) for the singular biomass level $B^*(t)$ is quite different in appearance from Eq. (8.15) for the singular solution to the forestry rotation problem, the behavior of the two solutions is markedly similar (see Exercise 5). In both cases an initial stage of undisturbed growth of the resource stock is followed by a continuous thinning that ultimately reduces the stock to a level of zero profitability. In both cases the harvest is completed by the time the natural growth rate of the stock begins to decline.

As before the case of zero discounting requires special attention. From Eq. (8.37) we can see that as $\delta \to 0$ the vertical asymptote $t = t_\delta$ moves to the right, approaching t_0. The singular curve $B^*(t)$ becomes progressively steeper, always passing through the point $(t_0, p^{-1}C)$. In the limit $B^*(t)$ reduces to the vertical line $t = t_0$ (see Figure 8.13a). Optimal fishing under zero discounting is an impulse control (if $F_{max} = \infty$) that harvests all profitable biomass at the single instant $t = t_0$. Obviously this corresponds to maximizing yield by employing an infinite fishing mortality, as we discuss in the previous section. If F_{max} is finite, there is a period $t_a \leq t \leq t_b$ in which fishing takes place at the maximum rate $F = F_{max}$. The solution of this blocked-interval problem is given in the next section.

High discount rates naturally induce heavier exploitation in earlier stages of the cohort's life. The reader can easily verify that Figure 8.13b properly describes the limiting situation as $\delta \to +\infty$. The optimal biomass curve $B^*(t)$ follows the zero-profit line $pB = C$, resulting as expected in the complete dissipation of rent.

For a numerical example we turn to the North Sea plaice population studied by Beverton and Holt (1957), who obtain the following parameter estimates:

natural mortality rate $M = 0.1$.
weight curve $w(t) = w_\infty(1 - e^{-K(t-\tau)})^3$,

where

$$w_\infty = 2867 \text{ g}$$
$$K = 0.095 \quad\quad\quad (8.40)$$
$$\tau = -0.815.$$

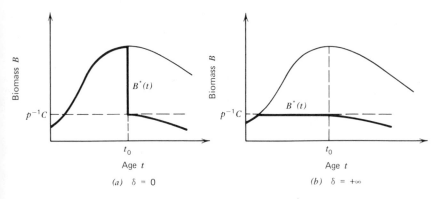

Figure 8.13 The cases $\delta = 0$ and $\delta = +\infty$.

If the cost–price ratio $p^{-1}C$ is low (i.e., if a large proportion of the maximum cohort biomass is of commercial value) then optimal fishing can be seen to capture most of the cohort soon after it reaches the age t_δ given by Eq. (8.38). This equation is readily solved for t_δ:

$$t_\delta = \tau + \frac{1}{K} \ln \left(1 + \frac{3K}{M + \delta} \right). \tag{8.41}$$

The relationship for plaice between the age t_δ and the discount rate δ is shown in Figure 8.14. For $\delta = 0$, t_δ is the age of maximum biomass $t_0 = 13.4$ yrs. A 5% discount rate reduces this to $t_{0.05} = 10.5$ yrs, whereas $\delta = 10\%$ yields $t_{0.10} = 8.5$ yrs.

For the moment assume that costs are negligible. Then

$$\rho_\delta = \frac{1}{R} B_0(t_\delta)$$

represents the total optimal yield from the cohort as a proportion of the number of recruits. This proportion is also shown in Figure 8.14. Although a discount rate of 5% decreases the optimal age t_δ by 22%, it produces a reduction in yield of only 5%; larger discount rates produce greater effects. When costs are considered the reduction in yield from discounting is less severe.

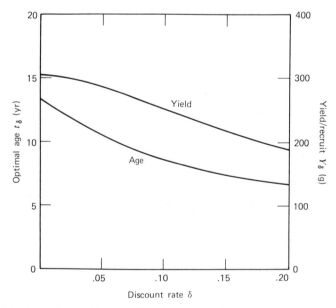

Figure 8.14 Optimal age of harvest T_δ and corresponding yield/recruit Y_δ for plaice.

Thus far our results apply only to the case of a single cohort, but they do suggest certain policy conclusions for the multicohort fishery as well. When time discounting is considered eumetric fishing may be suboptimal, the eumetric mesh size normally being larger than optimal size. Consider, for example, the problem of introducing regulations to control an over-exploited fishery. As Beverton and Holt (1957, Section 19.2.2) observe, the beneficial effects of any reduction in fishing effort or of any increase in mesh size are only fully realized subsequent to a transitional phase of significant duration. During the initial stages actual fish landings are reduced, as they must be to produce any ultimate improvement. Because they failed to specify a particular objective functional, Beverton and Holt naturally could not determine an optimal transitional phase. This problem perplexes later authors as well.

These results suggest that under the assumptions adopted in this section:

1. The optimal recovery policy consists of a complete moratorium that permits the fish population to recover to an optimal biomass level.

2. The optimal biomass level is not the eumetric (maximum-rent) level, but a lower "discounted" level; it can be reached more quickly, but it ultimately results in a somewhat reduced yield.

The first assertion clearly depends strongly on our assumption that price is constant. If fish is an essential or an extremely important item of consumption, as reflected by a low elasticity of demand, then clearly an asymptotic, gradual approach to the optimal biomass level is required.

It is less certain, however, that a gradual recovery policy is justified merely due to a consideration of the incomes of fishermen or fisheries. Only in cases where neither alternative employment opportunities nor temporary unemployment benefits are available does it seem advisable to continue fishing during the recovery stage. A deliberate decision to allow fishing to continue is thus in effect a subsidization of the fishing industry, the costs of which should always be recognized.

We conclude this section by rewriting the basic optimality equation [Eq. (8.37)] in a form that is more transparent from the economic viewpoint. Let

$$V(t) = pB(t) - C = pN(t)w(t) - C. \qquad (8.42)$$

We refer to $V(t)$ as the *net biovalue* of the cohort at age t; this biovalue changes as the result of natural growth and of both natural and fishing mortality. We also define

$$V^{\#}(t) = \frac{dV}{dt}\bigg|_{F=0} = p(N\dot{w} + w\dot{N}) = pB(t)\left(\frac{\dot{w}}{w} - M\right). \qquad (8.43)$$

Thus $V^{\#}(t)$ is the rate of increase in net biovalue in the absence of fishing: it is the potential or natural rate of increase in biovalue.

Equation (8.36) can be written

$$pN^*w\left(\frac{\dot{w}}{w} - M\right) = \delta(pN^*w - C),$$

which reduces simply to

$$\frac{V^{\#}(t)}{V(t)} = \delta. \qquad (8.44)$$

Optimal fishing is therefore adjusted to maintain equality between the proportional increase in potential net biovalue and the instantaneous discount rate. This makes our optimal fishing rule a rather simple marginality condition of standard form.

8.5 THE CASE $F_{max} < \infty$.

The optimal policy in the preceding section may necessitate a large fishing effort (mortality) over a short time period. In the case in which $\delta = 0$ an infinite effort (impulse control) is required to follow the singular path. [It is easy to see that impulse control is also required in the case in which the cost c is zero for any $\delta \geq 0$, because the singular path then consists of a vertical segment at the age $t = t_\delta$, given by Eq. (8.38). This case is important in our study of the multi-cohort problem in the following section.] What is the optimal policy in this case if we face the constraint

$$F(t) \leq F_{max} < \infty?$$

From our general understanding of linear control problems we know that the optimal control is a bang-bang control, with $F(t) = F_{max}$ on some blocked interval $t_1 \leq t \leq t_2$ containing t_0 and with $F(t) = 0$ otherwise. Given $\delta = 0$ the present value therefore equals

$$J(t_1, t_2) = \int_{t_1}^{t_2} [pN(t)w(t) - C]F_{max} \, dt$$

$$= \int_{t_1}^{t_2} \phi(t, t_1) \, dt, \qquad (8.45)$$

where we have

$$N(t) = Re^{-Mt_1}e^{-(M+F_{max})(t-t_1)} \qquad (8.46)$$

and hence

$$\phi(t, t_1) = \{pRe^{F_{max}t_1}e^{-(M+F_{max})t}w(t) - C\}F_{max}. \qquad (8.47)$$

By elementary calculus maximization of Eq. (8.45) requires that

$$\frac{\partial J}{\partial t_2} = \phi(t_2, t_1) = 0, \tag{8.48}$$

$$\frac{\partial J}{\partial t_1} = \int_{t_1}^{t_2} \frac{\partial \phi}{\partial t_1}(t, t_1)\, dt - \phi(t_1, t_1) = 0. \tag{8.49}$$

The condition given by Eq. (8.48) requires simply that $V(t_2) = 0$; that is, that harvesting continues until the biovalue is reduced to zero. It follows from Eqs. (8.47) and (8.48) that

$$\phi(t_1, t_1) = pRF_{max}e^{F_{max}t_1}\{e^{-(M+F_{max})t_1}w(t_1) - e^{-(M+F_{max})t_2}w(t_2)\}. \tag{8.50}$$

After slight manipulation Eq. (8.49) becomes

$$Re^{-Mt_1}w(t_1) - Re^{-Mt_1}e^{-(M+F_{max})(t_2-t_1)}w(t_2)$$
$$= RF_{max}\int_{t_1}^{t_2} e^{-Mt_1}e^{-(M+F_{max})(t-t_1)}w(t)\, dt.$$

This in turn can be written simply

$$B(t_1) - B(t_2) = F_{max}\int_{t_1}^{t_2} B(t)\, dt. \tag{8.51}$$

Equation (8.51) asserts that as a necessary condition for optimal yield,

change in biomass = total yield in biomass.

Because

$$\frac{dB}{dt} = \frac{d(Nw)}{dt} = -(M + F_{max})Nw + N\frac{dw}{dt},$$

we have in general

$$\Delta B = B(t_2) - B(t_1) = \int_{t_1}^{t_2} \frac{dB}{dt}\, dt = -(M + F_{max})\int_{t_1}^{t_2} B(t)\, dt + \int_{t_1}^{t_2} N\frac{dw}{dt}\, dt.$$

Thus

$$\Delta B = G - Y - Z, \tag{8.52}$$

where

$\Delta B =$ change in biomass during fishing phase,

$G = \int_{t_1}^{t_2} N\frac{dw}{dt}\, dt =$ natural growth in biomass,

$Y = F_{max}\int_{t_1}^{t_2} B(t)\, dt =$ total yield in biomass, and

$$Z = M \int_{t_1}^{t_2} B(t) \, dt = \text{natural mortality in biomass.}$$

Equation (8.52) merely states the obvious: the net change in biomass during the interval $t_1 \le t \le t_2$ equals the natural growth minus the losses due to mortality and fishing. This equality holds for *any* fishing program whether it is optimal or not.

The condition for optimal fishing given by Eq. (8.51) requires that

$$\Delta B = Y,$$

or equivalently from Eq. (8.52) that

$$G = Z. \tag{8.53}$$

This interesting conclusion does not appear to be self-evident, although Eq. (8.52) is often discussed (e.g., Graham, 1952; Beverton and Holt, 1957, p. 26). If an optimal fishing policy is followed for a single cohort in terms of maximizing the undiscounted net economic yield, then natural growth and natural mortality during the exploitation phase exactly cancel one another. Note that this conclusion is completely independent of the economic parameters. The parameters are brought in by the additional condition given by Eq. (8.48), stating that

$$pB(t_2) = C.$$

The foregoing results may be used to devise a simple numerical method for determining the optimum fishing interval $[t_1, t_2]$. Equation (8.53) implies that

$$\int_{t_1}^{t_2} \left(N \frac{dw}{dt} - MNw \right) dt = 0,$$

so that by Eq. (8.46)

$$\int_{t_1}^{t_2} \psi(t) \, dt = 0, \tag{8.54}$$

where

$$\psi(t) = e^{-(M+F_{max})t} \left(\frac{dw}{dt} - Mw \right). \tag{8.55}$$

This function is shown in Figure 8.15.

Equation (8.54) requires t_1 and t_2 to be chosen on opposite sides of t_0, so that the two areas are equal, as shown in Figure 8.15. Also t_1 and t_2

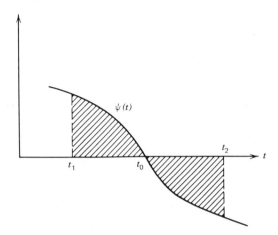

Figure 8.15   Calculation of the optimal fishing interval $[t_1, t_2]$.

must satisfy Eq. (8.48); that is,

$$e^{F_{\max}t_1} = \frac{Ce^{(M+F_{\max})t_2}}{pRw(t_2)}.\qquad (8.56)$$

A numerical method for computing t_1 and t_2 would be to choose an initial estimate for $t_2 > t_0$, and calculate t_1 using Eq. (8.56). We can then calculate the integral I in Eq. (8.54). If I is positive, our original estimate of t_2 was too small. We then increase t_2 by some specific amount (say, εI) and repeat the procedure.

8.6 MULTIPLE COHORTS: NONSELECTIVE GEAR

We come now to the dynamic optimization problem for a multicohort fish population. Even with the simplifying assumption to be made here that recruitment is independent of stock size, an analytic solution for the general problem seems completely unattainable. We therefore make an additional assumption that *the costs of fishing are negligible*; that is,

$$C = 0.\qquad (8.57)$$

The possible effects of relaxing this assumption are discussed in Section 8.8.

Let $N_k(t)$ denote the number of fish belonging to the kth cohort at time t. We assume that the kth cohort enters the fishery (recruits) at time $t = k$, and that the recruitment R is a constant that is the same for all cohorts. [This highly restrictive assumption is adopted here to produce a definite (deterministic) model for the purpose of analysis. The effect of varying cohort size is discussed later in the chapter.] Units of time are now specified as the time between successive recruitments, which are assumed to occur at regular intervals. If fishing mortality is zero we have

$$\frac{dN_k}{dt} = -MN_k, \quad N_k(k) = R \quad (t \geq k). \tag{8.58}$$

Let $w(a)$ now denote the weight of one fish of age a, where age is measured from the age of recruitment. Then the biomass of the kth cohort equals

$$B_k(t) = \begin{cases} N_k(t)w(t-k) & (t \geq k) \\ 0 & (t < k). \end{cases} \tag{8.59}$$

Next let $F = F(t)$ denote the coefficient of fishing mortality. Again assume that the fishing gear possesses knife-edge selectivity in the sense that no fishes below a certain age t_μ are captured, but all fishes of age $\geq t_\mu$ are equally subject to capture. We therefore have

$$\frac{dN_k}{dt} = \begin{cases} -MN_k & k \leq t < k + t_\mu \\ -[M + F(t)]N_k & t \geq k + t_\mu. \end{cases} \tag{8.60}$$

The value t_μ is to be considered a parameter subject to determination in an optimal manner.

It may not be necessary to draw the reader's attention to the large number of restrictions built into the above model. Although many of these restrictions are clearly unrealistic, there is probably no general agreement as to what constitutes the minimum "reasonable" dynamic model of a multicohort fishery. We shall attempt later to discuss the effect of relaxing some of these restrictions; this subject is largely unexplored, however, and there are many interesting unanswered questions for further research.

Here we begin by assuming that selectivity is nonexistent, so that all fishes are subject to capture as soon as they are recruited. (This assumption is relaxed in Section 8.8. In the fishery literature a distinction is made between the age of recruitment and the age of first liability to capture. When gear selectivity is fixed, there is no need to distinguish between these two ages in the theory.) We can then express the present value of revenues from fishing

$$PV = p \int_0^\infty e^{-\delta t} B(t) F(t) \, dt, \tag{8.61}$$

where $B(t)$ is the *total fishable biomass*

$$B(t) = \sum_{k=-\infty}^{\infty} B_k(t) = \sum_{k=-\infty}^{\infty} N_k(t)w(t-k). \tag{8.62}$$

The doubly infinite sum in Eq. (8.62) includes contributions from all cohorts recruited in the past. The upper limit of summation is actually not $k = +\infty$ but $k = [t] =$ greatest integer $\le t$, because $N_k(t) = 0$ for $k > t$. For any $t > 0$ we have

$$B_k(t) \le ce^{-M(t-k)} \quad \text{for} \quad t \ge k,$$

where $c = Rw_\infty =$ constant, so it follows that

$$B(t) \le c \sum_{k=-\infty}^{[t]} e^{-M(t-k)} \le \frac{c'}{1-e^{-M}} = c'',$$

where c' and c'' are also constants.

Thus the total biomass of the fish population is bounded by a fixed constant. [This important feature of the Beverton–Holt model does not hold for many multicohort models subjected to "optimization" analysis in the theoretical literature (cf. Beddington, 1974).] Both the infinite sum in Eq. (8.62) and the integral in Eq. (8.61) are therefore convergent (assuming $\delta > 0$).

The control-theoretic problem is to maximize Eq. (8.61) subject to the differential equations

$$\frac{dN_k}{dt} = -[M + F(t)]N_k \quad (t \ge k), \quad N_k(k) = R, \tag{8.63}$$

the single control variable $F(t)$ being subject to

$$F(t) \ge 0. \tag{8.64}$$

For mathematical simplicity we now allow impulse controls, remembering the results in the preceding section.

The Hamiltonian of this problem is

$$\mathcal{H} = e^{-\delta t} \sum N_k w(t-k)F - \sum \lambda_k (M + F)N_k, \tag{8.65}$$

with switching function

$$\sigma(t) = e^{-\delta t} \sum N_k w(t-k) - \sum \lambda_k N_k, \tag{8.66}$$

where $\{\lambda_k\}$ is an infinite system of adjoint variables. The adjoint equations are

$$\frac{d\lambda_k}{dt} = -\frac{\partial \mathcal{H}}{\partial N_k} = -e^{-\delta t} w(t-k)F(t) + \lambda_k(t)[M + F(t)].$$

Setting $\sigma(t) \equiv 0$ for singular control and differentiating, we obtain after simplification

$$\delta \sum N_k w_k = \sum N_k(\dot{w}_k - M w_k),$$

where

$$w_k(t) = w(t - k).$$

This can be written simply

$$\frac{V^{\#}(t)}{V(t)} = \delta, \tag{8.67}$$

where

$$V(t) = pB(t) = p \sum N_k(t) w(t - k)$$

again denotes the biovalue of the fish population, and

$$V^{\#}(t) = \frac{dV}{dt}\bigg|_{F=0} = \sum N_k(\dot{w}_k - M w_k).$$

From Eq. (8.67) we can see that there is no singular control $0 < F(t) < \infty$; this is also true for the single-cohort model with $c = 0$. To see this, note that the ratio

$$\frac{V^{\#}(t)}{V(t)} = \frac{\sum N_k(\dot{w}_k - M w_k)}{\sum N_k w_k}$$

is not changed by fishing, which is assumed to reduce all cohorts N_k by the same proportion. Thus there is no choice of $F(t)$ that can cause Eq. (8.67) to hold over a time interval $t_1 < t < t_2$.

It follows that the optimal pattern of fishing consists of a sequence of impulse controls at specific times $t = T_1, T_2, T_3, \ldots$, with the entire fish population being harvested each time. [It is important to remember the hypotheses under which this conclusion has been derived. Costs of fishing are zero (or at least do not depend on the biomass level), so that harvesting the entire biomass is feasible. More important, recruitment is completely stock independent, so that even reducing the population to zero does not affect future recruitment.] At first it may seem that the optimal harvest times T_1 should be determined from Eq. (8.67), but this is not correct. In fact the times T_j (which occur at regular intervals: $T_j = j \cdot T$) must be determined from the Faustmann formula

$$\frac{V_1^{\#}(T)}{V_1(T)} = \frac{\delta}{1 - e^{-\delta T}}, \tag{8.68}$$

where $V_1(T)$ is defined in Eq. (8.69) below. This can be proved by the same argument we apply to the forestry rotation problem. First, the harvests must occur at regular intervals, because following the first harvest at $t = T_1$ we are faced with the same problem we faced at $t = 0$ (assuming $t = 0$ corresponds to an initial harvest). Hence we know that $T_2 = 2T_1$, and so on.

Next let $V_1(T)$ denote the value of the population at a time T years after a complete harvest:

$$V_1(T) = pRe^{-MT}w(T) + pRe^{-M(T-1)}w(T-1) + \cdots + pRe^{-M}w(1)$$

$$= pR \sum_{k=1}^{T} e^{-kM}w(k). \qquad (8.69)$$

(In this equation and throughout the remainder of this section, we restrict T to integer values.) The total present value of all future harvests is therefore

$$PV = \sum_{j=1}^{\infty} e^{-j\delta T}V_1(T) = \frac{V_1(T)}{e^{\delta T} - 1},$$

just as it is in the forest rotation problem [Eq. (8.4)]. Consequently the optimal rotation T maximizing PV must be given by Eq. (8.68) as asserted earlier.

The solution can thus be described in the following terms. The biomass starts at zero at $t = 0$, following an initial harvest of the entire population. The biomass grows more and more rapidly as new cohorts are recruited to the fishery. Eventually as cohorts begin to die, an equilibrium is achieved and maintained with only small annual variations in biomass. Long before such an equilibrium is reached, however, harvesting again reduces the biomass to the zero level, and the process is repeated.

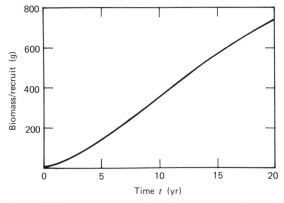

Figure 8.16 Biomass curve for multicohort stock (plaice).

We now apply these results to the data for North Sea plaice given in Eqs. (8.40). The age of recruitment of this species is given by Beverton and Holt (1957, p. 310) as

$$t_{\rho'} = 3.72 \text{ years.}$$

Combining this with Eqs. (8.40), we have for fish of age k years beyond recruitment age

$$w(k) = w_\infty \{1 - \exp\left[-K(k + 4.53)\right]\}^3 \text{ g.}$$

As above we suppose that nets are completely nonselective and all recruited fish in the population are equally liable to be captured. Values of the biomass per recruit

$$W(T) = \frac{V_1(T)}{pR} = \sum_{k=1}^{T} e^{-kM} w(k), \qquad (8.70)$$

TABLE 8.3. BIOMASS PER RECRUIT $W(T)$, PROPORTIONAL GROWTH RATES $\Delta W/W$ AND AVERAGE ANNUAL YIELDS $W(T)/T$ FOR NORTH SEA PLAICE: NONSELECTIVE GEAR.

Time T (yr)	Biomass per Recruit $W(T)$ (g)	$\dfrac{\Delta W(T)}{W(T)}$	$\dfrac{W(T)}{T}$
1	177.0	1.000	177.0
2	408.9	0.567	204.4
3	692.2	0.409	230.7
4	1021.3	0.322	255.3
5	1388.7	0.265	277.7
6	1786.4	0.223	297.7
7	2206.2	0.190	315.2
8	2640.3	0.164	330.0
9	3081.6	0.143	342.4
10	3523.9	0.126	352.4
11	3961.8	0.111	360.2
12	4390.9	0.098	365.9
13	4807.4	0.087	369.8
14	5208.8	0.077	372.1
15	5592.9	0.069	372.9
16	5958.2	0.061	372.4
17	6303.9	0.055	370.8
18	6629.5	0.049	368.3
19	6934.9	0.044	365.0
20	7220.3	0.040	361.0

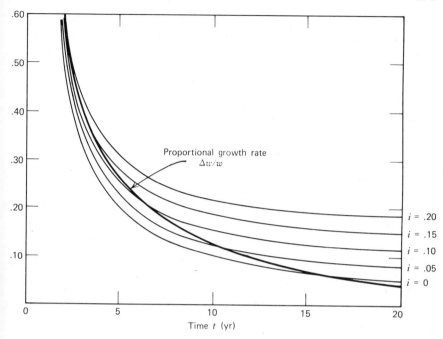

Figure 8.17 Graphical determination of optimal rotation for multicohort plaice fishery (nonselective gear); see Tables 8.3 and 8.4.

are given in Table 8.3, which also lists the corresponding proportional growth rate $\Delta W/W = \Delta V_1/V_1$ and the average annual yield $W(T)/T$ when the population is harvested at T years. Note that the biomass W is still increasing after 20 years, but at an annual rate of only 4%.

To obtain the optimal rotation period T, as determined by the Faustmann formula [Eq. (8.68)], we can plot the curve $\Delta W(T)/W(T)$ on the graph in Figure 8.2 (see Figure 8.17). The results are shown in Table 8.4. Average annual yield is maximized ($\delta = 0$) at 372.8 g/recruit, with a

TABLE 8.4. OPTIMAL "ROTATION" FOR PLAICE: NONSELECTIVE GEAR.

Discount rate δ	0	0.01	0.03	0.05	0.10	0.15	0.20	∞
Rotation T (yr)	15.0	14.4	11.5	10.2	6.4	3.0	2.5	1
Average Annual Yield per Recruit (g)	372.8	372.3	363.0	353.9	304.7	230.7	217.0	177.0

rotation of $T = 15.0$ yr. This may be compared with the maximum sustained yield of 257.2 g/recruit for the case of nonselective gear calculated by Beverton and Holt (1957, p. 310).

The preceding calculations can be repeated for various alternative assumptions regarding the selectivity of fishing nets. If we assume knife-edge selectivity at age $k = s$ yr beyond recruitment, Eq. (8.69) must be replaced by

$$V_1(T) = pR \sum_{k=s+1}^{T} e^{-Mk} w(k). \qquad (8.71)$$

The results for $s = 5$ and $s = 9$ are given in Table 8.5. For these cases average yield is maximized by rotations of length $T = 8$ and $T = 2$ yr, respectively. Discounting by 5% per annum reduces these optimal rotation periods to $T = 2$ yr and $T = 1$ yr, respectively. Note that increasing s increases both the maximum and the optimal average yield figures. Increasing s beyond 10 yr, however, reverses the yield increase. This is to be expected, because the age of maximum cohort biomass is 10 yr above recruitment age.

The results of this section may be summarized as follows. Whenever gear selectivity is fixed at a suboptimal level, optimal harvesting may require a periodic fishing effort. As gear selectivity decreases i.e., the proportion of immature fish in the catch increases, periodic harvesting becomes more effective. We study this question further in the following section.

TABLE 8.5. BIOMASS PER RECRUIT $W(T)$, PROPORTIONAL GROWTH RATES $\Delta W/W$, AND AVERAGE ANNUAL YIELDS $W(T)/T$ FOR NORTH SEA PLAICE: SELECTIVE GEAR.

Selectivity Parameter = 5 yr				Selectivity Parameter = 9 yr			
Time T (yr)	$W(T)$ (g)	$\dfrac{\Delta W(T)}{W(T)}$	$\dfrac{W(T)}{T}$	Time T (yr)	$W(T)$ (g)	$\dfrac{\Delta W(T)}{W(T)}$	$\dfrac{W(T)}{T}$
1	367.4	1.000	367.4	1	441.3	1.000	441.0
2	765.1	0.520	382.5	2	883.7	0.501	441.8
3	1184.9	0.354	394.9	3	1321.6	0.331	440.5
4	1619.0	0.268	404.7	4	1750.6	0.245	437.6
5	2060.3	0.214	412.1	5	2167.2	0.192	433.4
6	2502.6	0.177	417.1	6	2568.6	0.156	428.1
7	2940.5	0.149	420.1	7	2952.6	0.130	421.8
8	3369.6	0.127	421.2	8	3317.9	0.110	414.7
9	3786.1	0.110	420.7	9	3663.6	0.094	407.1
10	4187.5	0.096	418.8	10	3989.2	0.082	398.9

8.7 PULSE FISHING

We know from earlier chapters (see Sections 5.4 and 7.7) that considerations of economic efficiency can require an oscillating harvest policy,
which we refer to as pulse fishing, rather than a sustained-yield policy.
This is the case whenever the most efficient rate of harvest exceeds the
optimal sustainable yield. The analysis in the preceding section suggests a
completely different situation in which pulse fishing may be desirable.
This situation arises in a multicohort fishery where the selectivity of
fishing gear is not optimally adjusted.

Let us suppose, for example, that the gear is completely nonselective.
Thus the fishery does not possess the option of harvesting an individual
cohort at its particular optimal age; at any time the fishery must either
harvest all cohorts at once or no cohorts.

In this section we present a simple, direct argument that is not based on
control-theoretic considerations to show analytically that pulse fishing
produces a greater average biomass yield than any possible level of
sustained-yield fishing whenever gear is nonselective.

Now consider a simple, discrete-time cohort model in which a_k denotes
the biomass of an unexploited cohort of age $k = 1, 2, 3, \ldots$. Suppose that

$$a_1 < a_2 < \cdots < a_N > a_{N+1} > \cdots, \qquad (8.72)$$

so that N is the age of maximum biomass.

If the population is harvested completely once every m yr, then the
average annual yield equals

$$Y_m = \frac{1}{m}(a_1 + a_2 + \cdots + a_m).$$

If $m < N$ we have

$$a_1 + a_2 + \cdots + a_m < m a_{m+1},$$

so that

$$(m+1)(a_1 + a_2 + \cdots + a_m) < m(a_1 + a_2 + \cdots + a_{m+1});$$

that is,

$$Y_m < Y_{m+1}.$$

Thus if we wish to maximize the average yield, pulse harvesting cannot
occur more than once every N years, where N is the age of maximum
biomass. The example $\{a_i\} = \{1, 3, 2.5, 0, 0, \ldots\}$ shows that in general the
average yield $Y_m = (a_1 + a_2 + \cdots + a_m)/m$ may achieve its maximum for
$m > N$.

Suppose, however, that we harvest at a sustained-yield level, harvesting
a fixed proportion λ $(0 < \lambda < 1)$ of each cohort in each year. (Thus the

fishing gear is completely nonselective.) When equilibrium is reached the sustained yield Y_s consists of λa_1 from the youngest cohort plus $\lambda (1-\lambda)a_2$ from the one-year-old cohort, and so on. Thus

$$Y_s = \lambda a_1 + \lambda(1-\lambda)a_2 + \cdots + \lambda(1-\lambda)^{n-1}a_n + \cdots.$$

If Y_p denotes the maximum average yield that can be obtained by pulse fishing, then it can be shown that

$$Y_s < Y_p. \tag{8.73}$$

Proof.   Write $x = 1 - \lambda$, so that $0 < x < 1$. Then Eq. (8.73) is equivalent to

$$\sum_1^\infty a_i x^{i-1} < \frac{Y_p}{1-x}.$$

After summation by parts with $s_k = \sum_1^k a_i$ we have

$$\sum_1^n a_i x^{i-1} = s_n x^n - \sum_1^n s_i(x^i - x^{i-1})$$

$$= s_n x^n + (1-x) \sum_1^n s_i x^{i-1}.$$

Now for every $i \neq p$ we have $s_i/i < Y_p = \max (s_i/i)$. Hence

$$\sum_1^n a_i x^{i-1} < \left[nx^n + (1-x) \sum_1^n i x^{i-1} \right] Y_p$$

$$= (1 + x + x^2 + \cdots + x^{n-1}) Y_p$$

$$< \frac{Y_p}{(1-x)}. \qquad\qquad \text{Q.E.D.}$$

In other words, optimal pulse fishing produces a greater average yield than any sustained-yield harvest.

A significant special case arises when the maximum biomass occurs in the first year of a cohort's life; that is, when

$$a_1 > a_2 > a_3 > \cdots.$$

In this case—and only in this case—optimal pulse fishing harvests the entire cohort during its first year. This is equivalent to sustained-yield harvesting, with $\lambda = 1$ (i.e., $Y_s = Y_p$ in this case).

These results can be summarized as follows. Pulse fishing in a cohort fishery model produces a greater average yield than sustained-yield fishing, whenever the following conditions hold:

1. Fishing gear is nonselective.
2. The cohort biomass reaches a maximum at some age N greater than one generation interval.

3. Recruitment is constant and is not affected by harvesting (no matter how intensive).

Since neither condition (1) nor condition (3) is likely to hold strictly in practice, the usefulness of this result is obviously limited. Indeed it seems unlikely that the phenomenon of pulse fishing observed in practice in certain fisheries can be explained on purely biological grounds. From numerical studies we discuss in the following section, the benefits obtained from pulse fishing, while greater than zero, generally appear to be of minor practical importance. It seems probable, therefore, that pulse fishing is primarily the result of the economic factors already described.

Gear Selectivity

Now suppose that the selectivity of the fishing gear insures that all fish above a certain size are captured and that all smaller fish escape. Continuing with our simple, discrete-time model, suppose that the gear selects all fishes of age $\geq s$. Thus only cohorts of weight $a_s, a_{s+1}, \ldots$, are available to the fishery. If pulse fishing occurs every $(m-s+1)$ yr the average annual yield equals

$$Y_{m,s} = \frac{1}{m-s+1}(a_s + a_{s+1} + \cdots + a_m).$$

If $m \leq N =$ age of maximum biomass, $Y_{m,s}$ obviously increases with s. For $s = m = N$ in particular we obtain

$$Y_{N,1} = a_N.$$

This is a self-evident result: if the gear selects fishes from the cohort of maximum biomass (as well as older fishes), and if the cohort is harvested completely every year, the maximum possible yield can be achieved. The same results are obtained numerically in Tables 8.3 and 8.5 in Section 8.6.

For the discounted case, optimal gear selectivity corresponds to the age m determined by

$$\frac{\Delta a_m}{a_m} = \frac{a_{m+1} - a_m}{a_m} = \delta. \qquad (8.74)$$

In particular $\delta = 0$ implies $m = N =$ age of maximum biomass. To prove this assertion, we return to the framework of the Beverton–Holt model.

8.8 MULTIPLE COHORTS: SELECTIVE GEAR

Again we consider the Beverton–Holt model in Section 8.6, assuming now that the age of first capture is subject to choice. Still assuming that

$C = 0$, we wish to show that optimal fishing is achieved if knife-edge selectivity permits the gear to select fish at the age t_δ, for which

$$\frac{\dot{w}(t_\delta)}{w(t_\delta)} = M + \delta, \tag{8.75}$$

and if each year's fishing reduces the cohort of age t_δ to zero.

Indeed this result is now self-evident, because we know from the single-cohort model (Section 8.4) that the above policy maximizes the present value of revenue from each cohort separately and hence a fortiori from all cohorts taken together.

Because Eq. (8.75) is equivalent to $B'(t_\delta)/B(t_\delta) = \delta$, it is clear that Eq. (8.74) is the discrete-time analog of Eq. (8.75). The explicit solution of Eq. (8.75) for t_δ appears in Eq. (8.41), and the relationship between t_δ and δ is plotted in Figure 8.14. (To obtain the optimal age in terms of the number of years beyond recruitment, we must subtract 3.72 yr from t_δ.)

The results for the cases of both selective and nonselective gear depend strongly on various restrictive assumptions, the most serious being the assumptions of costless fishing and of recruitment that is entirely independent of stock levels. [The model discussed here with zero costs can be easily modified to allow for variable recruitment R_k. By Eq. (8.75), the optimal age of capture is in fact independent of the recruitment level R_k.] Relaxing either of these unrealistic assumptions leads to severe theoretical difficulties that have yet to be resolved.

On the basis of earlier results in this book and the work of other authors, it may be reasonable to make certain conjectures, however. The first conjecture is that pulse fishing will continue to be theoretically optimal in cases where gear selectivity is imperfect, but not in other cases, unless caused by cost nonlinearities (decreasing marginal costs). The second conjecture is that when recruitment is stock-dependent, optimal fishing will often consist of a *bimodal* harvest policy, with each cohort reduced to some optimal level and subsequently harvested more fully at some optimal age. Such bimodal harvest policies have recently been identified in a study of optimal deer harvesting (Beddington and Taylor, 1973) and in a study of optimal levels of cattle stocking (Jarvis, 1974).

Another important generalization of the Beverton–Holt model would allow for stock-dependent growth and mortality. These effects could also produce bimodal harvesting policies. The case of a single cohort in Exercise 5 is similar to the thinning problem in Section 8.2.

A numerical study of the Beverton–Holt model applied to the North Atlantic cod population has recently been conducted by Hannesson (1975). Figure 8.18 shows the optimal annual effort sequence $F = F_k$ from one of Hannesson's optimization runs. The following assumptions

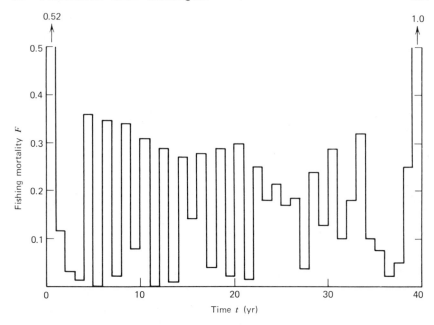

Figure 8.18 An example of pulse fishing (Hannesson, 1975).

hold: constant recruitment, positive costs and discount rate, and fixed gear selectivity (not the knife-edge variety, however). The solution shows a strong pulse-fishing effect with a two-year period. (The large initial and terminal values are bang-bang adjustments.)

Although the optimal policy in Hannesson's study is a pulse-fishing policy, a sustained-yield policy is almost as effective, producing more than 90% of maximum present value from pulse fishing. If gear selectivity could have been improved, pulse fishing would probably have enjoyed an even smaller advantage.

8.9 SUMMARY AND CRITIQUE

The results in this chapter consist primarily of elaborations of the basic capital-theoretic formula $F'(x) = \delta - \dot{p}/p$ (see also Exercise 5). An important modification is introduced by the rotation aspect of forest management; the same modification applies to replacement problems in general (see Exercise 1). Our forest management models, while clearly superficial from the practical viewpoint, show that the basic bioeconomic principles in previous chapters remain valid mutatis mutandi for this industry. In particular, because forests are typically slow-growing assets, time-preference and discounting effects are especially strong in this field.

Similarly discounting is seen to have potentially significant effects in fishery management, especially for long-lived species. The currently accepted management principles based on eumetric yield therefore suffer from the same shortcomings as rent-maximization policies in general. Such policies, which overlook time-preference phenomena, may encounter strong resistance from the exploiting industry.

We have obtained partial results here concerning the optimal control of cohort fisheries. Both practical and theoretical difficulties are to be anticipated in attempting to extend this analysis. It is no doubt a tactical error to expect dynamic optimization models to be extended to the development of practical management policies to any great degree. On the other hand it is equally dangerous to overlook these dynamic aspects entirely, and simple analytic models are valuable for their suggestive power.

EXERCISES

1. *The orchard replanting problem:* Let $\phi(t)$ denote the rate of production of fruit by an orchard of age t. Assume a constant price p and a constant cost c of replanting the orchard when production begins to decline. Show that the optimal age T of replanting satisfies

$$\frac{p\phi(T)}{p\int_0^T e^{-\delta t}\phi(t)\,dt - c} = \frac{\delta}{1-e^{-\delta T}}.$$

2. Again consider the problem of optimal aging, $V(t)$ denoting the sale value of an asset of age t. Suppose that a continuing cost is associated with caring for the asset, equal to a fixed amount c per unit time. The optimal age T for disposing of the asset is then given by

$$\frac{V'(T)-c}{V(T)} = \delta.$$

Show that, under normal assumptions, an increase in the value of the asset from $V(t)$ to $\lambda V(T)$, $\lambda > 1$, that is not accompanied by change in cost c, causes the owner to retain the asset for a longer period of time. [When many such assets of various ages are being held, it follows that the short-term response to an increase in price is a decrease in output. (Jarvis, 1974, asserts that such negative short-term price responses are typical in the Argentine cattle industry.)]

Also show that this phenomenon does *not* arise in the corresponding optimal rotation problem.

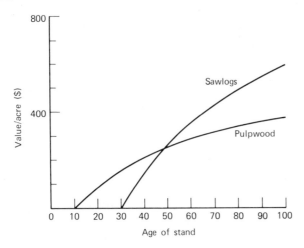

Figure 8.19 Net stumpage values for pulpwood and sawlogs; see Exercise 3.

3. The trees on a given site may be harvested either for pulpwood or for sawlogs. Net stumpage values for both products are shown in Figure 8.19. Using the Faustmann formula (with $c = 0$), prove that sawlogs are the preferred product at low discount rates, but that pulpwood is the preferred product if the discount rate is sufficiently high. What features of the graph in Figure 8.16 produce this result? Approximately what percentage rise in the price of pulpwood relative to sawlogs would make pulpwood the preferred product for all δ? Under what conditions on the value curves would sawlogs be preferred for all δ?

4. A given stand of trees is to be clearcut at some fixed age T. Fertilizer can be applied at a variable rate $Q = Q(t)$; the resulting growth equation is then

$$\frac{dx}{dt} = r \cdot (K - x) + F(Q, t), \quad x(0) = x_0,$$

where x is the volume of timber and r and K are positive constants. Letting p and c denote the constant price of timber and the cost of fertilization, respectively, we have

$$PV = e^{-\delta T} p x(T) - c \int_0^T e^{-\delta t} Q(t) \, dt.$$

If there are no binding constraints, show that the optimal fertilization rate $Q = Q(t)$ can be determined from the equation

$$\frac{\partial F(Q, t)}{\partial Q} = \frac{c}{p} \cdot e^{-(r+\delta)(t-T)}.$$

Discuss the properties of this solution, assuming $\partial F/\partial Q > 0$, $\partial^2 F/\partial Q^2 < 0$. (Adapted from Näslund, 1969.)

5. The growth equation for the Beverton–Holt model can be written in the form

$$\frac{dB}{dt} = \left[\frac{\dot{w}(t)}{w(t)} - M\right]B = g(t)B,$$

so that $g(t) = \dot{w}/w - M$ represents a biomass growth coefficient. In this model the net proportional growth of biomass is therefore density independent. Now consider the more general case in which net growth is density dependent, so that

$$\frac{dB}{dt} = g(t)F(B).$$

Solve the optimal harvesting problem for a single cohort in this case. Show that the singular biomass curve $B^*(t)$ satisfies the same equation [Eq. (8.15)] as it does in the optimal thinning model. Thus verify that, mathematically speaking, the thinning model contains both the Schaefer model and the single-cohort Beverton–Holt model as special cases.

6. (a) Prove that if $\phi(t)$ is a continuous, bounded function $(t \geq 0)$, then

$$\lim_{K \to +\infty} K \int_0^\infty e^{-Kt}\phi(t)\, dt = \phi(0).$$

(Consider the graph of the function Ke^{-Kt} for large K.)
(b) Use the result in (a) to establish Eq. (8.32).

BIBLIOGRAPHICAL NOTES

The Faustmann model (M. Faustmann, 1849) is frequently discussed by economists see Gaffney, 1960; Pearse, 1967; and Scott, 1972. More thorough studies of the economics of forestry appear in Schreuder (1968) and Gregory (1972). Control-theoretic models of optimal forest thinning are described by Näslund (1969), Kilkki and Vaisanen (1969), and Clark and de Pree (1975).

The multicohort fishery model in this chapter is due to Beverton and Holt (1957), whose definitive work is a standard reference in fisheries biology; many other aspects of fish-stock exploitation and much statistical data are presented in this work. Beverton and Holt also recognize the problem of transitional phases in the recovery of an overexploited fishery (Beverton and Holt, 1957, Chapter 19), but do not attempt any analysis based on criteria involving present values.

Although the Beverton–Holt model occasionally receives attention in the economic literature (Turvey, 1964; Smith, 1969), the dynamic complexities involved do not seem to be clearly recognized. Smith (1969), for example, introduces a mesh-size parameter into a lumped-parameter model, but is unable to say how a homogeneous population would be affected by alterations in mesh sizes. A recent economic study of the New England Yellowtail Flounder industry by Gates and Norton (1974) utilizes the Beverton–Holt model, but in a purely static form.

The dynamic optimization problem for the Beverton–Holt model is studied by Clark, Edwards, and Friedlaender (1973); Hannesson (1975) extends and applies these results to the North Atlantic cod fishery. (See also Waugh and Calvo, 1974.)

The Leslie matrix model is used extensively by ecologists (see Emlen, 1973, Section 9.3) as the standard model of an age-structured population. Because of its prediction of infinite exponential growth, the Leslie model is rarely applied to practical resource-harvesting problems. The model has nevertheless been studied from the point of view of optimal harvesting by Beddington (1974) and by Beddington and Taylor (1973). Reed (1975) has developed a nonlinear version of the Leslie matrix model that exhibits stable population growth.

A recent model of cattle production due to Jarvis (1974) has some contacts with the material in this chapter (see Exercise 2).

A special case of the result obtained in Section 8.5 for optimal harvesting at a limited rate $F_{max} < \infty$ was originally obtained by Goh (1973).

Computer simulation studies of multicohort fisheries by Walters (1969), Allen and Basasibwaki (1974), and Pope (1973) all indicate that pulse fishing sometimes leads to greater average yields than sustained-level fishing.

The optimal equilibrium solution (under discounting) for the Beverton–Holt model with nonzero costs, has recently been derived by the author; the result will appear elsewhere.

9

MULTISPECIES PROBLEMS

A given biological population is merely one component of a complex ecological system that contains predators, prey, competitors, disease organisms, and other living things. Modeling the dynamics of a population by means of a single differential or difference equation implies a neglect of these ecological interrelations, or at best some sort of generalized assumption of density-dependent mortality that may be caused by biologically related factors. [Our metered-model approach (Chapter 7) permits the inclusion of certain ecological relationships, but only in the form of given external factors.] This neglect may be justifiable in some cases, particularly where only one species of an ecosystem is subject to exploitation.

Today, however, the number of such isolated resource stocks appears to be rapidly diminishing. Most fisheries exploit more than one species, although in some instances part of the catch may be incidental. The International Commission for Northwest Atlantic Fisheries (ICNAF), for example, collects data on 43 species of fishes captured in the treaty area. The International Whaling Commission (IWC) establishes quotas for more than a dozen species subject to capture by the whaling fleets. Even specialized agencies such as the Inter-American Tropical Tuna Commission (IATTC) or the International Pacific Salmon Commission (IPSFC) are concerned with managing more than a single species. Moreover a single species often consists of several subpopulations, each of which may require separate control.

Interactions between exploited populations can be divided into two classes: biological interactions and economic interactions. Thus a single fishery that exploits several species may severely affect the dynamics and

302

stability of the corresponding ecosystem. On the other hand different components of the same ecosystem may be exploited independently, resulting in mutual externalities between the exploiters. The mathematical models in this chapter reflect this distinction quite clearly.

Problems related to the optimal exploitation of multispecies systems are obviously much more difficult both theoretically and practically than problems pertaining to the single-species case. Few ecosystems have been sufficiently studied to provide quantitatively valid estimates of interaction coefficients. Hopefully this situation is improving as a consequence of the many highly sophisticated ecosystem models currently under development [for example, see S. Levin (Ed.), 1974]. But practical resource-management decisions cannot be delayed until all the "scientific evidence" is complete; decisions must be made on the basis of available knowledge.

Our purpose in this chapter, as in the rest of the book, is to describe some of the more significant qualitative results that may be expected to arise from the interplay of biological and economic forces. As mentioned above, the theory now becomes much more difficult, and our progress is correspondingly limited. Indeed we cannot completely solve any of the dynamic multispecies optimization models proposed here. In those cases, however, in which two or more ecologically interacting species can be harvested independently, our theory (given in Section 9.3) seems adequate for most practical purposes. The case in which selective harvesting is not feasible is much more complex. As we show in Section 9.1, severe overexploitation of some species seems almost inevitable in many such cases.

9.1   DIFFERENTIAL PRODUCTIVITY

First we study the combined harvesting of two ecologically independent populations, once more focusing our attention on the fishery model. For simplicity we assume that each population is subject to logistic growth. If E denotes the effort devoted to combined harvesting, then we may write

$$\left. \begin{aligned} \frac{dx}{dt} &= rx\left(1 - \frac{x}{K}\right) - q_1 Ex \\ \frac{dy}{dt} &= sy\left(1 - \frac{y}{L}\right) - q_2 Ey \end{aligned} \right\}, \tag{9.1}$$

where q_1 and q_2 are the catchability coefficients for the two populations. Equations (9.1) constitute an extension of the Schaefer model to the case in which two independent populations are captured by the same gear.

If we also assume that the respective prices p_1 and p_2 are constant and that fishing costs are proportional to effort, we obtain the following expression for net revenue (economic rent):

$$\pi(x, y, E) = p_1 q_1 xE + p_2 q_2 yE - cE. \tag{9.2}$$

Bionomic Equilibrium

Equilibrium solutions $\dot{x} = \dot{y} = 0$ for Eqs. (9.1) can only occur on the coordinate axes ($x = 0$, or $y = 0$) or at a point (x, y) on the line segment

$$\frac{r}{q_1}\left(1 - \frac{x}{K}\right) = \frac{s}{q_2}\left(1 - \frac{y}{L}\right), \quad 0 \le x \le K, \quad 0 \le y \le L. \tag{9.3}$$

Let us suppose that

$$\frac{r}{q_1} < \frac{s}{q_2}. \tag{9.4}$$

Then the above equilibrium line intersects the y- axis at

$$\tilde{y} = L\left(1 - \frac{rq_2}{sq_1}\right) \tag{9.5}$$

(see Figure 9.1). [The ·case in which (9.4) is reversed is similar, with the intersection occurring at a point $\tilde{x} > 0$ on the x axis instead.]

Bionomic equilibrium of the open-access fishery is characterized by Eqs. (9.3), together with the condition

$$\pi = (p_1 q_1 x + p_2 q_2 y - c)E = 0. \tag{9.6}$$

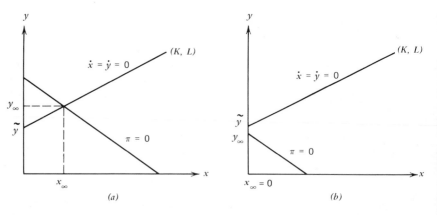

Figure 9.1 Bionomic equilibrium for the two-species fishery: (*a*) nonextinction; (*b*) extinction of the *x* population.

If $(p_1 q_1 x + p_2 q_2 y - c)$ is negative for all points (x, y) on the equilibrium line segment [Eqs. (9.3)], the fishery is incapable of yielding economic rent and therefore remains unexploited $(E = 0)$. Otherwise there are two cases, as indicated in Figure 9.1. In the first case (Figure 9.1a), the zero-profit line $\pi = 0$ intersects the equilibrium line at a point (x_∞, y_∞), where both x_∞ and y_∞ are positive. In the second case (Figure 9.1b), the two lines do not intersect, and bionomic equilibrium occurs at the point $(0, y_\infty)$ (i.e., $x_\infty = 0$). From Eqs. (9.5) and (9.6) we see that the necessary and sufficient condition for x_∞ to be positive is that

$$\frac{c}{p_2 q_2} > \tilde{y}. \tag{9.7}$$

We conclude that when Eq. (9.4) and the reverse inequality to Eq. (9.7) both hold, the open-access fishery leads to the ultimate extinction of the x population. [The model can be extended by adding a dynamic E reaction of the form $\dot{E} = k\pi$ (cf. Section 6.5, Example 9). In any case the equilibrium (x_∞, y_∞) is approached asymptotically because E remains bounded.] In the single-species Gordon–Schaefer model extinction cannot occur, because as $x \to 0$ the unit harvesting cost eventually exceeds the price. The present analysis shows that when two populations are exploited jointly, one population may be driven to extinction, whereas the other population continues to support the fishery in bionomic (one-species) equilibrium. The argument clearly extends to the case where n populations are harvested jointly: some populations may be eliminated, whereas others continue to survive. Before discussing the conditions under which elimination takes place, we consider the following example.

Blue and Fin Whales

To illustrate the preceding analysis, let us consider the Antarctic baleen whale fishery. For simplicity we suppose that only two species, blue whales and fin whales, are exploited. We wish to try to predict the outcome of the unregulated joint exploitation of these two populations. Let x and y denote the *numbers* of blue whales and fin whales, respectively. In Section 2.6 we use the Schaefer logistic model, with parameter values

$$s = 0.08, \quad L = 400,000$$

for the Antarctic fin-whale population. What evidence exists seems to suggest both a lower intrinsic growth rate and a smaller carrying capacity for the blue-whale population. For illustrative purposes here we use the values

$$r = 0.05, \quad K = 150,000.$$

Finally it seems reasonable to assume equal catchability for the two species, so that (with units of effort suitably specified) we may set

$$q_1 = q_2 = 1.$$

Equation (9.4) then holds, and we see that

$$\bar{y} = 400,000\left(1 - \frac{0.05}{0.08}\right) = 150,000.$$

If our model is correct, we can conclude that if the unregulated Antarctic whale fishery achieves bionomic equilibrium with fewer than 150,000 fin whales, then the equilibrium of the blue-whale population equals zero. In view of the fact that by 1975 the Antarctic fin-whale population had been reduced to approximately 75,000 whales, it seems fair to assume that the blue whale's existence in the Antarctic can be endangered by the continuing fin-whale fishery in the absence of regulation. The IWC seems to have recognized this danger in 1965, at which time a moratorium on blue whales was established.

It is perhaps unnecessary to review all of the important hypotheses underlying the above argument. But it should be noted that the estimate $\bar{y} = 150,000$ is critically dependent on the relative growth rate r/s. If, for example, the two whale species happen to have equal growth rates $r = s$, then $\bar{y} = 0$ and the model does not predict the extinction of either species. (Likewise if in fact $r > s$, then the fin whale could become the endangered species.)

On the other hand if other whale species, such as the sei or Minke whales now being captured, possess growth rates in excess of 8% per annum, our model could predict the progressive elimination of blue whales, fin whales, and other species. It is interesting that this argument does not depend on the relative *values* of the different species, but only on their relative growth rates. [Our present simple model does not explain why the Antarctic blue-whale population was severely depleted *before* the fin-whale stocks were heavily exploited. But we cover this point in our discussion of the quality of resource pools (see Section 5.2). In this regard it should be noted that in fact the two populations of blue whales and fin whales normally do not intermingle completely, because fin whales prefer to feed in more northerly waters than blue whales. Whether this separation would *ensure* the survival of a remnantal blue-whale population seems dubious.] However, if larger whales are easier to catch than smaller species, this factor affects survival rates. For example, in the blue-fin whale model suppose that $r = s$, but that $q_1 > q_2$. Then $\bar{y} > 0$, so that the blue whale becomes endangered if $y_\infty < \bar{y}$.

Biotechnical Productivity

Many other examples involving the elimination of certain populations under a regime of combined harvesting can be mentioned. In the ICNAF area severe overfishing of haddock (*Gadus aeglefinus*) and other major species has occurred in recent years, possibly as the result of modern methods of trawl fishing. The Alaskan salmon fisheries have experienced startling productivity losses since the beginning of the twentieth century, and ecologists suggest that this is primarily due to the progressive elimination of less productive subraces. In the case of waterfowl, wildlife biologists have expressed concern that heavy hunting pressures and the absence of species quotas may lead to the overexploitation of some of the slower breeding species such as the canvasback (*Aythya valisineria*).

At this point we review the conditions under which our present model predicts the open-access elimination of the x population. If $q_1 = q_2$ the basic requirement is that

$$r < s.$$

Under this condition x will be eliminated if the cost–price ratio for y is sufficiently low; that is, if

$$\frac{c}{p_2 q_2} < \bar{y} = L \frac{1-r}{s}.$$

Note that elimination of x in this model depends on the price p_2 of species y, and not on the price p_1 of species x.

If $q_1 \neq q_2$ the basic condition then becomes

$$\frac{r}{q_1} < \frac{s}{q_2}.$$

We refer to the ratio $\beta_1 = r/q_1$ as the *biotechnical productivity* (btp) of the x population. Thus a population possesses low btp if its intrinsic growth rate (i.e., its biotic potential) r is low or if its catchability is high. Our rule then becomes: populations with relatively low btp are subject to elimination under joint harvesting conditions, provided that the cost–price ratios of other species are sufficiently low. Although this result is derived on the assumption of logistic growth, it can readily be extended to more general growth functions (see Exercise 1 at the end of this chapter).

"Yield-effort" Curves

Our examination of the open-access model is complete, and we turn to optimal harvest policies. We begin with the static yield-effort diagram, keeping in mind that "yield" now refers to total economic revenue. Solving the equilibrium equations $\dot{x} = \dot{y} = 0$ for x and y in terms of E (Section 1.1), we have

$$TR = TR_x + TR_y$$

$$= p_1 q_1 KE\left(1 - \frac{q_1 E}{r}\right) + p_2 q_2 LE\left(1 - \frac{q_2 E}{s}\right). \tag{9.8}$$

Thus the total revenue curve TR is merely the sum of the two individual (parabolic) revenue curves. In the examples shown in Figure 9.2 it is assumed that the x population has a lower btp than the y population, so that elimination of x is feasible. Indeed with the total cost curve TC in the position shown in Figures 9.2a and b, bionomic equilibrium at $E = E_\infty$ does lead to the extinction of the x population.

Next we consider the rent-maximizing level E_0. In Figure 9.2a the x population is capable of producing much greater economic benefits than the y population, and rent maximization for the joint fishery is primarily determined by the contribution of the x population. The y population is in a sense only an incidental benefit to the fishery. Under these circumstances open-access exploitation $E = E_\infty$ is particularly disastrous, because it leads to the destruction of the valuable resource stock and reaches an equilibrium in which only the less valuable population remains.

In Figure 9.2b the situation is reversed, and the value of the y

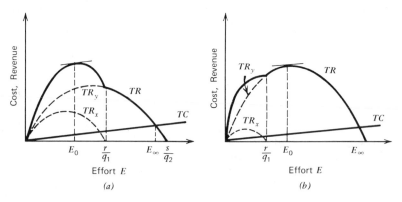

Figure 9.2 Sustained revenue and cost curves for the two-species model: (a) x population of greater value; (b) y population of greater value.

population is dominant. In this case maximization of sustained economic rent leads to the elimination of the x population. Thus even with zero discounting the extinction of one of the two populations is an economically optimal result. This perhaps unexpected conclusion depends strongly on the assumption that separate harvesting of the two populations is not feasible. To achieve maximum economic benefits for the fishery, it is necessary to exterminate one of the populations. Of course in actual practice it may be possible to separate the harvests, although this may greatly increase costs. Thus conservation of all populations belonging to an exploited ecosystem may not always be economically desirable.

Problems of this kind affect many existing fisheries. An interesting example concerns the Pacific yellowfin tuna (*Thunnus albacares*) fishery, in which large numbers of porpoises are inadvertently captured in the same nets as the tuna. In fact boats are able to locate schools of tuna by observing concentrations of porpoises, which associate with tuna. Because of its service to tuna fishermen, the porpoise population clearly has a positive preservation value to the tuna industry. However, releasing captured porpoises before hauling in the tuna nets is apparently expensive and time-consuming. The individual tuna fisherman probably does not consider it in his own interest to "conserve" the porpoises caught in his net, because he himself is unlikely to benefit significantly from the porpoises he saves. Thus it falls to a regulatory agency—the IATTC—to adopt rules regarding the conservation of porpoises. Indeed the IATTC is now in the process of trying to establish such regulations.

Many other examples of "incidental" catches can be quoted. In some cases valuable species such as Pacific salmon, halibut, and Atlantic haddock are captured by trawlers that primarily harvest other species. More complex are those cases in which one harvested species constitutes a food source for other species. This situation is discussed in Section 9.3.

Optimal Harvest Policies

We now attempt, with limited success, to solve the optimal control problem associated with the model given by Eqs. (9.1). Our objective is to maximize the present-value integral

$$PV = \int_0^\infty e^{-\delta t}[p_1 q_1 x + p_2 q_2 y - c]E(t)\, dt, \qquad (9.9)$$

subject to Eqs. (9.1) and the additional control constraint

$$0 \le E(t) \le E_{max}. \qquad (9.10)$$

The Hamiltonian is then

$$\mathcal{H} = e^{-\delta t}[p_1q_1x + p_2q_2y - c]E + \lambda_1(t)[F(x) - q_1Ex] + \lambda_2(t)[G(y) - q_2Ey]$$
$$= \sigma(t)E + \lambda_1F(x) + \lambda_2G(y), \tag{9.11}$$

where $\lambda_1(t)$ and $\lambda_2(t)$ are the adjoint variables. The adjoint equations are

$$\frac{d\lambda_1}{dt} = -\frac{\partial\mathcal{H}}{\partial x} = -e^{-\delta t}p_1q_1E - \lambda_1[F'(x) - q_1E] \tag{9.12}$$

$$\frac{d\lambda_2}{dt} = -\frac{\partial\mathcal{H}}{\partial y} = -e^{-\delta t}p_2q_2E - \lambda_2[G'(y) - q_2E]. \tag{9.13}$$

First we consider an optimal equilibrium solution, so that

$$E = \frac{F(x)}{q_1x} = \frac{G(y)}{q_2y}. \tag{9.14}$$

Hence Eqs. (9.12) and (9.13) become

$$\frac{d\lambda_1}{dt} - \gamma_1\lambda_1 = -p_1q_1Ee^{-\delta t}$$

$$\frac{d\lambda_2}{dt} - \gamma_2\lambda_2 = -p_2q_2Ee^{-\delta t},$$

where

$$-\gamma_1 = F'(x) - \frac{F(x)}{x} = \frac{rx}{K} \quad \text{and} \quad -\gamma_2 = \frac{sy}{L}.$$

These equations are easily solved:

$$\left.\begin{array}{l} e^{\delta t}\lambda_1(t) = \dfrac{p_1q_1E}{\gamma_1 + \delta} = \text{constant} \\[4mm] e^{\delta t}\lambda_2(t) = \dfrac{p_2q_2E}{\gamma_2 + \delta} = \text{constant} \end{array}\right\} \tag{9.15}$$

Thus the shadow prices $e^{\delta t}\lambda_i(t)$ of the two populations remain constant in equilibrium. [This involves a tacit assumption that these shadow prices remain bounded as $t \to \infty$, a condition that is often referred to as the *transversality condition at* ∞. No mathematical justification for this assumption appears to be known (see Arrow and Kurz, 1970, p. 46; and Halkin, 1974).]

We also know that the Hamiltonian given in Eq. (9.11) must be maximized for $E \in [0, E_{\max}]$. Assuming that optimal equilibrium does not occur either at $E = 0$ or $E = E_{\max}$, we must therefore have singular control:

$$\frac{\partial \mathcal{H}}{\partial E} = e^{-\delta t}(p_1 q_1 x + p_2 q_2 y - c) - \lambda_1 q_1 x - \lambda_2 q_2 y = 0.$$

From Eqs. (9.15) we then use Eq. (9.14) again to obtain

$$p_1 q_1 \left[x - \frac{F(x)}{\gamma_1 + \delta} \right] + p_2 q_2 \left[y - \frac{G(y)}{\gamma_2 + \delta} \right] = c. \qquad (9.16)$$

Taken with Eq. (9.14), this determines the optimal equilibrium populations $x = x_\delta$, $y = y_\delta$.

As expected, the limiting case $\delta = +\infty$ corresponds to the dissipation of economic rent

$$p_1 q_1 x_\infty + p_2 q_2 y_\infty = c.$$

The reader may also verify that the case $\delta = 0$ corresponds to the maximization of sustainable rent. In addition both x_δ and y_δ decrease with increasing δ toward x_∞ and y_∞, respectively. Thus if $x_\infty = 0$, then we also have $x_\delta = 0$ for sufficiently large finite δ. Up to this point these are straightforward generalizations of the single-species model in Chapter 2.

Our problem therefore possesses an equilibrium solution that satisfies the necessary conditions of the maximum principle. However, economic interpretation of this solution does not seem obvious. Moreover it appears to be extremely difficult to determine the optimal approach path, which must consist of some combination of bang-bang controls and nonequilibrium singular controls. (Unlike the one-dimensional model, singular solutions here need not be equilibrium solutions.) We make no further attempt to obtain the complete solution to this problem. [The difficulty in this problem seems to be associated with the noncontrollability of the system given by Eqs. (9.1) in the sense that $\dot{x}$ and $\dot{y}$ cannot be controlled independently. Harvesting therefore produces complex long-term effects on the dynamics of the system that influence the optimal harvest policy. This noncontrollability is somewhat similar to the case of nonselective gear in the multicohort fishery appearing in Section 8.6.]

9.2 HARVESTING COMPETING POPULATIONS

In this section we examine Gause's model of interspecific competition, based on the equations

$$\left. \begin{array}{l} \dfrac{dx}{dt} = F(x, y) = rx\left(1 - \dfrac{x}{K}\right) - \alpha xy \\[2mm] \dfrac{dy}{dt} = G(x, y) = sy\left(1 - \dfrac{y}{L}\right) - \beta xy \end{array} \right\}. \qquad (9.17)$$

The reader should review briefly the discussion of this system in Section 6.6. Recall that depending on the relative positions of the isoclines $\dot{x} = 0$ and $\dot{y} = 0$, there are various possibilities regarding equilibria and their stability (see Figure 6.17).

It is important not to interpret the Gause model as a definitive description of a natural system of competing populations. In certain cases, of interest in this section, the model predicts the complete exclusion of either population x or population y. In the natural environment, however, populations are distributed over space, and space is strongly inhomogeneous. A population that is completely out-competed by another population may find various refuges where it can continue to survive, at least in small numbers. [Gause (1935) demonstrated this possibility in his famous flour-beetle experiments, in which the introduction of small lengths of glass tubing into the flour permitted one beetle species to survive, rather than be exterminated by a competitive second species.]

Thus even where the model given by Eqs. (9.17) predicts an equilibrium of the form $x = x_0 > 0$, $y = 0$, we admit the existence of a "refugee" y population that, if x were suddenly eliminated (by harvesting, for example), would be in a position to grow according to the second of Eqs. (9.17).

Now suppose that the isoclines of Eqs. (9.17) are as shown in Figure 9.3a. The only stable equilibrium is at $(K, 0)$, so that only a refugee y population exists. Now let the x population be subject to harvesting, so that Eqs. (9.17) become

$$\left. \begin{array}{l} \dfrac{dx}{dt} = F(x, y) - qEx \\[2mm] \dfrac{dy}{dt} = G(x, y) \end{array} \right\} . \qquad (9.18)$$

The effort E is considered to be a parameter.

Figures 9.3b and c show the outcome of increasingly large effort levels. The x isocline

$$y = \frac{r}{\alpha}\left(1 - \frac{x}{K}\right) - \frac{q}{\alpha} E$$

moves downward, parallel to itself, as the value of the parameter E increases. For small values of E a single equilibrium exists at $(x_E, 0)$, where

$$x_E = K\left(1 - \frac{qE}{r}\right). \qquad (9.19)$$

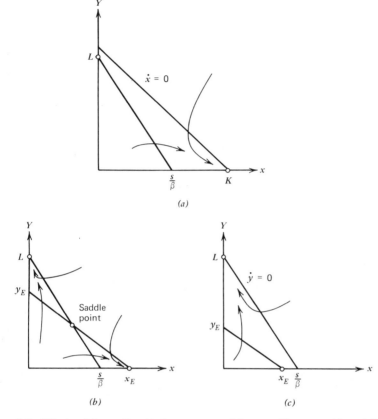

Figure 9.3 Effects of harvesting in the presence of interspecific competition: (a) zero fishing effort $E = 0$; (b) positive E; (c) large E.

A bifurcation occurs at $E = E_1$, for which the two isoclines intersect at (0, L). Larger values of E lead to the situation depicted in Figure 9.3b, where two stable equilibria at $(x_E, 0)$ and $(0, L)$ are separated by a saddle-point equilibrium. Another bifurcation occurs at $E = E_2$, where $x_{E_2} = s/\beta$. For $E > E_2$, the x population is eliminated (not directly by harvesting, but by the competitive interaction). Figure 9.4a shows the equilibrium population x as a function of effort level E. The corresponding yield-effort curve $Y = qEx_E$ appears in Figure 9.4b or c, depending on the relationship between E_2 and the value $E = r/2q$. In the case in which $E_2 > r/2q$, MSY occurs at the peak of the yield-effort parabola and corresponds to a stable equilibrium for the system given by Eqs. (9.18). However, if $E_2 < r/2q$, then MSY occurs at the bifurcation point E_2 itself and is therefore unstable.

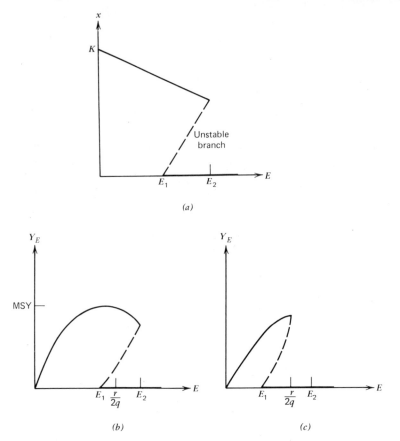

Figure 9.4 Effects of harvesting in the presence of interspecific competition: (a) equilibrium x population; (b) sustainable yield $E_{MSY} < E_2$; (c) sustainable yield $E_{MSY} > E_2$.

The practical implications of the latter case are clear. Suppose that the existence of a competitive interaction is not suspected during the developmental stages of the fishery for the x population. Using the Schaefer model biologists predict MSY to occur at the level $E = r/2q$. Actually the fishery collapses before effort increases to this level. This is particularly surprising because the yield-effort curve is still rising when the collapse occurs. There may be a tendency to attribute the collapse to changes in environmental conditions, rather than to purely biological mechanisms.

The Pacific Sardine

Biologists offer the preceding model as an explanation for the collapse of the Pacific sardine (*Sardinops caerula*) fishery in the late 1940s and early 1950s (see Figure 9.5). The sardines were replaced by an anchovy population (*Engraulis mordax*), which now appears to be dominant, and is inhibiting the recovery of the sardines. Under natural conditions the system may possess a saddle-point equilibrium similar to the one in Figure 9.3*b*, so that either sardines or anchovies are capable of dominating the system. In this case changing environmental conditions could cause dominance to shift from one species to another. Evidence from sea-bed core samples suggests that such a shift has a cycle of several hundred years. [See Culley (1971) for a description of the California sardine fishery and its political aspects. As usual our biological model is a severely oversimplified description of the collapse of the Pacific sardine fishery. In practice heavy fishing leads to the removal of most of the older cohorts, rendering the fishery especially vulnerable to one or two recruitment failures. Murphy (1967) suggests that two consecutive recruitment failures were indeed the direct cause of the Pacific sardine collapse.]

Effort as a Dynamic Variable

We now extend the model given by Eqs. (9.18), assuming that in the open-access fishery effort itself is a dynamic variable that satisfies

$$\frac{dE}{dt} = kE(x - x_\infty), \qquad (9.20)$$

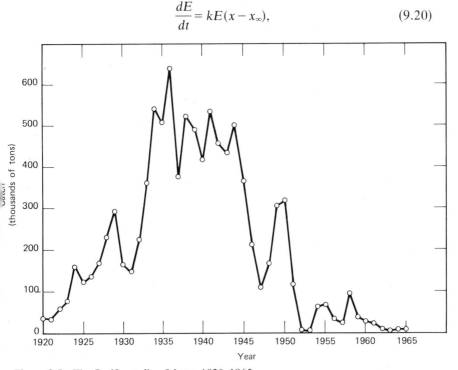

Figure 9.5 The Pacific sardine fishery, 1920–1965.

where $x_\infty = c/p$ is the zero-rent population level. Thus we obtain a three-dimensional dynamical system that is correspondingly more complex than the two-dimensional system in Section 6.5. The behavior of the model can be studied by means of computer simulations, but an alternative approach is to assume that the E reaction occurs much more *slowly* than the x–y reaction. [This is related to the *fast–slow dynamics hypothesis* used by Zeeman (1974).]

We therefore imagine that the system given by Eqs. (9.18) achieves a "moving equilibrium" that is continually adjusted according to the "slower" Eq. (9.20). We then construct a pseudo x isocline in the x–E plane by eliminating y from the equations $\dot{x} = \dot{y} = 0$. This is the curve shown in Figure 9.4a. The E isocline is simply the line $x = x_\infty$. There are two cases to consider, depending on whether the equilibrium point is on the stable or the unstable branch of the x isocline (cf. Figure 6.16). In the case shown in Figure 9.6a the equilibrium (x_∞, E_∞) is stable, whereas in the case shown in Figure 9.6b the unstable equilibrium at (x_∞, E_∞) gives rise to a pseudo limit-cycle behavior. [Figure 9.6b is only an approximation of the x–E behavior of the three-dimensional system given by Eqs. (9.18) and (9.20); this system does not necessarily undergo precise limit-cycle oscillations.]

We do not pursue the question of optimal harvesting for the present model, because the noncontrollability we discuss in the preceding section arises again owing to the assumption that the y population cannot be harvested. (Noncontrollability is a real problem in the California sardine fishery. The California state legislature has prohibited the harvesting of anchovies as a result of lobbying by sport fishermen who believe that the

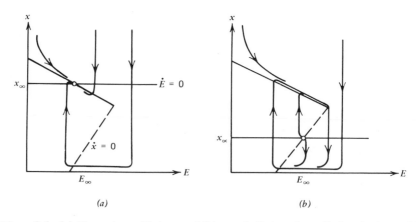

(a)   (b)

Figure 9.6 (a) Bionomic equilibrium, and (b) pseudo limit-cycle oscillations in the model of interspecific competition.

anchovy is an essential food source for sport-fish species. It is conceivable that the development of an anchovy fishery could lead to the recovery of the sardine population.) Controllable models are examined in Section 9.3.

In the foregoing discussion we assume that the harvested system [Eqs. (9.18)] never possesses a stable equilibrium in which both x and y are positive. The opposite case in which competitive coexistence is possible for certain levels of effort E is easily dealt with (see Exercise 2). The results can also be extended to more general forms of the growth functions $F(x, y)$, $G(x, y)$ (see Yodzis, 1975).

9.3   SELECTIVE HARVESTING

In this section we discuss the following model, in which both populations of a two variable system can be harvested independently:

$$\left.\begin{aligned}\frac{dx}{dt} &= F(x, y) - h_1(t) \\ \frac{dy}{dt} &= G(x, y) - h_2(t)\end{aligned}\right\}. \tag{9.21}$$

For simplicity we continue to adopt the specific functional forms

$$\begin{aligned}F(x, y) &= rx\left(1 - \frac{x}{K}\right) + \alpha xy, \\ G(x, y) &= sy\left(1 - \frac{y}{L}\right) + \beta xy.\end{aligned} \tag{9.22}$$

The case $\alpha < 0$, $\beta < 0$ is the Gause model we use in Section 9.2 [see Eqs. (9.17)]. The alternative case $\alpha < 0$, $\beta > 0$ gives rise to a predator–prey model in which the predator y feeds on the prey x (see Larkin, 1966). (The case of mutualism, or symbiosis $\alpha > 0$, $\beta > 0$ is not examined here.) It is important to note that the predator–prey model given by Eqs. (9.22) is structurally stable, unlike the Lotka–Volterra predator–prey model (see Section 6.4). A simple isocline sketch shows that when $\alpha < 0$, $\beta > 0$ the positive first-quadrant equilibrium if it exists is a stable node or focus, and du Lac's test confirms that there are no limit cycles (see Exercise 5, Chapter 6).

We now consider the optimal harvesting problem with the objective functional

$$J\{h_1, h_2\} = \int_0^\infty e^{-\delta t}\{[p_1 - c_1(x)]h_1(t) + [p_2 - c_2(y)]h_2(t)\}\, dt. \tag{9.23}$$

This is to be maximized subject to the state equations [Eqs. (9.21)] as well as the usual constraints, including

$$0 \leq h_i(t) \leq h_i^{\max}, \quad i = 1, 2. \tag{9.24}$$

Neglecting the abnormal case the Hamiltonian is

$$\mathcal{H} = e^{-\delta t}\{[p_1 - c_1(x)]h_1 + [p_2 - c_2(y)]h_2\} + \lambda_1[F(x, y) - h_1] + \lambda_2[G(x, y) - h_2]$$

$$= \{e^{-\delta t}[p_1 - c_1(x)] - \lambda_1\}h_1 + \{e^{-\delta t}[p_2 - c_2(y)] - \lambda_2\}h_2$$

$$+ \lambda_1 F(x, y) + \lambda_2 G(x, y).$$

First we consider the case of "doubly singular" control, where the coefficients of both h_1 and h_2 vanish identically:

$$\lambda_1(t) \equiv e^{-\delta t}\{p_1 - c_1[x(t)]\}$$
$$\lambda_2(t) \equiv e^{-\delta t}\{p_2 - c_2[y(t)]\}. \tag{9.25}$$

The adjoint equations are

$$\frac{d\lambda_1}{dt} = -\frac{\partial \mathcal{H}}{\partial x} = e^{-\delta t}c_1'(x)h_1 - \lambda_1 F_x - \lambda_2 G_x$$

$$\frac{d\lambda_2}{dt} = -\frac{\partial \mathcal{H}}{\partial y} = e^{-\delta t}c_2'(y)h_2 - \lambda_1 F_y - \lambda_2 G_y. \tag{9.26}$$

Differentiating Eqs. (9.25), equating with Eqs. (9.26), and simplifying, we obtain

$$[p_1 - c_1(x)]F_x + [p_2 - c_2(y)]G_x - c_1'(x)F(x, y) = \delta[p_1 - c_1(x)]$$
$$[p_1 - c_1(x)]F_y + [p_2 - c_2(y)]G_y - c_2'(y)G(x, y) = \delta[p_2 - c_2(y)]. \tag{9.27}$$

Thus the case of doubly singular control in the two-species problem corresponds to the case of singular control in the one-species model [see Eq. (2.16)]. In particular Eqs. (9.27) yield an optimal equilibrium solution $x = x^*$, $y = y^*$. The economic interpretation of these equations is also the same as before; the additional terms $[p_2 - c_2(y)]G_x$ and $[p_1 - c_1(x)]F_y$ reflect the addition to the marginal value product of the y (or x) population afforded by means of an increase in x (or y). These cross-dependencies are clearly an essential feature of the two-species coupled dynamical system.

Introducing the expression

$$R(x, y) = [p_1 - c_1(x)]F(x, y) + [p_2 - c_2(y)]G(x, y) \tag{9.28}$$

for the sustainable economic rent, we see that Eqs. (9.27) can be rewritten

$$\frac{\partial R}{\partial x} = \delta[p_1 - c_1(x)]$$

$$\frac{\partial R}{\partial y} = \delta[p_2 - c_2(y)].$$

(9.29)

When $\delta = 0$, Eqs. (9.29) reduce to the condition for maximum sustained rent

$$\frac{\partial R}{\partial x} = \frac{\partial R}{\partial y} = 0.$$

More generally we define

$$z(x) = \int_{x_0}^{x} [p_1 - c_1(\xi)] \, d\xi$$

$$w(y) = \int_{y_0}^{y} [p_2 - c_2(\eta)] \, d\eta,$$

(9.30)

where $x_0 = x(0)$ and $y_0 = y(0)$. Also let

$$R_\delta(x, y) = R(x, y) - \delta\{z(x) + w(y)\}.$$ (9.31)

Then we see that Eqs. (9.29) represent a necessary condition for maximization of the function $R_\delta(x, y)$. [As written, the condition given by Eqs. (9.29) neglects the state-variable constraints $x \geq 0$, $y \geq 0$. One of these constraints may well be binding for the optimal solution. Thus Eqs. (9.29) should actually be replaced by the standard Kuhn–Tucker conditions for constrained maximization of the function $R_\delta(x, y)$.] The fact is useful in the discussion of optimal approach paths.

Optimal Approach Paths*

Replacing h_1 by $F(x, y) - \dot{x}$ and h_2 by $G(x, y) - \dot{y}$, we can write the objective functional in Eq. (9.23)

$$J = \int_0^\infty e^{-\delta t} \{[p_1 - c_1(x)][F(x, y) - \dot{x}] + [p_2 - c_2(y)][G(x, y) - \dot{y}]\} \, dt.$$

Given Eqs. (9.30) we have

$$\dot{z} = [p_1 - c(x)]\dot{x} \quad \text{and} \quad \dot{w} = [p_2 - c_2(y)]\dot{y}.$$

* This section is related to some unpublished work of R. M. Solow and M. Spence.

Integrating the above expression for J by parts we obtain

$$J = \int_0^\infty e^{-\delta t}\{[p_1 - c_1(x)]F(x, y) - \delta z + [p_2 - c_2(y)]G(x, y) - \delta w\}\, dt$$

$$= \int_0^\infty e^{-\delta t} R_\delta[x(t), y(t)]\, dt. \qquad (9.32)$$

With this transformation it is apparent why Eqs. (9.29) determine the optimal equilibrium solution (x^*, y^*). Once we reach (x^*, y^*) we obviously want to stay there indefinitely, since the integrand given in Eq. (9.32) is as large as possible at that point. The only question left is to determine the best way of moving from some initial point (x_0, y_0) to point (x^*, y^*). Graphically this problem is illustrated in Figure 9.7, which shows the *objective surface* $R = R_\delta(x, y)$. Our problem is to find functions $x = x(t)$, $y = y(t)$ that satisfy the constraints on $\dot{x}$, $\dot{y}$ induced by Eqs. (9.24), so that the corresponding path on the objective surface is as high as possible, for all time t.

The problem would be solved if we could determine an approach path $[x^*(t), y^*(t)]$ with the property that

$$R_\delta[x^*(t), y^*(t)] \ge R_\delta[x(t), y(t)] \qquad (9.33)$$

for all feasible paths $(x(t), y(t))$. (It is easy to verify that this is precisely the condition satisfied by the most-rapid approach paths of the one-dimensional problem.) Unfortunately it is by no means clear that such a path exists, because of the complex form of the constraints given in Eqs.

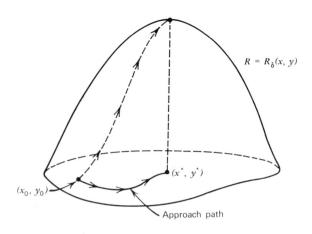

Figure 9.7 The objective surface $R = R_\delta(x, y)$.

(9.24), which can now be written

$$F(x, y) - h_1^{max} \le \dot{x} \le F(x, y)$$
$$G(x, y) - h_2^{max} \le \dot{y} \le G(x, y). \tag{9.34}$$

To appreciate the difficulty, we first solve a simpler problem in which the constraints are

$$A_1 \le \dot{x} \le A_2$$
$$B_1 \le \dot{y} \le B_2, \tag{9.35}$$

where A_i and B_i are constants.

Our aim is to choose at each point (x, y) a velocity vector $(\dot{x}, \dot{y})$ that satisfies these constraints, so that the function $R(x, y) = R_\delta(x, y)$ increases as rapidly as possible. [For simplicity we drop the subscript δ; confusion with the previous function $R(x, y)$ should not arise.] We have

$$\frac{dR}{dt} = R_x \dot{x} + R_y \dot{y}.$$

To maximize this, clearly we must have

$$\dot{x} = \begin{cases} A_1 & \text{if } R_x < 0 \\ A_2 & \text{if } R_x > 0 \end{cases}$$

$$\dot{y} = \begin{cases} B_1 & \text{if } R_y < 0 \\ B_2 & \text{if } R_y > 0. \end{cases}$$

Consider the curves $R_x = 0$ and $R_y = 0$ in the x–y plane; they intersect at the optimal equilibrium point (x^*, y^*) (see Figure 9.8). These curves may

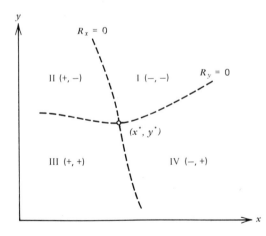

Figure 9.8 Because (x^*, y^*) is a maximum of $R(x, y)$, the signs of R_x and R_y are as shown.

be thought of as ridge-lines on the surface shown in Figure 9.7. Clearly the optimal control $(\dot{x}, \dot{y})$ must be given by

$$(\dot{x}, \dot{y}) = \begin{cases} (A_1, B_1) & \text{in region I} \\ (A_2, B_1) & \text{in region II} \\ (A_2, B_2) & \text{in region III} \\ (A_1, B_2) & \text{in region IV.} \end{cases}$$

The resulting optimal approach paths are shown in Figure 9.9. The rule is simple: follow the straight-line approach path from a given initial point (x_0, y_0) until this path meets a ridge; then follow the ridge up to the point (x^*, y^*).

In fact we can employ the maximum principle to justify this rather intuitive argument. To maximize Eq. (9.32) subject to the constraints given by Eqs. (9.35), we introduce the new control variables

$$\dot{x} = u, \quad \dot{y} = v.$$

The corresponding Hamiltonian is

$$\mathcal{H} = e^{-\delta t} R(x, y) + \lambda_1 u + \lambda_2 v.$$

If, for example, u is not a bang-bang control ($u \neq A_1$ or A_2), then it must be singular, so that $\lambda_1 = 0$. By the adjoint equation, $\lambda = -e^{-\delta t} R_x = 0$ also.

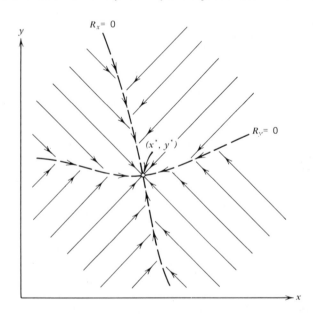

Figure 9.9 Optimal approach paths for the simplified constraints given by Eqs. (9.34).

Thus the ridge-line $R_x = 0$ corresponds to a *semisingular control* in which u is singular but v is a bang-bang control. Similarly $R_y = 0$ is the semisingular trajectory when v is singular. Except on these ridge-lines, both controls must be bang-bang controls. Therefore our solution is obviously correct.

Now what happens if the constraints A_i and B_i are not fixed, but depend on position, as in Eqs. (9.34)? The preceding application of the maximum principle is not valid, because in the maximum principle (Section 4.4) *the control set is not permitted to depend on the state variables.* In fact it is easy to show that the ridge-lines are *not* optimal for the constraints given in Eqs. (9.24) by applying the maximum principle to the original problem. This is also intuitively reasonable, because it may be possible to increase the effectiveness of the velocity vector $(\dot{x}, \dot{y})$ by moving off the ridge-line, but only if constraints A_i and B_i vary with the position of (x, y).

The exact solution of the original problem is not known. It is possible to determine unique semisingular optimal trajectories passing through (x^*, y^*), and it may be conjectured that these trajectories are the desired substitutes for the ridge-lines. The author has not succeeded in proving this, however.

Practical Approach Paths

Again consider the objective surface shown in Figure 9.7. Clearly given any initial point (x_0, y_0), it is better to move up the surface toward (x^*, y^*) than to remain at (x_0, y_0). [For simplicity we are assuming that the objective surface has a unique local maximum (x^*, y^*). Otherwise we face the usual problem of multiple equilibria in which the optimal equilibrium may depend on the initial position.] It is not important that we do not know the "ideal" approach path corresponding to the particular set of assumed constraints (that are more or less arbitrary in themselves). Even the simplistic rule

$$h_1 = \begin{cases} h_{max} & \text{if } x > x^* \\ 0 & \text{if } x < x^*, \end{cases}$$

similar for h_2, although suboptimal is surely a practically acceptable approach. The conclusion: if you know where you want to be and if many feasible approach paths are available, do not be concerned if the ideal path is not apparent.

Examples

For all practical purposes, then, the optimal harvesting problem for our selective harvesting model reduces to the rather simple question of maximizing the function $R_\delta(x, y)$ over $x \geq 0$, $y \geq 0$. Of course in general this can be readily accomplished on a computer by using standard maximization routines. To illustrate, we consider some special cases.

We begin with the old stand-by MSY, where "yield" now refers to a weighted sum

$$Y = p_1 h_1 + p_2 h_2 = p_1 F(x, y) + p_2 G(x, y).$$

Here p_1 and p_2 can be any desired non-negative numbers that are not necessarily market prices. This is the special case $\delta = c_1 = c_2 = 0$ of our general optimization model. With F and G given by Eqs. (9.22) we therefore have

$$Y = R(x, y) = p_1 f(x) + p_2 g(y) + (p_1 \alpha + p_2 \beta) xy, \qquad (9.36)$$

where $f(x)$ and $g(y)$ are logistic functions. Figure 9.10a shows the function $p_1 f(x) + p_2 g(y)$, to which must be added the term $(p_1 \alpha + p_2 \beta) xy$ [see Figure 9.10b].

Given $\alpha < 0$, in the competition model $\beta < 0$, whereas in predator–prey model $\beta > 0$. Thus the coefficient $(p_1 \alpha + p_2 \beta)$ is negative for the competition model, but may be positive for the predator–prey model. If this coefficient is positive, it is clear from Figure 9.10 that at the maximizing point (x^*, y^*), both x^* and $y^* > 0$. But when the coefficient is negative, either x^* or y^* may vanish. An obvious example is the case $p_2 = 0$. Clearly $Y = p_1[f(x) - \alpha xy]$ is maximized for $y^* = 0$ (no competitors or no predators, as the case may be), and $x^* = K/2$.

The general case

$$R_\delta(x, y) = [p_1 - c_1(x)]F(x, y) + [p_2 - c_2(y)]G(x, y) - \delta[z(x) + w(y)]$$

is difficult to analyze completely. For example, we could have $y^* < y_\infty$,

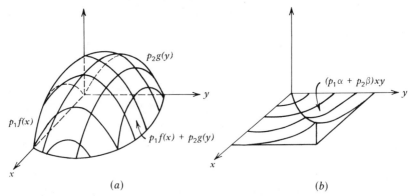

Figure 9.10 Components of the objective surface given by Eq. (9.36).

where $p_2 = c_2(y_\infty)$; if the x population is particularly valuable, it may be worthwhile to remove competitors or predators, even at a loss. (This should not be construed as blanket approval of predator-control programs, especially those based on the objective of complete extermination—an objective that is almost never successfully achieved. However, properly conceived control programs are obviously worthwhile.)

9.4 A DIFFUSION MODEL: THE INSHORE–OFFSHORE FISHERY

In this section we consider a rather stylized fishery model in which the spatial distribution and movement of the fish are taken into account. We imagine a straight shoreline, and let x denote the distance from shore. The density $u = u(x, t)$ of a fish population at time t is assumed to depend on the distance x. We adopt the partial differential equation

$$\frac{\partial u}{\partial t} = \sigma^2 \frac{\partial^2 u}{\partial x^2} + F(x, u) \quad (0 \le x \le S) \tag{9.37}$$

as a model of the growth and diffusion of the population where $F(x, u)$ represents the natural growth rate of the population at distance x from shore, and S denotes the outer limit of the population's habitat. We suppose that

$$F(x, u) > 0 \quad \text{for } 0 < u < K(x);$$

that is, $K(x)$ is the carrying capacity at location x.

The term $\sigma^2 u_{xx}$ in Eq. (9.37) is the *diffusion term*, and σ^2 is the *diffusion coefficient*. The physical interpretation of the diffusion term is straightforward. Consider a fixed location x. The rate of diffusion of fish is assumed to be proportional to the gradient of the density. Thus the rate of diffusion of fish toward x from the population further offshore is approximately proportional to

$$\frac{u(x + h) - u(x)}{h}.$$

Similarly the rate of diffusion from x toward the shore is proportional to

$$\frac{u(x) - u(x - h)}{h}.$$

Thus the net rate of increase of u at x is given by

$$\frac{1}{h}\left\{\frac{u(x + h) - u(x)}{h} - \frac{u(x) - u(x - h)}{h}\right\} = \frac{u(x + h) - 2u(x) + u(x - h)}{h^2}.$$

As $h \to 0$, this difference approaches u_{xx}, as in Eq. (9.37).

Equation (9.37) is a nonlinear partial differential equation of parabolic type. The discussion of this equation here is limited to equilibrium solutions. As an example, consider the case in which

$$F(x,\bar{u}) = \begin{cases} ru(1 - u/k) & \text{for } 0 < x < a \\ -su & \text{for } a < x < \infty. \end{cases} \tag{9.38}$$

Thus the inshore area $0 < x < a$ can be considered the breeding area of the population, which also diffuses offshore where it experiences a fixed mortality rate s. The corresponding equilibrium solution of Eq. (9.37) is shown in Figure 9.11. The dashed curve in this figure represents the equilibrium population density for no diffusion ($\gamma = 0$); the solid curve indicates the result of diffusion. For $x > a$, the solution is easily seen to be

$$u(x) = ce^{-\sqrt{s}x/\sigma}, \quad c = \text{constant}.$$

In other words the population density decreases exponentially as $x \to \infty$ at a rate that is dependent on the ratio $\sqrt{s}/\sigma$ involving the mortality rate and the diffusion constant. For $x < a$, we have $u_{xx} = \sigma^{-2}F(x, u) > 0$, so that the curve $u(x)$ is concave downward as shown in Figure 9.11. [The solution $u(x)$ and its derivative are assumed to be continuous at $x = a$. By prescribing $u_x(0) = 0$ (i.e., no flux at the shoreline), we specify a unique equilibrium solution $u(x)$.]

The Open-access Fishery

Next let $E = E(x)$ denote fishing effort, which we assume to depend on the distance x from shore. Then Eq. (9.37) is replaced by

$$\frac{\partial u}{\partial t} = \sigma^2 \frac{\partial^2 u}{\partial x^2} + F(x, u) - E(x)u. \tag{9.39}$$

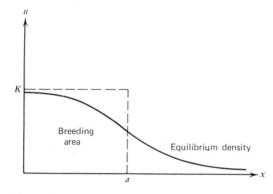

Figure 9.11 A simple diffusion model.

(The catchability coefficient q can be set equal to 1 without loss of generality.) We suppose that the unit effort cost $c = c(x)$ also depends on x, and in particular that $c(x)$ is an *increasing* (or at least a nondecreasing) function of x. What can be said about the equilibrium distribution of effort in the open-access fishery?

If price p is constant we have the expression

$$R = [pu(x) - c(x)]E(x) \tag{9.40}$$

for the rent earned by vessels fishing at a distance x from shore. (More precisely R denotes the *rent density*, or rent per unit distance.) Bionomic equilibrium is therefore characterized by the conditions

$$R \equiv 0, \quad \frac{\partial u}{\partial t} \equiv 0. \tag{9.41}$$

From Eq. (9.40) we must then have for all x

$$\textit{either} \quad u(x) = \frac{c(x)}{p} \quad \textit{or} \quad E(x) = 0. \tag{9.42}$$

First consider the case of no diffusion $\sigma = 0$. The fishery can then be imagined as a continuous distribution of independent populations, each having its own growth rate $F(x, u)$. Clearly the bionomic equilibrium in this case is given by

$$u(x) = \min \left[\frac{c(x)}{p}, \, K(x) \right]. \tag{9.43}$$

This solution is shown in Figure 9.12. The fishery is viable out to a certain distance L from shore, at which fishing costs become too high to allow profitability. The inshore fishery is heavily exploited, with effort $E(x) = F[x, u(x)]/u(x)$ falling off steadily as x approaches L. If the cost–price ratio $c(x)/p$ subsequently decreases, the fishery becomes more·heavily exploited at each location, and the outer limit L moves further offshore. Many marine fisheries actually follow this pattern of development.

What is the effect of diffusion on this simple model? The conditions of bionomic equilibrium are now given by Eqs. (9.42), and

$$\sigma^2 u_{xx} + F(x, u) - E(x)u = 0. \tag{9.44}$$

At this stage a further simplification of the model may clarify the exposition. For the moment assume that $c(x) = mx + b$ is linear ($m, b > 0$) and that $F(x, u) = F(u)$ is independent of x (i.e., $K(x) = K = $ constant). We then note that the previous solution [Eq. (9.43)] satisfies Eq. (9.44) except at the single point $x = L$, because $u_{xx} \equiv 0$ except at $x = L$. This exceptional point is crucial, however, because u_{xx} does not exist at $x = L$. [In fact $u_{xx}(L)$ can

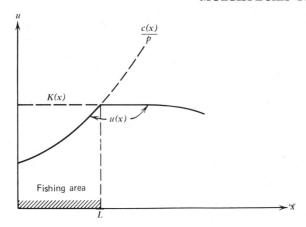

Figure 9.12   Bionomic equlibrium in the inshore–offshore fishery ($\delta = 0$).

be identified with $-m\delta_L(x)$, where $\delta_L(x)$ denotes the Dirac delta function concentrated at $x = L$. Such a solution is physically impossible, because the corresponding effort $E(x)$ would have to equal $-\infty$ at $x = L$, whereas clearly $E(x)$ must be ≥ 0.]

The correct solution is shown in Figure 9.13. We no longer have $u(x) = c(x)/p$ for all $x < L$, but only for $x < x_1$ where $x_1 < L$. Diffusion affects the equilibrium position on the interval $[x_1, S]$. The equilibrium solution $u(x)$ on this interval is determined from Eq. (9.44) with $E(x) = 0$, or

$$\sigma^2 u_{xx} + F(x, u) = 0, \tag{9.45}$$

and from the boundary conditions

$$\left.\begin{array}{l} pu(x_1) = c(x_1) \\ pu_x(x_1) = c'(x_1) \\ u_x(S) = 0 \end{array}\right\}. \tag{9.46}$$

Equations (9.45) and (9.46) constitute a free *boundary-value problem*, with undetermined end point x_1. Geometrically we are simply required to fit a solution of Eq. (9.45) into the interval $[x_1, S]$ that joins smoothly with the curve $u = c(x)/p$ and that has zero slope at $x = S$.

The interpretation of this solution is straightforward. Because of the depletion of the stock on the inshore fishing ground $(0 \leq x \leq x_1)$, the density further offshore $(x_1 \leq x \leq S)$ is reduced as a result of diffusion toward the inshore area. In equilibrium no fishing occurs beyond x_1, because diffusion reduces the density below the zero-profit level $c(x)/p$. [It is easy to see that the equilibrium effort level $E(x)$ in this model has a

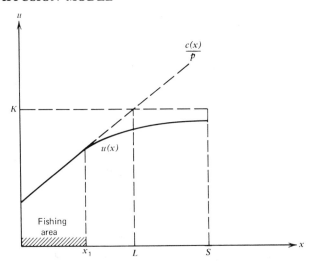

Figure 9.13 Bionomic equilibrium in the inshore–offshore fishery ($\sigma > 0$).

discontinuity at $x = x_1$; this is in contrast to the nondiffusion case, where $E(x) \to 0$ continuously as $x \to L$. In other words the limits of the fishing area under bionomic equilibrium would be clearly discernible (in terms of effort distribution) in the diffusion case, but not in the nondiffusion case.]

It should be clear that similar results can be derived for a more general model. The model can also be applied to other situations. For example, suppose that the bionomic equilibrium shown in Figure 9.13 has become established, but that subsequently a distant-water fleet begins to exploit the offshore fishery. Diffusion to the inshore fishery is then reduced, possibly to a level at which the inshore fishery becomes completely unprofitable. Canada's maritime fisheries, for example, appear to have become severly impoverished as the result of offshore trawling operations.

Optimization

First we treat the optimal yield problem under the assumption of zero diffusion. In this case, because all locations are independent, our basic rule [Eq. (2.16)] applies (with suitable changes in notation):

$$\frac{\partial R_s}{\partial u} = \delta \left[p - \frac{c(x)}{u} \right]. \tag{9.47}$$

Here $R_s = R_s(x, u)$ denotes sustainable economic rent:

$$R_s(x, u) = \left[p - \frac{c(x)}{u} \right] F(x, u).$$

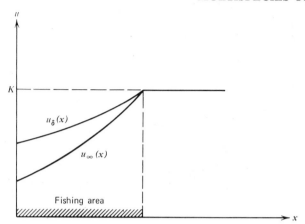

Figure 9.14 Bionomic open-access equilibrium density $u_\infty(x)$ and optimal equilibrium density $u_\delta(x)$.

The resulting optimal population level $u_\delta = u_\delta(x)$ is shown in Figure 9.14; the figure also shows the open-access bionomic equilibrium $u_\infty(x)$. We note that $u_\delta(x) > u_\infty(x)$ for all $x < L$, and also that the fishing area is the same for open-access and for optimal exploitation.

Figure 9.14 can be interpreted in terms of Ricardian rent, similar to the original analysis of fishery economics given by H. S. Gordon (1954). The marginal fishing ground at distance $x = L$ from shore provides no economic rent—not because productivity is low [although this could be the case if $K(x)$ decreases with x], but because fishing costs are too great. Fishing grounds closer to shore can yield economic rent, but fail to do so under open-access conditions of exploitation; competitive fishermen tend to equate price with average cost $c(x)/u(x)$, thus dissipating the available rent.

What happens when diffusion is reintroduced into the model? The general control problem is then to maximize

$$J\{E\} = \int_0^\infty e^{-\delta t} \int_0^S [pu(x, t) - c(x)]E(x, t)\, dx\, dt,$$

subject to the state equation [Eq. (9.39)] and to the usual control constraints. The solution to this *distributed-parameter control problem* (Lions, 1971) is not attempted here. Instead we introduce a drastically simplified (and probably much more realistic) model, in which the fish stock consists of two geographical subpopulations (one inshore, one offshore) with diffusion between the two.

An Explicit Inshore–Offshore Model

Returning to a more customary notation, let x_1 and x_2 now denote the respective biomasses of inshore and offshore subpopulations of the same fishery. The dynamical system of the exploited fishery is modeled by

$$\frac{dx_1}{dt} = F_1(x_1) + \sigma(x_2 - x_1) - E_1 x_1$$

$$\frac{dx_2}{dt} = F_2(x_2) + \sigma(x_1 - x_2) - E_2 x_2. \tag{9.48}$$

Thus if $x_2 > x_1$, the offshore population diffuses into the inshore area at a rate that is proportional to $x_2 - x_1$. This diffusion provides the only coupling between the two populations. For simplicity the two subpopulations are presumed to be equally catchable.

Assume that the price p is the same for both populations, but that the inshore fishery supplies effort at a lower cost c_1 than the offshore fishery cost c_2. The objective functional for the combined fishery is then given by

$$J[E_1, E_2] = \int_0^\infty e^{-\delta t} \{(px_1 - c_1)E_1 + (px_2 - c_2)E_2\}\, dt. \tag{9.49}$$

This problem takes the general form of the selective two-species harvesting model in the previous section. Carrying out the same integration-by-parts procedure as before, we have

$$J = \int_0^\infty e^{-\delta t}\left\{\left(p - \frac{c_1}{x_1}\right)[F_1(x_1) + \sigma(x_2 - x_1) - \dot{x}_1]\right.$$

$$\left. + \left(p - \frac{c_2}{x_2}\right)[F_2(x_2) + \sigma(x_1 - x_2) - \dot{x}_2]\right\} dt$$

$$= \int_0^\infty e^{-\delta t}\left\{R_1(x_1) + R_2(x_2) - \delta[z_1(x_1) + z_2(x_2)] + \sigma(x_2 - x_1)\left(\frac{c_2}{x_2} - \frac{c_1}{x_1}\right)\right\} dt,$$

where

$$R_i(x_i) = \left(p - \frac{c_i}{x_i}\right)F_i(x_i)$$

represents sustainable rent *neglecting diffusion* and where

$$z_i(x_i) = \int_{x_{i\infty}}^{x_i}\left(p - \frac{c_i}{x_i}\right) dx_i.$$

Optimal equilibrium corresponds to maximization of the integrand in the preceding expression for J. After slight simplification, the first-order conditions become

$$R_1'(x_1) - \delta\left(p - \frac{c_i}{x_1}\right) = \frac{\sigma}{x_1^2 x_2}(c_2 x_1^2 - c_1 x_2^2)$$

$$R_2'(x_2) - \delta\left(p - \frac{c_2}{x_2}\right) = \frac{\sigma}{x_1 x_2^2}(c_1 x_2^2 - c_2 x_1^2).$$

$$(9.50)$$

In other words we obtain our usual rule [Eq. (2.16)] for each separate subpopulation, except for a modification involving the diffusion constant σ. What can be said about the effect of this modification?

Let us assume that the nondiffusion optimality equations

$$R_i'(x_i) - \delta\left(p - \frac{c_i}{x_i}\right) = 0, \quad i = 1, 2 \qquad (9.51)$$

possess unique solutions $x_i^* > 0$. Then let

$$Q(x_1, x_2) = c_2 x_1^2 - c_1 x_2^2.$$

In Figure 9.15a and b are shown the three lines $Q = 0$, $x_1 = x_1^*$, and $x_2 = x_2^*$; these lines divide the x_1,x_2-plane into seven regions. The various expressions in Eqs. (9.50) vanish on these lines. By examining the signs of these expressions in each of the seven regions, it is easy to see that any solution $(x_1^\dagger, x_2^\dagger)$ of Eqs. (9.50) must lie within the shaded triangle in Figure 9.15a and b. A simple continuity argument then shows that such a solution does exist.

The relation between the diffusion optima $(x_1^\dagger, x_2^\dagger)$ and the optima without diffusion (x_1^*, x_2^*) can be described as follows. First if

$$\frac{c_1}{(x_1^*)^2} < \frac{c_2}{(x_2^*)^2},$$

as in Figure 9.15a, then

$$x_1^\dagger < x_1^*$$
$$x_2^\dagger > x_2^*. \qquad (9.52)$$

In other words if the marginal harvest cost $\partial C/\partial x_1 = c_1/(x_1^*)^2$ for the nondiffusion inshore fishery is lower than the marginal cost $\partial C/\partial x_2$ for the offshore fishery, then diffusion has the effect of decreasing the optimal population level inshore and increasing this level offshore. To a certain extent it is worthwhile to allow some of the offshore population to swim inshore, where it can be captured at a lower cost. If the marginal cost situation is reversed, as in Figure 9.15b, then the effect of diffusion is also reversed.

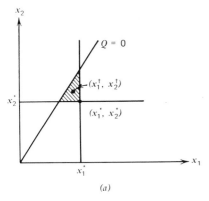
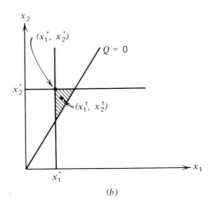

Figure 9.15 Solutions to Eqs. (9.50).

For the solutions $x_i^\dagger$ obtained from Eqs. (9.50) to be feasible, the corresponding effort levels obtained from the equilibrium equations $\dot{x}_i = 0$ must be non-negative. If this is not the case, then the calculation must be modified by setting the appropriate E_i equal to zero a priori.

9.5 SUMMARY AND CRITIQUE

In this chapter we study some typical examples of multispecies harvesting models. Most of the results we examine here are fairly straightforward. When selective harvesting of each species is not feasible, for example, those species or subpopulations whose biotechnical productivity is relatively low may be eliminated by the harvesting process. In some cases this may even be an economically optimal result, although the eliminated population in itself is valuable. Open-access exploitation, however, is much more likely to result in the elimination of highly desirable species than optimally controlled exploitation. Where many different species or subpopulations are involved, the outcome may be a long history of progressively declining productivity. As in other cases the greater the demand, the more severe will be the overexploitation.

In this chapter we also observe how mechanisms of ecological competition can result in extremely unstable harvested populations. In some cases the concept of maximum sustainable yield loses all meaning.

Another phenomenon of practical significance is the difficulty of achieving—or even delineating—optimal harvest policies for multipopulation ecosystems subject to nonselective harvesting. This problem requires further study; quite possibly there is no completely satisfactory solution. In cases where selective harvesting is feasible, on the other hand, we can see how the problem of determining optimal equilibrium population levels (x^*, y^*) is reduced to a simple calculus problem. [This solution readily extends to the case of n species (see Exercise 4).]

Finally in this chapter we study the effects of diffusion within the confines of an inshore–offshore fishery model. Under open-access conditions the inshore fishery tends to be more heavily exploited (due to lower costs) than the offshore fishery. The optimal distribution of effort between inshore and offshore fisheries can be determined in a two-population model by the same method we use in the selective harvesting problem.

From a practical viewpoint the principal criticism related to the kind of models we discuss in this chapter is the problem of "identification"; that is, the problem of estimating parameters so that the resulting model has useful predictive power. No marine ecosystem has probably been studied yet in sufficient depth to warrant the use of any sophisticated multispecies model. Nevertheless the type of general results we obtain here at least comprises preliminary research that is essential to an understanding of the exploited ecosystem. Current work in practical systems ecology seems to have little economic content; this may be the time for a more closely coordinated multidisciplinary approach.

Another shortcoming of our analysis is the failure to specify simple rules for the optimal transitional harvest policy. We argue that this may be only a minor failure for the case where selective harvesting is feasible. However, the nonselective case in particular is poorly understood. Problems of this type are surely among the most difficult facing fishery scientists today. It is at least agreed that unregulated open-access exploitation can have devastating effects on both the biological and the economic performances of multispecies fisheries. Unfortunately we still seem to be far from establishing optimal regulatory management institutions and policies, although many worthwhile steps are being taken. It seems likely that some general concept of ecosystem conservation will eventually emerge, as the benefits of maintaining healthy ecosystems become more apparent.

EXERCISES

1. Let $F(x)$ and $G(y)$ be growth curves with no depensation (i.e., $F'' < 0$, $G'' < 0$).

    (a) Show that the equilibrium curve

    $$\dot{x} = F(x) - q_1 Ex = 0, \quad \dot{y} = G(y) - q_2 Ey = 0$$

    is increasing and that it meets the y axis at the point $\bar{y}$ given by

    $$\frac{G(\bar{y})}{\bar{y}} = \frac{q_2 r}{q_1},$$

    provided that $r/q_1 < s/q_2$. Here $r = F'(0)$, and $s = G'(0)$.

(b) Now show that population x is eliminated under open-access exploitation, if $\bar{y} > c/p_2q_2$.

2. Show how the curves in Figure 9.4 are modified when the equilibrium point (see Figure 9.3b) is a node rather than a saddle point. (Recall that this depends on the relative slopes of the isoclines; see Figures 6.17a and b.) Discuss the implications of this change.

3. Equations (9.18) assume that only the x population is subject to harvesting. Discuss the case in which both populations are jointly harvested:

$$\frac{dx}{dt} = F(x, y) - q_1 Ex$$

$$\frac{dy}{dt} = G(x, y) - q_2 Ey.$$

Sketch the yield-effort curves $Y = TR$.

4. Extend the theory of Section 9.3 to the case of n species. In particular show that the problem of determining optimal equilibrium population levels $x_i^*\,(1 \le i \le n)$ can be reduced to the problem of maximizing a certain function of n variables.

BIBLIOGRAPHICAL NOTES

Multispecies fisheries models are studied by Larkin (1963, 1966), who observes that an MSY policy in a trawling fishery can lead to the ultimate extinction of some species. Several optimal control problems based on the Lotka–Volterra model of a predator–prey system are examined by Goh, Leitmann, and Vincent (1974); however, this work does not consider economically based objectives. The Hamiltonian formalism of multispecies fisheries appears in Quirk and Smith (1970). A discrete-time model for the optimal harvesting of a population in which males and females are treated separately is given by Mann (1970). The optimal equilibrium rule [Eqs. 9.29] has been extended to the case of a nonlinear objective by Silvert and Smith (1976).

SUPPLEMENTARY READING

Under each topical heading in this section, list A should be considered essential background material to be studied by anyone who is seriously interested in renewable-resource management. List B contains specialized, definitive works that the reader may wish to consult. All of the references in this section are detailed in the bibliography that follows.

Capital and Interest

A. J. Hirshleifer (1970).
B. K. J. Arrow and M. Kurz (1970); F. A. Lutz (1966).

Cost-benefit Analysis

B. R. Layard, ed. (1972); E. J. Mishan (1971); A. R. Prest and R. Turvey (1965).

Differential Equations and Dynamical Systems

B. A. A. Andronov et al. (1973); G. Birkhoff and G.-C. Rota (1969); M. Hirsch and S. Smale (1974).

Ecological Models

B. R. M. May (1973); J. Maynard Smith (1974); E. C. Pielou (1969); L. B. Slobodkin (1961); K. E. F. Watt (1968).

Economic Theory

A. P. A. Samuelson (1973); or any introductory text.
B. E. Burmeister and A. R. Dobell (1970); P. A. Samuelson (1965).

Fishery Biology

A. R. J. H. Beverton and S. J. Holt (1957); M. Graham (1952); M. B. Schaefer (1957).
B. D. H. Cushing (1968); J. A. Gulland (1974); G. V. Nikolskii (1969).

Optimization and Optimal Control

A. M. D. Intriligator (1971).

B.   L. D. Berkovitz (1974); M. D. Cannon, C. D. Cullum, and E. Polak (1970); E. B. Lee and L. Markus (1968); L. S. Pontryagin et al. (1962).

Resource Economics

A.   S. V. Ciriacy-Wantrup (1972); C. W. Clark (1973a); C. W. Clark and G. R. Munro (1975); H. S. Gordon (1954); H. Hotelling (1931); A. D. Scott (1972); R. M. Solow (1974).
B.   F. T. Christy, Jr., and A. D. Scott (1965); J. A. Crutchfield and A. Zellner (1962); G. Gregory (1972); Herfindahl, O. C., and A. V. Kneese (1974).

BIBLIOGRAPHY

Allee, W. C., *Animal Aggregations*, University of Chicago Press (Chicago) 1931.

Allen, K. R., Analysis of the stock-recruitment relation in Antarctic fin whales, in B. Parrish (Ed.), Fish Stocks and Recruitment, *Journal du Conseil International pour l'Exploration de la Mer, Rapports et Procès-Verbaux de Réunions* **164** (1973), 132–137.

Allen, R. L., and P. Basasibwaki, Properties of age-structured models for fish populations, *Journal of the Fisheries Research Board of Canada* **31** (1974), 1119–1125.

Anderson, L. G., Optimum economic yield of a fishery given a variable price of output, *Journal of the Fisheries Research Board of Canada* **30** (1973), 509–518.

Andronov, A. A. et al., *Qualitative Theory of Second-Order Dynamical Systems*, Wiley (New York, 1973).

Arrow, K. J., Optimal capital policy, the cost of capital, and myopic decision rules, *Annals of the Institute of Statistical Mathematics* **16** (1964), 21–30.

Arrow, K. J., Optimal capital policy with irreversible investment, in J. N. Wolfe (Ed.), *Value, Capital and Growth, Papers in Honour of Sir John Hicks*, Edinburgh University Press (Edinburgh, 1968), pp. 1–20.

Arrow, K. J., and M. Kurz, *Public Investment, the Rate of Return and Optimal Fiscal Policy*, Johns Hopkins Press (Baltimore, 1970).

Aström, K. J., *Introduction to Stochastic Control Theory*, Academic (New York, 1970).

Bachmura, F. T., The economics of vanishing species, *Natural Resources Journal* **11** (1971), 674–692.

Baumol W. J., On the social rate of discount, *American Economic Review* **58** (1968), 788–802.

Beddington, J. R., Age structure, sex ratio, and population density in the harvesting of natural animal populations, *Journal of Applied Ecology* **11** (1974), 915–924.

Beddington, J. R., and D. B. Taylor, Optimal age-specific harvesting of a population, *Biometrics* **29** (1973), 801–809.

Beddington, J. R., C. M. K. Watts, and W. D. C. Wright, Optimal cropping of self-reproducible natural resources, *Econometrica* **43** (1975), 789–802.

Bell, F. W., et al., The future of the world's fishery resources, National Marine Fisheries Service, Division of Economic Research, Working Paper 71-1 (1970), pp. 75–112.

Bellman, R., *Dynamic Programming*, Princeton University Press (Princeton, 1957).

338

Bellman, R., and R. Kalaba, *Dynamic Programming and Modern Control Theory*, Academic (New York, 1965).

Berkovitz, L. D., *Optimal Control Theory*, Springer-Verlag (New York, 1974).

Beverton, R. J. H., and S. J. Holt, *On the Dynamics of Exploited Fish Populations*, Ministry of Agriculture, Fisheries and Food (London) Fisheries Investigations Series 2(19) (1957).

Birkhoff, G., and G.-C. Rota, *Ordinary Differential Equations*, Blaisdell (Waltham, Mass., 1969).

Bradley, P. G., Some seasonal models of the fishing industry, in A. D. Scott (Ed.), *Economics of Fisheries Management—A Symposium*, University of British Columbia, Institute of Animal Resource Ecology (Vancouver, 1970), pp. 33–43.

Brauer, F., and D. A. Sanchez, Constant rate population harvesting: equilibrium and stability, *Theoretical Population Biology* **8** (1975), 12–30.

Breder, C. M., Jr., On the survival value of fish schools, *Zoologica* **52** (1967), 25–40.

Brock, V. E., and R. H. Riffenburgh, Fish schooling: a possible factor in reducing predation, *Journal du Conseil International pour l'Exploration de la Mer* **25** (1963), 307–317.

Brock, W. A., Some results on the uniqueness of steady states in multisector models when future utilities are discounted, *International Economic Review* **14** (1973), 535–559.

Brown, G., Jr., An optimal program for managing common property resources with congestion externalities, *Journal of Political Economy* **82** (1974), 163–174.

Bryson, A., and Ho, Y. C., *Applied Optimal Control*, Ginn (Waltham, Mass., 1969).

Burmeister, E., and A. R. Dobell, *Mathematical Theories of Economic Growth*, Macmillan (New York, 1970).

Burt, O. R., and R. G. Cummings, Production and investment in natural resource industries, *American Economic Review* **60** (1970), 576–590.

Bushaw, D., Optimal discontinuous forcing terms, in S. Lefschetz (Ed.), *Contributions to the Theory of Nonlinear Oscillations IV*, Annals of Mathematics Studies **41**, Princeton University Press (Princeton, 1958), pp. 29–52.

Cannon, M. D., C. D. Cullum, and E. Polak, *Theory of Optimal Control and Mathematical Programming*, McGraw-Hill (New York, 1970).

Cheung, S. N. S., Contractual arrangements and resource allocation in marine fisheries, in A. D. Scott (Ed.), *Economics of Fisheries Management—A Symposium*, University of British Columbia, Institute of Animal Resource Ecology (Vancouver, 1970), pp. 97–108.

Christy, F. T., Jr., *Alternative arrangements for Marine Fisheries: An Overview*, Resources for the Future (Washington, D. C., 1973).

Christy, F. T., Jr., and A. D. Scott, *The Common Wealth in Ocean Fisheries*, Johns Hopkins Press (Baltimore, 1965).

Ciriacy-Wantrup, S. V., The economics of environmental policy, *Land Economics* **47** (1971), 36–45.

Ciriacy-Wantrup, S. V., *Resource Conservation: Economics and Policies*, 2nd ed., University of California Press (Berkeley, 1972).

Clark, C. W., Economically optimal policies for the utilization of biologically renewable resources, *Mathematical Biosciences* **12** (1971), 245–260.

Clark, C. W., The dynamics of commercially exploited animal populations, *Mathematical Biosciences* **13** (1972), 149–164.

Clark, C. W., The economics of overexploitation, *Science* **181** (1973a), 630–634.

Clark, C. W., Profit maximization and the extinction of animal species, *Journal of Political Economy* **81** (1973b), 950–961.

Clark, C. W., Possible effects of schooling on the dynamics of exploited fish populations, *Journal du Conseil International pour l'Exploration de la Mer* **36** (1974a), 7–14.

Clark, C. W., Mathematical bioeconomics, in P. van den Driessche (Ed.), *Mathematical Problems in Biology (Victoria Conference)*, Lecture Notes in Biomathematics **2**, Springer-Verlag (New York, 1974b), pp. 29–45.

Clark, C. W., A delayed-recruitment model of population dynamics, with an application to baleen whale populations, *Journal of Mathematical Biology* (to appear).

Clark, C. W., and J. de Pree, A simple linear model for the optimal exploitation of renewable resources (unpublished, 1975).

Clark, C. W., G. Edwards, and M. Friedlaender, Beverton–Holt model of a commercial fishery: optimal dynamics, *Journal of the Fisheries Research Board of Canada* **30** (1973), 1629–1640.

Clark, C. W., and G. R. Munro, Economics of fishing and modern capital theory: a simplified approach, *Journal of Environmental Economics and Management* **2** (1975), 92–106.

Copes, P., The backward-bending supply curve of the fishing industry, *Scottish Journal of Political Economy* **17** (1970), 69–77.

Copes, P., Factor rents, sole ownership, and the optimum level of fisheries exploitation, *The Manchester School of Social and Economic Studies* **40** (1972a), 145–163.

Copes, P., The resettlement of fishing communities in Newfoundland, Canadian Council of Rural Development (Ottawa, 1972b).

Council on Environmental Quality, 5th Annual Report, U. S. Government Printing Office (Washington, D. C., 1974).

Crutchfield, J. A., Management of the North Pacific fisheries: economic objectives and issues, *Washington Law Review* **43** (1967), 283–307.

Crutchfield, J. A., Economic aspects of international fishing conventions, in A. D. Scott (Ed.), *Economics of Fisheries Management—A Symposium*, University of British Columbia, Institute of Animal Resource Ecology (Vancouver, 1970), pp. 63–78.

Crutchfield, J. A., An economic view of maximum sustainable yield, in P. M. Roedel (Ed.), *Optimum Sustainable Yield as a Concept in Fisheries Management*, American Fisheries Society Special Publication No. 9 (Washington, D.C., 1975).

Crutchfield, J. A., and A. Zellner, Economic aspects of the Pacific halibut fishery, *Fishery Industrial Research*, Vol. 1, No. 1, U. S. Department of the Interior (Washington, D.C., 1962).

Culley, M., *The Pilchard*, Pergamon Press (Oxford, 1971).

Cummings, R. G., Some extensions of the economic theory of exhaustible resources, *Western Economic Journal* **7** (1969), 201–210.

Cushing, D. H., *Fisheries Biology: A Study in Population Dynamics*, University of Wisconsin Press (Madison, 1968).

Danø, S., *Nonlinear and Dynamic Programming*, Springer-Verlag (New York, 1975).

de Lury, D. B., On the estimation of biological populations, *Biometrics* **3**(4) (1947), 145–167.

Dorfman, R., An economic interpretation of optimal control theory, *American Economic Review* **59** (1969), 817–831.

Emlen, J. M., *Ecology: An Evolutionary Approach*, Addison-Wesley (Reading, Mass., 1973).

Fan, L.-T., and C.-S. Wang, *The Discrete Maximum Principle*, Wiley (New York, 1974).

Faustmann, M., Berechnung des Werthes, welchen Waldboden sowie nach nicht haubare Holzbestande für die Weldwirtschaft besitzen, *Allgemeine Forst und Jagd Zeitung* **25** (1849), 441.

Feldstein, M. S., The social time-preference rate in cost-benefit analysis, *Economic Journal* **74** (1964), 360–379.

Feller, W., On the logistic law of growth and its empirical verification in biology, *Acta Biotheoretica* **5** (1940), 51–66.

Fisher, A. C., J. V. Krutilla, and C. J. Cicchetti, The economics of environmental preservation: a theoretical and empirical analysis, *American Economic Review* **62** (1972), 605–619.

✗Fisher, I., *The Theory of Interest*, MacMillan (New York, 1930).

Gaffney, M. M., *Concepts of Financial Maturity of Timber and Other Assets*, North Carolina State College, Department of Agricultural Economics (Raleigh, 1960).

Gaffney, M. M., (Ed.), *Extractive Resources and Taxation*, University of Wisconsin Press (Madison, 1967).

Garrod, D. J., Management of multiple resources, *Journal of the Fisheries Research Board of Canada* **30**(12, Part 2) (1973), 1977–1985.

Gates, J. M., and V. J. Norton, The benefits of fisheries regulation: a case study of the New England yellowtail flounder fishery, University of Rhode Island Technical report No. 21 (1974).

Gause, G. F., *La Théorie Mathématique de la Lutte Pour la Vie*, Hermann (Paris, 1935).

Goh, B. S., Optimal control of renewable resources and pest populations, *Proceedings of Sixth Hawaii International Conference of Systems Sciences* (1973), 26–28.

Goh, B. S., G. Leitmann, and T. L. Vincent, Optimal control of a prey–predator system, *Mathematical Biosciences* **19** (1974), 263–286.

Goldsmith, O. S., Market allocation of exhaustible resources, *Journal of Political Economy* **82** (1974), 1035–1040.

Gordon, H. S., Economic theory of a common-property resource: the fishery, *Journal of Political Economy* **62** (1954), 124–142.

Gordon, R. L., A reinterpretation of the pure theory of exhaustion, *Journal of Political Economy* **75** (1967), 274–286.

Gould, J. R., Extinction of a fishery by commercial exploitation: a note, *Journal of Political Economy* **80** (1972), 1031–1038.

Goundry, G. K., Forest management and the theory of capital, *Canadian Journal of Economics and Political Science* **26** (1960), 439–451.

Graham, M., Overfishing and optimum fishing, *Conseil International pour l'Exploration de la Mer, Rapports et Procès-Verbaux des Réunions* **132** (1952), 72–78.

Gray, L. C., Rent under the assumption of exhaustibility, *Quarterly Journal of Economics* **28** (1914), 466–489.

Gregory, G., *Forest Resource Economics*, Ronald (New York, 1972).

Gulland, J. A., *Manual of Methods for Fish-Stock Assessment*, United Nations Food and Agriculture Organization (Rome, 1969).

Gulland, J. A., *The Management of Marine Fisheries*, University of Washington Press (Seattle, 1974).

Halkin, H., Necessary conditions for optimal control problems with infinite horizons, *Econometrica* **42** (1974), 267–272.

Hamilton, W. D., Geometry for the selfish herd, *Journal of Theoretical Biology* **31** (1971), 295–311.

Hannesson, R., Fishery dynamics: a North Atlantic cod fishery, *Canadian Journal of Economics* **8** (1975), 151–173.

Hardin, G., The tragedy of the commons, *Science* **162** (1968), 1243–1247.

Herfindahl, O. C., and A. V. Kneese, *Economic Theory of Natural Resources*, Bobbs Merrill (Columbus, Ohio, 1974).

Hermes, H., and J. P. La Salle, *Functional Analysis and Time Optimal Control*, Academic (New York, 1969).

Hirsch, M., and S. Smale, *Differential Equations, Dynamical Systems, and Linear Algebra*, Academic (New York, 1974).

Hirshleifer, J., *Investment, Interest, and Capital*, Prentice-Hall (Englewood Cliffs, N. J., 1970).

Holling, C. S., Resilience and stability of ecological systems, *Annual Review of Ecology and Systematics* **4** (1973), 1–24.

Hoppensteadt, F. C., and J. M. Hyman, Periodic solutions of a logistic difference equation, *Journal on Applied Mathematics* (to appear).

Hotelling, H., The economics exhaustible resources, *Journal of Political Economy* **39** (1931), 137–175.

Idyll, C. P., The anchovy crisis, *Scientific American* **228** (June 1973), 22–29.

Institut del Mar de Peru, *Report of the Fourth Session of the Panel of Experts on Stock Assessment of the Peruvian Anchoveta*, Boletin, Vol. 2, No. 10, Callao, Peru (1974).

Intriligator, M. D., *Mathematical Optimization and Economic Theory*, Prentice-Hall (Englewood Cliffs, N. J., 1971).

Jaquette, D. L., A discrete-time population-control model, *Mathematical Biosciences* **15** (1972), 231–252.

Jaquette, D. L., A discrete-time population-control model with setup cost, *Operations Research* **22** (1974), 298–303.

Jarvis, L. S., Cattle as capital goods and ranchers as portfolio managers: an application to the Argentine cattle sector, *Journal of Political Economy* **82** (1974), 489–520.

Jones, R., and W. B. Hall, A simulation model for studying the population dynamics of some fish species, in M. S. Bartlett and R. W. Hiorns (Eds.), *The Mathematical Theory of the Dynamics of Biological Populations*, Academic (New York, 1973).

Keyfitz, N., *Introduction to the Mathematics of Population*, Addison-Wesley (Reading, Mass, 1968).

Kilkki, P., and U. Vaisanen, Determination of the optimal policy for forest stands by means of dynamic programming, *Acta Forestalia Fennica* **102** (1969), 100–112.

Kurz, M., Optimal economic growth and wealth effects, *International Economic Review* **9** (1968), 348–357.

Kushner, H. J., *Introduction to Stochastic Control*, Holt, Rinehart & Winston (New York, 1971).

Larkin, P. A., Interspecific competition and exploitation, *Journal of the Fisheries Research Board of Canada* **20** (1963), 647–678.

Larkin, P. A., Exploitation in a type of predator–prey relationship, *Journal of the Fisheries Research Board of Canada* **23** (1966), 349–356.

Larkin, P. A., R. F. Raleigh, and N. J. Wilimovsky, Some alternative premises for constructing theoretical production curves, *Journal of the Fisheries Research Board of Canada* **21** (1964), 477–484.

La Salle, J. P., The time optimal control problem, in S. Lefschetz (Ed.), *Contributions to the Theory of Nonlinear Oscillations* V, *Annals of Mathematical Studies* **45,** Princeton University Press (Princeton, 1960), pp. 1–24.

Lawler, G. H., Fluctuation in the success of year-classes of whitefish populations with special reference to Lake Erie, *Journal of the Fisheries Research Board of Canada* **22** (1965), 1197–1227.

Layard, R. (Ed.), *Cost-Benefit Analysis,* Penguin Books (Harmondsworth, England, 1972).

Lee, E. B., and L. Markus, *Foundations of Optimal Control Theory,* John Wiley (New York, 1968).

Leighton, W., *Ordinary Differential Equations,* Wadsworth (Belmont, Calif., 1963).

Leslie, P. H., Some further notes on the use of matrices in population mathematics, *Biometrics* **35** (1948), 213–245.

Leviatan, N., and P. A. Samuelson, Notes on turnpikes: stable and unstable, *Journal of Economic Theory* **1** (1969), 454–475.

Levin, S. (Ed.), *Ecosystem Analysis and Prediction,* Society for Industrial and Applied Mathematics (Philadelphia, 1975).

Lewis, T. R., and R. Schmalensee, Nonconvexities and the theory of renewable resource management, *Journal of Environmental Economics and Management* (to appear).

Lions, J. L., *Optimal Control of Systems Governed by Partial Differential Equations,* Springer-Verlag (New York, 1971).

Lotka, A. J., *Elements of Physical Biology,* Williams and Wilkins (Baltimore, 1925).

Lutz, F. A., *The Theory of Interest,* Aldine (Chicago, 1968).

Mann, S. H., A mathematical theory for the harvest of natural animal populations when birth rates are dependent on total population size, *Mathematical Biosciences* **7** (1970), 97–110.

Massé, P., *Optimal Investment Decisions,* Prentice-Hall (Englewood Cliffs, N. J., 1962).

May, R. M., *Stability and Complexity in Model Ecosystems,* Monographs in Population Biology VI, Princeton University Press (Princeton, N. J., 1973).

May, R. M., Biological populations with nonoverlapping generations: stable points, stable cycles, and chaos, *Science* **186** (1974), 645–647.

May, R. M., and G. F. Oster, Bifurcation and dynamic complexity in simple ecological models, *American Naturalist* (to appear).

Maynard Smith, J., *Mathematical Ideas in Biology,* Cambridge University Press (Cambridge, England, 1971).

Maynard Smith, J., *Models in Ecology,* Cambridge University Press (Cambridge, England., 1974).

Miele, A., Flight mechanics and variational problems of a linear type, *Journal of Aero-Space Sciences* **25** (1958), 581–590.

Mishan, E. J., *Cost-Benefit Analysis,* Unwin (London, 1971).

Mohring, H. S., The costs of inefficient fishery regulation: a partial study of the Pacific Coast halibut industry, *MS,* University of Minnesota, Minneapolis, 1973.

Murphy, G. I., Vital statistics of the Pacific sardine (*Sardinops caerulea*) and the population consequences, *Ecology* **48** (1967), 731–736.

Murphy, G. I., Population biology of the Pacific sardine (*Sardinops caerula*), *Proceedings of the California Academy of Science* (4th series), **34** (1966), 1–84.

Näslund, B., Optimal rotation and thinning, *Forest Science* **15** (1969), 446–451.

Nicholson, A. J., and V. A. Bailey, The balance of animal populations, *Proceedings of the Zoological Society of London* (1935), 551–598.

Nikolskii, G. V., *Theory of Fish Population Dynamics*, Oliver and Boyd (Edinburgh, 1969).

Oguztörelli, M. N., *Time-Lag Control Systems*, Academic (New York, 1966).

Oster, G., Stochastic behavior of deterministic models, in S. A. Levin (Ed.), *Ecosystem Analysis and Prediction*, SIAM–SIMS Conference Proceedings (Alta, Utah, 1975), pp. 24–37.

Parrish, B. B., (Ed.), Fish stocks and recruitment, *Conseil International pour l'Exploration de la Mer, Rapports et Procès-Verbaux des Réunions* **164** (1973).

Paulik, G. J., Anchovies, birds, and fishermen in the Peru current, in W. W. Murdoch (Ed.), *Environment: Resources, Pollution, and Society*, Sinauer Associates (Stamford, Conn., 1971), pp. 156–185.

Pearl, R., *The Biology of Population Growth*, Knopf (New York, 1930).

Pearse, P., The optimal forest rotation, *Forestry Chronicle* **43** (1967), 178–195.

Pielou, E. C., *An Introduction to Mathematical Ecology*, Wiley-Interscience (New York, 1969).

Plourde, C. G., A simple model of replenishable resource exploitation, *American Economic Review* **60** (1970), 518–522.

Plourde, C. G., Exploitation of common-property replenishable resources, *Western Economic Journal* **9** (1971), 256–266.

Plourde, C. G., Diagrammatic representation of dynamic exploitation of common-property resources (unpublished, 1974).

Pontryagin, L. S., V. S. Boltyanskii, R. V. Gamkrelidze, and E. F. Mishchenko, *The Mathematical Theory of Optimal Processes*, Wiley-Interscience (New York, 1962).

Pope, J. G., An investigation into the effects of variable rates of the exploitation of fishery resources, in M. S. Bartlett and R. W. Hiorns (Eds.), *The Mathematical Theory of Biological Populations*, Academic (New York, 1973), pp. 23–34.

Prest, A. R., and R. Turvey, Cost-benefit analysis: a survey, *Economic Journal* **75** (1965), 683–735.

Quirk, J. P., and V. L. Smith, Dynamic economic models of fishing, in A. D. Scott, (Ed.), *Economics of Fisheries Management—A Symposium*, University of British Columbia, Institute of Animal Resource Ecology (Vancouver, 1970), pp. 3–32.

Reed, W. J., A stochastic model for the economic management of a renewable animal resource, *Mathematical Biosciences* **22** (1974), 313–337.

Reed, W. J., Some stochastic models in animal resource management (Ph.D. Thesis), University of British Columbia (Vancouver, 1975).

Ricker, W. E., Stock and recruitment, *Journal of the Fisheries Research Board of Canada* **11** (1954), 559–623.

Ricker, W. E., Handbook of computations for biological statistics of fish populations, *Bulletin of the Fisheries Research Board of Canada*, **119** (1958).

Rothschild, B. J., An exposition on the definition of fishing effort, *Fishery Bulletin* **70** (1972), 671–679.

Samuelson, P. A., A catenary turnpike theorem involving consumption and the golden rule, *American Economic Review* **55** (1965), 486–496.

Samuelson, P. A., *Foundations of Economic Analysis*, Harvard University Press (Cambridge, Mass., 1965).

Samuelson, P. A., *Economics*, 9th ed., McGraw-Hill (New York, 1973).

Schaefer, M. B., Some considerations of population dynamics and economics in relation to the management of marine fisheries, *Journal of the Fisheries Research Board of Canada* **14** (1957), 669–681.

Schaefer, M. B., Fishery dynamics and the present status of the yellowfin tuna population of the Eastern Pacific ocean, *Bulletin of the Inter-American Tropical Tuna Commission* **12**(3) (1967).

Schoener, T. W., Population growth regulated by intraspecific competition for energy or time: some simple representations, *Theoretical Population Biology* **4** (1973), 56–84.

Schreuder, G. F., Optimal forest investment decisions through dynamic programming, *Yale University School of Forestry Bulletin* **72** (1968).

Schwartz, L., *La Théorie des Distributions*, Hermann et Cie. (Paris, 1966).

Scott, A. D., The fishery: the objectives of sole ownership, *Journal of Political Economy* **63** (1955), 116–124.

Scott, A. D., *Natural Resources: The Economics of Conservation*, 2nd ed., McLelland-Stewart (Toronto, 1972).

Shaw, E., Schooling in fishes: critique and review, in L. R. Aronson (Ed.), *Development and Evolution of Behavior*, W. H. Freeman (San Francisco, 1970), 452–480.

Silverberg, R., *The Auk, the Dodo, and the Oryx: Vanished and Vanishing Creatures*, T. Y. Crowell (New York, 1967).

Silvert, W., and W. R. Smith, Optimal exploitation of a multi-species community, *Mathematical Biosciences* (to appear).

Slobodkin, L. B., *Growth and Regulation of Animal Populations*, Holt, Rinehart & Winston (New York, 1961).

Small, G., *The Blue Whale*, Columbia University Press (New York, 1971).

Smith, V. L., Economics of production from natural resources, *American Economic Review* **58** (1968), 409–431.

Smith, V. L., On models of commercial fishing, *Journal of Political Economy* **77** (1969), 181–198.

Solow, R. M., The economics of resources or the resources of economics, *American Economic Review* **64** (1974), 1–14.

Solow, R. M., Intergenerational equity and exhaustible resources, in G. M. Heal (Ed.), *Symposium on the Economics of Exhaustible Resources, Review of Economic Studies* (1975: Special Issue), pp. 29–46.

Southey, C., Policy prescriptions in bionomic models: the case of the fishery, *Journal of Political Economy* **80** (1972), 769–775.

Spence, M., *Blue Whales and Applied Control Theory*, Technical Report No. 108, Stanford University, Institute for Mathematical Studies in the Social Sciences, 1973.

Spence, M. and D. Starrett, Most rapid approach paths in accumulation problems, *International Economic Review* **16** (1975), 388–403.

Thom, R., *Stabilité Structurelle et Morphogènese*, W. A. Benjamin (New York, 1972).

Turvey, R., Optimization and suboptimization in fishery regulation, *American Economic Review* **54** (1964), 64–76.

Verhulst, P. F., Notice sur la loi que la population suit dans son accroissement, *Correspondance Mathématique et Physique* **10** (1838), 113–121.

Volterra, V., *Leçons sur la Théorie Mathématique de la Lutte pour la Vie*, Gauthier-Villars (Paris, 1931).

Vousden, N., Basic theoretical issues of resource depletion, *Journal of Economic Theory* **6** (1973), 126–143.

Walters, C. J., A generalized computer simulation model for fish population studies, *Transactions of the American Fisheries Society* **98** (1969), 505–512.

Wangersky, P. J., and W. J. Cunningham, Time lag in predator–prey population models, *Ecology* **38** (1957), 136–139.

Warga, J., Relaxed variational problems I, II, *Journal of Mathematical Analysis and Applications* **4** (1962), 111–128 and 129–145.

Watt, K. E. F., *Ecology and Resource Management*, McGraw-Hill (New York, 1968).

Waugh, G., and P. Calvo, Economics of exhaustible resources: the fishery, *Economic Record* **50** (1974), 423–429.

Wiegert, R. C., Competition: a theory based on realistic, general equations of population growth, *Science* **185** (1974), 539–541.

Wright, C., Some political aspects of pollution control, *Journal of Environmental Economics and Management* **1** (1974), 173–186.

Yodzis, P., The effects of harvesting on competitive systems, *Bulletin of Mathematical Biology* (to appear).

Young, L. C., *Lectures on the Calculus of Variations and Optimal Control Theory*, W. B. Saunders (Philadelphia, 1969).

Zeeman, E. C., Levels of structure in catastrophe theory, in R. D. James (Ed.), *Proceedings of the International Congress of Mathematicians*, (Vancouver, 1974), Vol. 2, pp. 533–546.

Index